Just Transformations

'A hugely important book, setting a radical agenda for societal transformation. Drawing on grassroots alternatives from across the world, the book offers a vital guide for both scholars and activists. Everyone committed to just transformations for sustainability should read this book now!'

—Ian Scoones, Professor, Institute of Development Studies, University of Sussex

'A fantastic collection that illustrates that just transformations are already being imagined and implemented on the ground. The authors offer an important, creative example of genuine scholar-activism keenly focused issues of justice, power, and the transformative potential of EJ.'

—David Schlosberg, Professor of Environmental Politics and Director, Sydney Environment Institute, University of Sydney

'A splendid collective book co-produced by an impressive international group of twenty-five socio-environmental academics and activists ... focusing both on the alternatives that are born from the resistance to extractivism or pollution, and on sustainable practices such as community textile production. Building on detailed knowledge of the local protagonists and issues, this optimistic, inspiring book jumps scales to national and international dimensions.'

—Joan Martinez-Alier, Institute of Environmental Science and Technology, Universitat Autònoma de Barcelona

Just Transformations

Grassroots Struggles for Alternative Futures

Edited by Iokiñe Rodríguez, Mariana Walter and Leah Temper

First published 2024 by Pluto Press
New Wing, Somerset House, Strand, London WC2R 1LA
and Pluto Press, Inc.
1930 Village Center Circle, 3-834, Las Vegas, NV 89134

www.plutobooks.com

This work is based on research supported through the Transformations to Sustainability (T2S) programme (2014–19) of the International Social Science Council (ISSC), which in 2018 became the International Science Council (ISC), with funding from the Swedish International Development Cooperation Agency (Sida).

British Library Cataloguing in Publication Data
A catalogue record for this book is available from the British Library

ISBN 978 0 7453 4477 5 Paperback
ISBN 978 0 7453 4481 2 PDF
ISBN 978 0 7453 4479 9 EPUB

This book is printed on paper suitable for recycling and made from fully managed and sustained forest sources. Logging, pulping and manufacturing processes are expected to conform to the environmental standards of the country of origin.

Typeset by Riverside Publishing Solutions, Salisbury, England

Simultaneously printed in the United Kingdom and United States of America

Contents

Introduction

Iokiñe Rodríguez, Mariana Walter and Leah Temper

Transition and transformation are buzzwords on everybody's lips in today's climate crisis-ridden world. If we are to save humanity and other species, something big must change. This has led to a wealth of literature aimed at understanding, managing and guiding society towards the needed transformation. Much of this has focused on the potential contribution of markets (the green economy), the state (green technocracy) and technology (the eco-modernist utopia/dystopia, depending on your perspective). Yet surprisingly absent from this conversation are activists, communities and movements on the ground who have been struggling and building social transformations, crafting alternatives and recreating their worlds from the ground up, as well as engaged activist academics who are working with movements in trying to make this change happen. This book seeks to contribute to this necessary conversation.

Environmental justice movements may not use terms like transformation, sustainability or environment, but they have a lot to teach about how just transformations happen. Such lessons become particularly relevant nowadays, when climate and pandemic crises are fostering accelerated top-down technocratic and authoritarian paths that are making invisible and sometimes disabling ongoing bottom-up transformative processes.

However, little is still known about how bottom-up just transformations to sustainability are taking place, what makes them possible and what sustains them over time. Understanding this is not merely of academic interest. Resistance movements and communities are also often interested in reflecting about this, and learning with others about how to make their strategies stronger or understand better what might be preventing them from achieving their objectives. Yet the urgency of their struggles often leaves little time for reflection, let alone for developing the methods for joint or cross-learning.

This book offers a window into learning from just transformation to sustainability through the lens of locally led processes, by sharing the results of a three-year international research project (ACKnowl-EJ: Academic-Activist Co-produced Knowledge for Environmental Justice) in which academics and resistance movements across the world worked together, analysing and learning from citizen-led transformations. We draw on a variety of case studies in Argentina, Belgium, Bolivia, Canada, India, Lebanon, Turkey and Venezuela (see Figure 0.1),

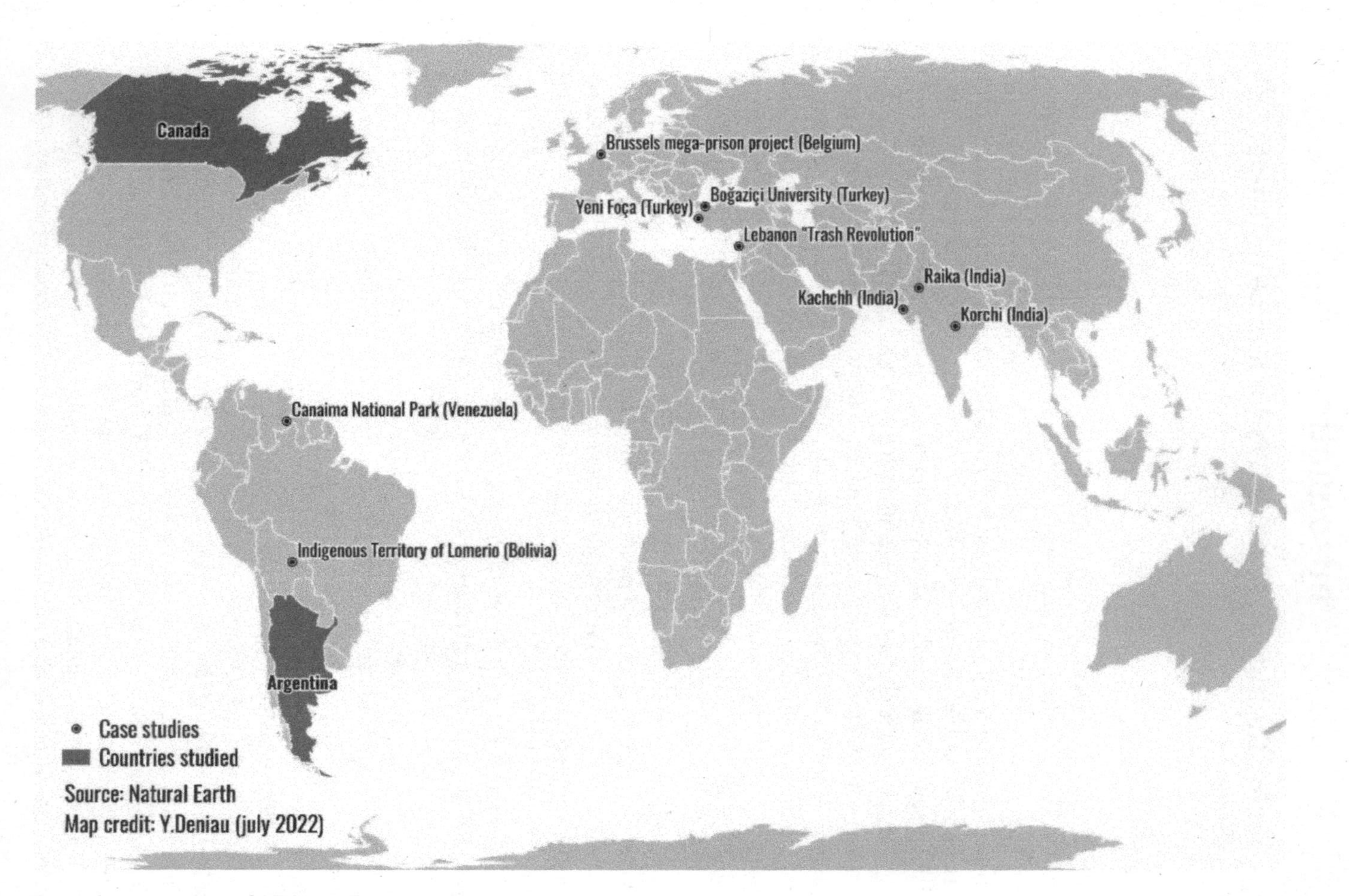

Figure 0.1 Locations of ACKnowl-EJ case studies

where resistance movements have been taking the lead, with different degrees of success and failure, in bringing about just sustainability transformations. These case studies range from resistance to the fossil fuel rush (Turkey), Indigenous peoples' struggles against extractivism (Venezuela), urban mobilizations around waste management (Lebanon) and against the building of a mega-prison (Belgium), to the search for alternative transformations such as new forms of political governance for managing commons (Lomerío in Bolivia, Korchi in India), the articulation of women's pastoralist vision for the future (the Raika in India) or a revival of communal and family-based economies (Kachchh in India). This book also draws out lessons from country-wide transformations against mining and for the defence of Indigenous lands (Argentina and Canada respectively) from a global atlas of resistances (the EJAtlas: www.ejatlas.org).

This book is also a toolkit for how to do engaged transformations research, which we hope will be of use both to environmental justice academics and activists. We share the ACKnowl-EJ approach to co-production of knowledge and for doing 'politically rigorous' research with activists on just transformations to sustainability. We also present two frameworks that we developed to assess how progress towards transformation is taking place, which communities, activists and academics can use to examine and learn from specific transformative processes. We do this, however, recognizing the immense challenges within engaged environmental justice research. Across the book we reflect both on the internal ethical, relational and epistemological challenges that we faced as a research collective in doing this type of research, and on the external political challenges we confronted arising from a world that is turning ever more violent towards environmental justice movements and activist scholars. This book shows, however, that despite these challenges, engaged research has an important role to play in supporting bottom-up just transformations to sustainability.

Finally, based on the lessons drawn from the case studies and from cross-country analysis from the EJAtlas, this book contributes towards the development of a theory of just transformations to sustainability from the perspectives of those fighting on the ground to achieve it. We hope this will help rethink how transformations to sustainability are currently being conceptualized, and inform a conversation between and across resistance movements about the changes they are trying to bring about for a more just and sustainable world.

This introductory chapter develops as follows. In the following second section we situate the ACKnowl-EJ project and ourselves, explaining how this work draws from several grounded and grassroots initiatives and aims to co-produce knowledge with and for communities and social movements. The third section discuss some key concepts that have guided this process: radical transformations, conflicts and alternatives. The fourth section lays out the conceptual framework that we developed to jointly study just transformations to sustainability with a focus on power relations, dimensions and scales of change. A framework that is multidimensional and intersectional, balancing ecological concerns with social, economic, cultural and democratic spheres. We conclude with a detailed description of the content of the book chapter by chapter.

The ACKnowl-EJ Project: Who Are We and What Brought Us Together?

ACKnowl-EJ was a three-year project funded by UNESCO's International Social Science Council as one of three Transformative Knowledge Networks within their Transformations to Sustainability programme and a broader Future Earth Science programme. ACKnowl-EJ aimed to engage in action and collaborative analysis of the transformative potential of community responses to environmental and social injustices, particularly those framed as extractivism, and alternatives born from resistance. The project was an experiment in co-producing knowledge that could answer the needs of social groups, advocates, citizens and social movements, while supporting communities and movements in their push for change. The network was made up of activist-scholars and activists with ties to academic institutions, non-governmental organizations, communities and social movements (see Figure 0.2), and drew on work that members were carrying out in India, Turkey, Bolivia, Canada, Belgium, Lebanon, Venezuela and Argentina. In all cases the research involved some degree of activist research, either through researchers: 1) being activists in the struggles themselves, 2) having a double role as researcher/activist or 3) developing action-research teams with communities or movements to document the case studies.

While most case studies were developed between 2016 and 2019, during the ACKnowl-EJ project, two cases (Turkey and Bolivia) are published with updated

Figure 0.2 Academic and community co-partners from the ACKnowl-EJ project, India 2018

perspectives since significant changes took place over these last two years, which we considered important to include.

This group of scholar-activists shared an interest in 1) understanding and supporting social transformation and resistance to extractive activities and imposed development, 2) creating linkages between academia and activists, and 3) helping to give visibility to communities, movements and initiatives that are putting transformative alternatives into practice. Thus, we came together to create a space for reflection and action on questions such as:

- What role do resistance movements against 'extractivism' play in shaping local and global transformations for sustainability from the ground up, and in dealing with the global environmental and social crisis?
- How are resistance strategies and the creation of development alternatives carried out?
- What is transformed in this process?
- What determines the success of resistance movements and of transformations over time?

In other words, we wanted to contribute to a better understanding of 'what needs to be transformed' for more sustainable futures from the perspective of resistance movements, as well as 'how it can be transformed' and 'what truly transformative alternatives are'. In this sense, ACKnowl-EJ subscribes to a 'right here, right now' approach for transformations and aims for its research to be transformative and to affect change that empowers others (Moser 2016, Temper and Del Bene 2016).

This ACKnowl-EJ project is grounded in three initiatives that aim to co-produce knowledge with and for communities – the EJAtlas, the Grupo Confluencias network and Vikalp Sangam, described below. All three initiatives are dedicated to capacities for action and practice-based research with co-production of knowledge, learning/teaching processes, reflexivity, and the creation of research outputs that answer to the scientific rigour of academia and to the political rigour of actors in environmental struggles. This transformative environmental justice (EJ) research agenda stresses the importance of engagement with critical scholars, scholar-activists and activists as well as the recognition of the epistemologies and ontologies of marginalized voices, for a co-production and reproduction of plural knowledges.

The EJAtlas

Research as part of the Environmental Justice Atlas (www.ejatlas.org) project over the past ten years has focused on producing bottom-up documentation and mapping of the numerous extraction conflicts taking place throughout the world, and has helped make visible the violence perpetrated by states and corporations against resisting populations (Temper et al. 2015). The EJAtlas's 3,880 cases (as of June 2023) provide a repository of diverse, radically challenging and overtly political agonistic forms of contestation of environmental inequality by subaltern social movements. It offers an opportunity to tune into the plurality of grassroots voices opposing specific economies, institutions, infrastructures and cultures that are at the root of the ecological crisis. It demonstrates the diversity in these movements as well as the commonalities that join them under a global and globalizing movement for environmental justice (Martinez-Alier et al. 2016).

While the EJAtlas was originally designed to emphasize, make visible and dissect processes of environmental injustice, in the ACKnowl-EJ project it was used as an empirical base for examining what EJ looks like in practice and for understanding the multiple and creative agency of EJ groups as 'altering' forces of the status quo. In many cases these struggles propose and put forward their own visions of transformations. The cases illuminate how and when democratic and transformative processes that arise in response to extractive processes move from the individual to the community level and then disseminate outwards.

Grupo Confluencias – conflict transformation practitioners in Latin America

Grupo Confluencias, a group of Latin American socio-environmental conflict transformation practitioners and researchers, have been working since 2005 as a platform for deliberation, joint research and capacity-building on this topic. Members of this network play a combination of roles in the transformation of socio-environmental conflicts: dialogue facilitation, peace-building, advice and capacity-building for Indigenous peoples and urban/rural communities, policy advice on environmental and sustainable development issues and action-research in their respective countries. The countries include Argentina, Bolivia, Chile, Ecuador, Costa Rica, Guatemala, Peru and Venezuela.

After years of deliberation around what a truly transformative process entails, one of the most important outcomes of the work of Grupo Confluencias was a framework to help assess if processes being developed on the ground to transform environmental conflicts were effectively contributing to reducing violence and increasing justice. The group merged conflict transformation theory (a concept that originated in peace studies in post-war scenarios) with decolonial thought, power theory and political ecology, and developed a Socio-environmental Conflict Transformation (SCT) Framework that seeks to guide conflict transformation work for greater environmental justice in the region.

The framework was largely informed by the work the group carries out supporting Indigenous peoples in environmental justice struggles. Because of this it places the development of intercultural relations (Walsh 2005a, 2005b) (*construcción de interculturalidad* in Spanish) at the core of the conflict transformation process. Two other key features of this framework are the attention paid to understanding the role that power dynamics and culture play in environmental conflicts and their transformation, and the development of indicators to help assess if the strategies used are effectively contributing to the reduction of violence, greater justice and ultimately the development of interculturality (Rodríguez et al. 2019, Rodríguez and Inturias 2018).

Although the framework was widely discussed among conflict transformation practitioners and Indigenous leaders in capacity-building meetings and workshops in Latin America, the ACKnowl-EJ project offered the first opportunity to apply it in case studies.

Vikalp Sangam (Alternatives Confluence), India

Vikalp Sangam ('Alternatives Confluence') (https://vikalpsangam.org/) is an Indian platform for networking between groups and individuals working on alternatives to the currently dominant model of development and governance, in various spheres of life (see Daga 2014, Kothari 2016, Thekaekara 2015). Its major activity is the convening of regional and thematic confluences across India (Kothari 2016), whereby people exchange experiences and ideas emerging from practice and thinking in a whole range of endeavours including sustainable agriculture and pastoralism, renewable energy, decentralized governance, community health, craft and art revival, multiple sexualities, inclusion of the differently abled, alternative learning and education, community-based conservation, decentralized water management, urban sustainability, gender and caste equality, and more.

Beyond the sharing of practical experiences and the documentation and dissemination of stories of transformation hosted on the website, one of the most important outputs of the Vikalp Sangam process is a conceptual framework of transformative alternatives. This framework aims to dissect the different spheres of transformation involved in radical alternatives. It is important to realize that while this framework has significant elements of 'ideology' in it, it is not based on or emanating from Marxist, Gandhian, Ambedkarite or other radical ideologies to which movements in India relate, but rather on the wisdom and concepts emerging from grassroots communities and groups (see Kothari 2016 for linkages between the concepts in the framework and actual alternative initiatives in India). It is constantly evolving, after discussions at each Sangam. Several hundred people from the range of sectors mentioned above have debated the various aspects of this framework.

The ACKnowl-EJ project offered the opportunity for these three networks to come together to conceptualize what an approach for analysing radical transformations to sustainability could look like. Before introducing the framework that we developed to analyse just transformations to sustainability, we discuss three core ideas that grounded our analysis: the need for radical transformations, conflicts as catalysts of change, and alternatives as concrete materializations of the needed changes.

Grounding Just Transformations to Sustainability

ACKnowl-EJ adopted an understanding of social transformation towards sustainability that focuses on the following three considerations:

- Social transformation towards more sustainable futures often occurs as a result of conflict. Oppositional consciousness and resistance to hegemonic structures are a key element in the creation of alternative ways of being and doing.

- A perspective of conflict as productive, rather than something to be avoided, suggests the usefulness of a 'conflict transformation' approach that can address the root issues of ecological conflicts as a path towards transformations to sustainability.
- Radical alternatives are a form of resistance that advances a vision of what sustainable transformative processes could look like.

We now turn to examining each one of these considerations.

The need for radical transformations

When we talk about transformation, what are we really talking about? When can we say that something has been transformed? Who are the agents of transformation? And what is it that needs to be transformed? Transformation is an amorphous term and recently somewhat of a buzzword. This has led to calls for a clearer definition of the term; and the need to differentiate transformation from transition. Further, we believe it is necessary to parse out and better define radical initiatives and alternatives as those that offer the clearest paths to transformation.

Transformation by definition needs to reconfigure the structures of development through changing an overarching global political economy dominated by neoliberal capitalism with, in our day, increasing authoritarian tendencies (Pelling 2011). It includes 'radical shifts, directional turns or step changes in normative and technical aspects of culture, development or risk management' (Pelling et al. 2015). From this perspective, transformation deals with the deeper and obscured roots of unsustainability, laden in social, cultural, economic and political spheres. These relatively invisible root causes often overlap and interact to produce uneven outcomes (Pelling 2012) including feedbacks. According to Scoones (2016), transformations to sustainability require a shift beyond scarcity discourses towards a politicized understanding of resources and sustainability. Thus, if transformation is to be achieved in an empowering and pro-poor way then a truly politicized view which exposes, problematizes and resists the ongoing reproduction of harmful power relations is inevitable (Gillard et al. 2016).

While there is broad acknowledgement that a transformation to sustainability requires a radical shift, including a shift in society's value-normative system and shifting relations across the personal (i.e. beliefs, values, worldviews), political (i.e. systems and structures) and practical (i.e. behaviours and technical responses) levels simultaneously (O'Brien and Sygna 2013), there is less consensus about what the 'radical' in radical transformations means. The word '*radicalis*' comes from the Latin 'of or having root' and refers to 'change at the root', with connotations to fundamental and revolutionary change of social systems. A radical social perspective inherently calls for addressing social justice and power issues, as well as environmental ones, in the transformation process.

Nancy Fraser's distinction between what she terms affirmative versus transformative change is illustrative. She argues that injustices may be resolved either affirmatively or transformatively. Affirmative redistributive remedies aim to correct existing income inequality by facilitating transfer of material resources to maligned groups, for example through the social welfare state. However, these remedies tend to leave intact the conditions, such as the capitalist mode of production, that were responsible for generating income inequality in the first place. In contrast, transformative redistributive remedies are aimed at eradicating the origins of economic injustice and eliminating the root causes of economic inequality, and would include 'redistributing income, reorganizing the division of labour, subjecting investment to democratic decision-making, or transforming other basic economic structures' (Fraser 1995: 73). Regarding recognition and identity conflicts, the transformative remedy, in contrast to affirmative action, entails the deconstruction of identities themselves and the transformation of the underlying cultural-valuational structure.

In this way, we believe it is important to differentiate initiatives by communities, civil society organizations, government agencies and businesses that are dealing *only* with the symptoms of the problem, and can be considered reformist initiatives, from those alternatives and movements which are confronting the basic structural reasons for unsustainability, inequity and injustice, such as capitalism, patriarchy, state-centrism, or other inequities in power resulting from caste, ethnic, racial and other social characteristics. We call these transformative or radical alternatives.

It should also be noted that there is no *necessary* contradiction between reform and transformation; many reform measures may well be contained within transformative processes, and some reforms if stretched far enough can also be transformative.

However, we may argue that a radical transformation needs to be based on attaining the impossible, rather than limiting itself to purely technical questions and narrowly constrained approaches based on questions of ecological sustainability such as energy production technologies and costs. David Harvey (2011) calls this a 'co-revolutionary theory', which picks up transformative steam from grassroots movements but without ignoring the reclamation of hegemonic state structures. The 'Initial point of entry for alternatives is less important than the need to infect and influence other domains', suggests Pelling (2012: 7), where societal 'shifts and movements are not minor historical events and most likely require energies both at the grassroots as well as momentum from above'. This, we argue, is the basis of a radical transformative agenda: flourishing rooted, local alternatives connected to wider political transformations, meanwhile paying utmost attention to historical, social and political specificities to build emancipatory sustainabilities (Scoones et al. 2017).

Because EJ movements propose that environmental problems are political issues that cannot be solved apart from social and economic justice, and that these call for a transformative approach and the restructuring of dominant social relations and institutional arrangements, we argue that EJ movements need to be at the core of sustainability transformations. EJ brings attention to both the multivalent aspects of justice – from distribution to cultural recognition to participation, capabilities, cognitive justice and beyond – as well as an intersectional approach to forms of difference across lines of class, race, gender, sexual preference, caste, ability, etc. This multidimensional and intersectional approach has been sorely lacking from transformation studies. Further, the EJ approach focuses on the interdependency of issues, seeing environmental devastation, ecological racism, poverty, crime, social despair, and alienation from community and family as aspects of a larger rooted systemic crisis. Finally, radical politics and alternatives and knowledge on how to confront hegemonic power and injustices is often created through processes of struggle.

For us, radical transformation implies one which refers to a transformation of power structures and relations, from a situation of domination, injustice, violence and unsustainability to one of reduced violence, increased equality and flourishing. It entails challenging the sources of domination and oppression, including capitalism, patriarchy, state-centrism and inequities along lines of race, caste, ethnicity, gender, ableism, sexuality and others – and it is thus multidimensional and intersectional, balancing ecological concerns with social, economic, cultural and democratic spheres.

Socio-environmental conflicts as catalysts of change

From this perspective, the manifestation of socio-environmental conflict is the first step of sustainability transformations. This is because conflicts express a questioning of the status quo and of a system where some have to be 'polluted', displaced and deprived while others benefit.

Yet EJ struggles go beyond demanding redistribution of environmental harms and resources, but rather contest the very economic, ecological, social and cultural principles behind particular uses of the environment (Gadgil and Guha 1993). In some cases, those resisting an extractivist project are often articulating an anti-systemic vision for societal transformation to sustainability within their resistance practices. Further, the organizing and collective action in which they engage, in defence of their lives and livelihoods, often inspires the quest for more localized and democratic forms of governing resources and commons and leads to new practices and alternative forms of provisioning and production. This highlights the productivity of conflicts in the creation of transformation and alternatives.

Conventional approaches to social and ecological conflicts generally adopt a perspective focused on conflict resolution/management, which aims on achieving a mutual satisfaction of interests among actors based on the maximization of individual gains – win–win solutions, through cooperation, negotiation and consensus seeking (Fisher and Ury 1981; Ury et al. 1988). Under this approach, conflicts tend to be seen as negative phenomena to be avoided and 'resolved' as quickly as possible. However, such approaches can lead environmental conflicts to become recurrent and cyclical because they offer little opportunity for developing solid democratic and sustainable agreements for the use and management of the environment and territories. Socio-environmental conflicts have complex and profound roots, in the majority of cases with important political, historical, social, environmental and cultural components and profound power asymmetries and institutional failures, which limit the possibility of them being successfully dealt with through conventional, facilitated conflict resolution methods.

In contrast, a *conflict transformation approach* sees conflicts as a natural and inevitable part of human interactions that can have constructive potential. The starting point of conflict transformation is that conflict is rooted in situations that are perceived as unjust, and by unearthing and making injustices visible, conflicts become catalysts for social change (Dukes 1996; Lederach 1995). While conflict resolution tends to focus on reaching agreements and overcoming a crisis situation, conflict transformation engages with a much bigger question: the pursuit of justice in society through restoration, rectification of wrongs and the creation of right relationships based on equity and fairness (Botes 2003; Lederach 1995). More specifically, Lederach (1995) defines conflict transformation as:

> the process that helps us visualise and answer to the flow and backflow of social conflict as life opportunities, that can create processes of constructive change, reduce violence, increase justice in interactions and social structures and respond to the real problems of human relations.

Thus, following Lederach, in this book we see socio-environmental conflicts as an opportunity for constructive change, and conflict transformation as the multiple processes that can make this change happen.

Alternatives as concrete materializations of the needed changes

Just as EJ conflicts signal the need for change, they also find expression in the form of counter-hegemonic alternative processes and narratives. Political ecologist Paul Robbins advocates what he terms a 'hatchet and seed' approach (Robbins 2004). This entails a dual task of deconstructing and discarding dominant narratives, while also identifying alternative practices and knowledges and bringing these positive examples and theoretical innovations, developed by and through social movements and community activists, to light.

While we are concerned with the role of conflict and resistance in transformation, an integral element of this resistance is the social movements that are not actively opposing particular projects such as those defined by the EJAtlas, but are engaged in practices that provide an alternative to a part or the whole of the currently dominant system, challenging one or more of the capitalist, statist, patriarchal, religious, casteist or other structures of power inequity. For instance, a group of women farmers transforming their agricultural systems away from one of dependence on chemicals, corporate seeds and government credit towards self-reliance for seeds, organic inputs, local exchange and collective credit, and local knowledge are not necessarily struggling against a particular project or company but rather against a global agro-industrial model of injustice.

Alternatives can be understood as practices, performances, systems, structures, policies, processes, technologies and concepts/frameworks, practised or proposed/propagated by any collective or individual, communities, social enterprises, etc. that usurp or challenge the capitalist mainstream and that reflect a diversity of exchange relations, social networks, forms of collective action and human experiences in different places and regions (Gibson-Graham 2006). Alternatives can be continuations from the past, reasserted in or modified for current times, or new ones; it is important to note that the term does not imply these are always 'marginal' or new, but that they adopt and operate with values and ideologies that overtly reject hegemonic economic and political practices. While they may position their activities in non-confrontational and potentially apolitical terms, their attempt to create alternatives to the hegemonic system is also often informed by an oppositional consciousness. This may include groups engaged in small-scale energy production, organic farming and permaculture, open-source software, and other forms of radical grassroots experimentation. While these groups are less likely to explicitly position themselves as EJ movements, through their embodied practices they can be said to be advancing a vision of what EJ could look like.

In summary, we see conflict and alternatives as intertwined processes. EJ struggles are spaces of reimagination, where one's and the other's ways of thinking, seeing the world and doing are disputed and reshaped in a dynamic and multi-scalar learning process. Moreover, alternatives can be both the result or the root of resistance processes. Communities can rebel against the de-legitimation of their values, worldviews and related practices. For instance, in the context of increased pressures and conflicts related to the mining activities in Latin America, social movements are formulating strategies to develop and strengthen local alternatives during, after and before the unfolding of conflicts. Alternatives are also fostered as a strategy to prevent, and oppose (e.g. Walter et al. 2016). Thus, social movements, resistance and alternatives are linked processes. People move across these spaces, protesting and engaging in rebuilding when they need to.

A Framework for Understanding Movements of Resistance as Agents of Transformative Change

Although EJ struggles/conflicts and alternatives are powerful processes where intended (and unintended) social transformations occur, the particularities of these processes remain largely under-examined. Furthermore, frameworks to analyse transformations with a focus on conflicts and alternatives from the perspectives of EJ movements are rare.

In ACKnowl-EJ we wanted to develop a framework that could help analyse bottom-up transformations based on two core ideas:

- A transformation to sustainability must entail transformation of power relations.
- Social transformation studies need to pay attention to such power relations across multiple dimensions and scales to fully capture how truly transformative processes and alternatives occur, and how to prevent inequalities and injustices from being created elsewhere or displaced.

Most importantly, we needed to develop a common point of reference that could help us look in detail at cases but also make comparison across different political systems. For instance, much of the content of Grupo Confluencias's framework, which focuses on the development of intercultural relations using a decolonial environmental justice approach (Rodríguez 2020), and which makes sense in the context of current plurinational politics prevalent in many Latin American countries, did not necessarily apply to other countries which function under liberal or consociational democracies (e.g. Lebanon).[1]

Thus, we developed a common framework for analysis that took elements from the three networks that could be applicable across all case studies. From Grupo Confluencias we adopted the work on power, from Vikalp Sangam we took the work on alternatives and from the EJAtlas the work on scales. We thus developed a framework that focuses on three core elements: strategies used to transform power, dimensions/spheres of transformation and scales of transformation.

Analysing transformative strategies

As we have argued, a radical perspective on transformation calls for an explicit engagement with the issue of power in environmental struggles. This requires however differentiating between hegemonic and transformative power.

It is by impacting on hegemonic power structures that EJ movements may manage to advance their vision of EJ. Yet in order to see how this process of change takes place or how it can be more effectively produced, it is necessary to dissect hegemonic power in its different forms (Gaventa 1980; Lukes 1974).

The notion of power as domination is the most commonly known definition of hegemonic power. It implies the idea of imposing a mandate or an idea (Bachrach and Baratz 1962). However, the power of domination is not always exercised coercively, but through subtle mechanisms. In this sense, domination can manifest in the form of visible (Weber 1971), hidden (Giddens 1984) and invisible/internalized forms of power (Foucault 1971).

In society, the 'visible' face of power is commonly manifested through decision-making bodies (institutions) where issues of public interest, such as legal frameworks, regulations and public policies, are decided (e.g. parliaments, legislative assemblies, formal advisory bodies). This is the public space where different actors display their strategies to assert their rights and interests. At a global scale this is also expressed in the use of Western/modern institutional forms of power (like the nation-state) on non-Western societies to organize and control labour, its resources and products. Visible power may also manifest through economic frameworks at a national and global scale that shape economic activities and productive systems in society, such as global capitalism. This type of political and economic power is also commonly known as structural power. It is important to stress that some dimensions of structural power may also be invisible, such as some governance processes of global capitalism that seem disconnected and dispersed, but in reality are linked through nodes or stages in commodity chains, extraction, transportation, production of goods and final disposal (Temper et al. 2018).

Much of the time, power is also exercised in a 'hidden' way by incumbent authorities attempting to maintain their privileged position in society, by creating barriers to participation, excluding issues from the public agenda or controlling political and economic decisions 'behind the scenes'. In other words, the power of domination is exercised also by people and power networks (Long and Van Der Ploeg 1989) that are organized to ensure that their interests and world views prevail over those of others. For instance, in socio-environmental conflicts, coalitions between particular sectors of the nation-state and private actors, national and international organizations, development banks and foreign governments are often the norm.

Thirdly, the power of domination also works in an 'invisible' way through discursive practices, normative rules, narratives, worldviews, knowledge, behaviours and thoughts that are assimilated by society as true without public questioning (Foucault 1971). This invisible, capillary, subtle form of power often takes the shape in practice (following Galtung 1990) of cultural violence, through the imposition of values and beliefs and knowledge systems that exclude or violate the physical, moral or cultural integrity of certain social groups by underestimating their own value, knowledge and belief systems. This impacts via mechanisms of subjectivation on the lives, bodies and minds of marginalized peoples, to the point of stripping them of their very essence and subjective, individual and collective identities. This is what some authors have termed epistemic or cognitive violence/injustice (de Sousa Santos 2008; Visvanathan 1997).

In this way, invisible and hidden power often act together, one controlling the world of ideas and the other controlling the world of decisions.

This distinction between powers concentrated in institutions, people and culture is very important for understanding relationships of power and domination in environmental struggles and in the perpetuation of environmental injustices. The challenge for overcoming violence and injustice in its different forms (direct, structural, cultural/epistemic/cognitive) and therefore for achieving conflict transformation is to generate strategies to impact on these three areas in which power is concentrated: 1) institutions, legal and economic frameworks, 2) people and their networks, and 3) discourses, narratives, knowledge and ways of seeing the world. The final outcome of the struggles, in terms of achieving the desired

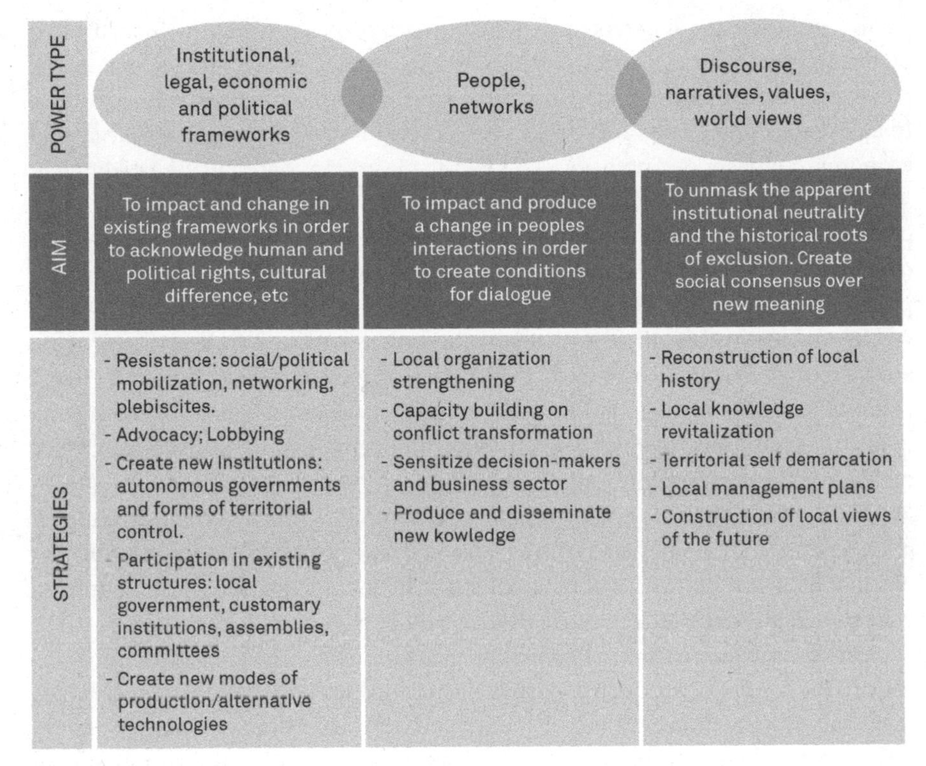

Figure 0.3 Strategies used by Indigenous people in Latin America to impact on the structural, personal and cultural dimensions of domination

Source: Rodríguez and Inturias 2018.

transformation, depends on knowing how and when to impact on each one of the types of hegemonic power.

An understanding of the transformative power strategies used by resistance movements to impact on the different types of hegemonic power, and their successes or limitations, is an essential part of a radical approach to the study of transformation to sustainability.

Drawing on the work of Grupo Confluencias, we provide examples of some transformative power strategies commonly used by Indigenous peoples in Latin America, which also find expression in other parts of the world (see Figure 0.3 for a summary). However, as we will see throughout this book, EJ movements across the world use a much wider array of transformative strategies.

Strategies to impacting on structural power

Resistance movements impact on structural power in different ways. One is through outright confrontation, impacting through political and social mobilization on laws, regulations and norms that have been created without consultation or that do not represent the differentiated rights of society. Another way is by activating democratic procedures, such as plebiscites/referenda (Walter and Urkidi 2017). Although effective in the short term, these strategies will not necessarily transform in a profound way institutional structures, unless they impact also macro legal and economic frameworks. Another way is by ensuring greater representation of different sectors of society in the formulation of public policy in existing institutions, or by creating new institutional arrangements where none exist, such as decision-making councils, co-management committees, round tables or processes of consultation/ prior informed consent. However, co-optation processes become a risk.

In contrast to this affirmative approach, in the context of Indigenous peoples, a transformative approach towards public participation processes should be intercultural. Here the focus is not to open up participation for marginalized sectors in already established institutions, but rather to strengthen and respect customary decision-making procedures and approaches for managing the commons, such as autonomous or customary forms of government. The Indigenous peoples' movement in Latin America has been very active in the last decades, pushing for plurinational and pluricultural national states in some Latin American countries, with successful results in Venezuela, Bolivia and Ecuador.

Other strategies to impact structural power include developing methods to control the means of production through new productive and economic activities that challenge the dominant global economic system and help sustain community life, such as solidarity economies or alternative technologies and forms of energy production.

Strategies to impact on people and networks

Resistance movements also impact on dominant networks in different ways. One is by creating and strengthening their own networks to advance political action and social mobilization strategies that can help them impact on existing laws, political systems and economic frameworks at a national and global scale. Examples of such networks include the CIDOB (the Confederation of Indigenous Peoples of Bolivia) in Bolivia and the COICA (the Confederation of Indigenous Organizations of the Amazon Basin) in South America, which have both taken a lead in the defence of Indigenous rights, the protection of their territories and the fight against extractivism in Latin America in the last three decades. Only last year, COICA launched a new initiative called 'Amazonia for Life' which aims at guaranteeing the protection of 80% of the Amazon Basin by 2025 (https://amazonia80x2025.earth/).

Often, movements and communities create alliances with academics and human rights and environmental justice activists to help them strengthen their social and political organization, local leadership, resistance and dialogue/negotiation tactics or make their struggles visible at a national and global scale. A good example of this is the ICCA Consortium (www.iccaconsortium.org/), an international membership organization dedicated to promoting the appropriate recognition and support of

Indigenous Peoples' and Community Conserved Areas and Territories (ICCAs). Other example is the EJAtlas, which apart from helping to make EJ-specific struggles visible through the online map is being increasingly used by movements and activists to produce featured maps that expose the actors involved in causing environmental conflicts and hold them accountable for the damage and violence (https://ejatlas.org/featured).

At a more local scale, the articulation of transformative knowledge networks is also often crucial for the generation of new knowledge in dealing with uncertainties inherent to socio-environmental conflicts. Environmental conflicts often arise out of social perceptions of risk generated by extractive activities, large-scale development or local natural resource use practices. These can range from the health risks related to mining to the environmental impacts of local subsistence activities such as slash and burn agriculture and savannah burning. In both cases, conflict is often perpetuated by the lack of reliable information to determine accurately the real impacts of certain activities.

Communities often generate knowledge about these risks themselves, for example through community participatory research or environmental monitoring projects that seek to assess the impact of their own livelihood practices or of mining and extraction activities in their territories. In other cases, new knowledge to help solve uncertainties is generated through alliances with sectors of the scientific community (Rodríguez et al. 2013). When the research is carried out jointly, additional to the value of knowledge networks helping reduce and clarify uncertainties regarding environmental change, this strategy has great value in the revitalization of local environmental knowledge. Communities armed with such knowledge can negotiate or discuss the effects of specific projects or activities on their lives with other actors in more equitable conditions (Capasso 2017). Similarly, public bodies can make decisions or modify environmental policies based on 'objective' information.

Strategies to impact on cultural power

The long-term challenge for many social groups whose worldviews are not represented equally in the dominant ways of knowing the world is to influence and impact on the realm of social representations in order to protect and defend their own identity, through the creation of new meanings, norms and values. If over time, a sufficient number of people confirm and reaffirm the new meanings through the creation of counter-narratives or counter-discourses, systemic changes in cultural power can take place.

We refer for example to dominant views of development, to the way nation-state models define citizenship rights, to dominant climate change or environmental change discourses. Many actors and social movements are creating new social meanings when they position themselves against mining or against infrastructure projects based on their own conceptions of the environment, the land and development. In other cases, it is often necessary to begin the process by strengthening local cultural power. This entails raising collective awareness of the problem through processes that can help strengthen local identity. The revitalization of local environmental knowledge and the reconstruction of local history are some of the actions that can help with this. Building visions of the future through community life plans, processes of self-demarcation or local territorial management can also contribute.

Analysing dimensions/spheres of transformation

What changes or what is transformed as a result of the strategies used by EJ movements? How just and sustainable are these transformations?

When redressing an injustice, there is always the potential threat of producing new problematic power relations and recreating new systems and structures of domination and oppression. In the processes of transformation, initiatives that focus on confronting one dimension of injustice can negatively impact other dimensions. For example, initiatives aiming to increase community control over natural resources through community management can lead to the entrenching of unfair gender relations by transferring power over resource use from women to men. Corporations use 'greenwashing' acclaiming how they improve their ecological impact at one scale, while continuing to oppress workers and force developmental visions that erase local cultures.

Thus, we argue that in analysing transformation, a holistic and integrated perspective on transformation and the multiple dimensions across which transformations occur can serve to support actors to undertake more comprehensive transformations and encourage greater reflexivity to impacts and outcomes of the changes being experienced. Such a comprehensive approach could also improve the way external actors (e.g. researchers, practitioners, governments, NGOs) address and approach social transformation processes. It can also bring attention to the paradox that those who are victims of oppression can also become agents of other forms of oppression.

The following five dimensions/spheres of alternative transformation have been developed in the Vikalp Sangam experience. It is proposed that alternatives are built on the following interrelated, interlocking dimensions/spheres, seen as an integrated whole.

1. Ecological integrity and resilience, which includes the conservation of the rest of nature (ecosystems, species, functions, cycles) and its resilience, and respect for ecological limits at various levels, local to global.
2. Social well-being and justice, including lives that are fulfilling and satisfactory physically, socially, culturally and spiritually; where there is equity between communities and individuals in socio-economic and political entitlements, benefits, rights and responsibilities; where there is communal and ethnic harmony.

3. Direct and delegated democracy, where decision-making starts at the smallest unit of human settlement, in which every human has the right, capacity and opportunity to take part, and builds up from this unit to larger levels of governance; and where decision-making is not simply on a 'one person one vote' basis but respectful of the needs and rights of those currently deprivileged, e.g. some minorities.
4. Economic democracy, in which local communities and individuals (including producers and consumers, often combined into one as 'prosumers') have control over the means of production, distribution, exchange and markets; where localization is a key principle, and larger trade and exchange is built on it.
5. Cultural diversity and knowledge democracy, in which pluralism of ways of living, ideas and ideologies are respected, and where the generation, transmission and use of knowledge (traditional/modern, including science and technology) are accessible to all.

These five spheres or dimensions overlap in significant ways. Many or most current initiatives may not fulfil all the above. The direction of the alternative transformation process and to what extent these different spheres/dimensions are taken into consideration can offer a valuable measure of how transformative and radical a certain alternative is. For instance, a producer company that achieves economic democracy but is ecologically unsustainable (and does not care about this), and is inequitable in governance and distribution of benefits (and does not care about this) may not be considered an alternative from a radical perspective. Similarly, a brilliant technology that cuts down power consumption but is affordable only by the ultra-rich would not qualify (though it may still be worth considering if it has potential to be transformed into a technology for the poor also).

It should be noted that these five spheres of transformation are based on, and in turn influence, the set of values that individuals and collectives hold, encompassed within their worldviews. These encompass spiritual and/or ethical positions on one's place in the universe, relations with other humans and the rest of nature, identity, and other aspects. For example, the Vikalp Sangam process in India has identified a set of values and principles as crucial parts of transformation, including self-governance/autonomy (*swashasan/swaraj*); cooperation, collectivity, solidarity and 'commons'; rights with responsibilities; the dignity of labour (*shram*) and livelihoods as ways of life (*jeevanshali*); respect for subsistence and self-reliance (*swavalamban*); simplicity and sufficiency (*aparigraha*); respect for all life forms (*vasudhaiv kutumbakam*); non-violence (*ahimsa*); reciprocity; and pluralism and diversity, just to take a few examples (Kothari 2016).

Analysing scales of transformation

Transformation processes entail complex scalar dynamics which structure political action and outcomes (Kurtz 2003; Staeheli 1994). Key questions when examining scalar considerations include: How do transformations at one scale impact others across scales? How do processes of transformation, the building of alternatives and the stitching together of new forms of governance/production/being diffuse and translate across space? Finally, can we consider something transformative if change is confined to the very local or small scale (even down to the family unit or to individual experience); or must transformation entail an increasing sphere of influence? In this section we refer to three key scalar dynamics at play when examining EJ movement- and community-led transformations processes: spatial scales, temporal scales and human/societal scales.

New repertoires of action and mobilization practices, such as those power strategies discussed above, are often diffused from location to location. The way such transformative experiences move across scales is complex and surprising.

Let us take for example the case of Argentina (discussed in Chapter 8), where after the successful experience of the Esquel community mobilization in 2002, nine provinces approved mining restricting regulations (although two bans were later reverted). The articulation of activist networks across different provinces was key for movements to achieve transformations at different scales (Walter and Wagner 2021).

In a similar vein, the Vikalp Sangam process described above is an initiative that aims to contribute to the sharing and replication of this type of transformative experience and local-scale institutional innovation. The emphasis is on out-scaling alternative initiatives, rather than upscaling them. In the latter, a single initiative attempts to become bigger and bigger, often leading to the replication of bureaucratic, top-down structures that defeat the principles of democracy that the initiative may have started with, whereas in out-scaling different actors and organizations and communities learn from each other, absorb the key principles and processes, and attempt transformations in their own areas and sectors while mindful of local/sectoral particularities. The focus of the multilevel perspective on vertical uptake can overlook this type of horizontal transmission of transformation.

Regarding temporal scales, the dynamic and contingent nature of transformation and the methodological challenges in capturing these non-linear processes must be highlighted. What initially might seem a radical transformative process might be lost in time due to both internal or external drivers, such as state co-optation and/or repression, or inter/intra-community conflicts. On the other hand, a transformative experience can be triggered as a result of a failure or a tragic circumstance.

A scalar perspective can restore agency to grassroots movements, emphasizing how movements of resistance from below act as agents for transformative change, something that has remained poorly understood until the present.

Finally, transformations can occur at the single individual level (as in a shift in worldviews), to the social movements, communities or societal levels and the interrelations between them. We refer to this as the human or societal scale of transformation. The transformation of human behaviour is considered an essential part of transitions and transformations to global sustainability (Gifford 2011;

Swim et al. 2011). O'Brien and Sygna (2013) highlight the relevance of what they define as the personal sphere in transformation processes. The personal sphere considers the individual beliefs, values and worldviews that shape the ways in which the systems and structures (the political institutions) are perceived, and affects what types of solutions are considered 'possible'.

In one way, we may say that the personal level of transformation is what provides the building blocks for group and societal transformation. However, it is common for those sharing personal experiences that transform them to talk about a realization that occurred through collective action. For example, protest is not only a lever in processes of transformation; it also develops transformative capacity, including solidarity, social capital and forms of collective identity and knowledge that are immensely productive and which create indispensable resources and relations towards further transformation (della Porta 2008). This has also been termed the 'political productivity' of conflicts (Merlinsky and Latta 2012).

Protests have cognitive, affective and relational impacts on the individuals and movements that carry them out. Meanwhile, street actions, blockades and occupations create arenas where communities are formed and where social, ecological and democratic experimentation is able to take place.

Application of the framework

The different chapters of this book explore how the application of this framework can:

1. help analyse and recognize the contribution of grassroots EJ movements to societal transformations to sustainability, and
2. support and aid radical transformation processes.

The focus on resistance, conflict and alternatives, and the dimensions and elements of transformation we have outlined (i.e. forms of power, dimensions of change and scales of transformation) can provide a useful framework for situating the agency of EJ activists and how these lead to alternatives. By combining the three elements we may characterize and map these movements and the dimensions, scales and power structures they are focused towards transforming.

For example, bringing our attention to 'people power' – relational and associational power – we can establish how social connections and the building of networks lead to transformative change. This includes 'connecting the dots' between disparate movements to form stronger alliances. It also includes increasing intersectionality and broadening of struggles through the integration of multiple dimensions, through combining ecological concerns with social, economic and cultural ones. For instance, Chapter 11 of this book by Jen Gobby and Leah Temper shows how Indigenous struggles against specific mining, windfarm, and oil and gas projects in Canada move to an understanding of the broader industrial energy system, climate justice and rethinking how energy can be produced and managed at local scales.

Regarding institutional (structural) power, some of the chapters of the book show how institutions for organizing and alternatives for commoning are transmitted across scales vertically and horizontally – this may include consultations/referendums, as well as new strategies for direct action or new local approaches to governing the commons.

Finally, a focus on discursive power elucidates how social movements create narratives and frames that disrupt the status quo, destabilize the system and eventually yield profound social, political and environmental change. This is reflected in our Venezuelan and Bolivian case studies (Chapters 4 and 9), which show how the struggles of the Pemon and Monkoxi Indigenous peoples respectively, against large-scale development and extractivism, contributed to the recent changes to pluricultural and plurinational nation-state models in these two countries.

By linking conflicts and alternatives, we can better understand the interconnections between these various ways of impacting on power and how movements move from defensive to proactive actions. For example, the Korchi case study in India (see Chapter 10) shows how new forms of direct democracy (institutional power) emerge through processes of organizing (relational power). Meanwhile, new and reclaimed cultural values are reaffirmed in contrast to those being opposed through collective action. These reclaimed cultural values and ways of being are alternatives to the dominant development model, even though they are not new.

As we will see in Chapters 3, 4 and 5, this framework also helps us to see the 'dark side of transformation' in action (Blythe et al. 2018), by unpacking and making visible the role of the state and corporate power in constraining transformative power by reinforcing their hegemonic strategies, while movements from below organize to overcome these political relations and cultural obstacles.

As said before, this conceptual framework was largely designed for communities and resistance movements to learn from the transformations they are trying to achieve. A holistic and integrated perspective to evaluate transformation can serve to support actors to undertake more comprehensive transformations and maintain greater reflexivity about the impacts and outcomes of the changes being experienced.

It is important to stress, however, that the framework can be adapted in different ways to facilitate a discussion about just transformations to sustainability. For example, not all case studies in this book used the framework in the same way. Some, using the Conflict Transformation Framework, preferred to focus on the systematization of the struggles by unpacking power, while others also paid attention to what was transformed as a result. Some cases, which were part of the Vikalp Sangam, prefer to focus on understanding alternatives transformations. Other cases with a cross-country focus were able to pay more attention to multi-scalar transformations (Chapters 8 and 11). Also, while some used the framework as a guide for systematization or reconstruction of the EJ struggles or alternatives, others used it in a much more reflective and participatory way.

As we will see in Chapter 2, this framework is still very much a work in progress, which was enriched through the ACKnowl-EJ experience and will no doubt continue to be enriched beyond the project as the paths of its members continue crossing. Hopefully others will also use it and adapt it in ways that they consider useful.

The Structure of the Book

This book is structured in three parts. Part I, 'Our Approaches and Methods for Engaging with Transformations', is developed in two chapters that provide more detail about the approaches and the methods we developed in ACKnowl-EJ to engage with transformations and to co-produce knowledge about environmental justice.

Chapter 1 defines what we understand as co-production for environmental justice, and introduces the reader to a series of participatory methods used and invented during the project to co-produce knowledge with communities and among the project members about environmental justice and transformation to sustainability. Chapter 2 presents the two frameworks (the Conflict Transformation Framework and the Alternatives Transformation Format) produced by two of our project institutional members, which were used in the case studies to assess the concrete contribution that resistance movements are making towards just transformations to sustainability. The chapter brings together both frameworks, with a focus on their development history and how they were used in the project.

Part II of the book, 'Analysing Transformations from and with Environmental Justice Movements', presents a series of case studies analysed to understand how resistance movements across the world are trying to contribute to just transformation to sustainability. We divide the chapters into three blocks that explore key outstanding aspects of the different case studies for understanding the dynamics of transformations from the ground up:

- the tension between hegemonic and counter-hegemonic power dynamics,
- different scales on which transformations take place, and
- the role that new forms of democracy, culture and alternatives to development play in winning struggles in the long term.

Section 1 of Part II, entitled 'Double Movements Against State and Market', presents cases from Turkey (transformations and resistance to the fossil fuel rush in Yeni Foça and to the neoliberalization of knowledge production), Venezuela (struggles against extractivism in Canaima National Park) and Lebanon (the 2015 trash crisis) to examine the structure and agency dynamics (double movement) within environmental justice struggles, using power analysis to illustrate the tension between hegemonic and counter-hegemonic power dynamics in transformation struggles. The focus is on the role of the state and corporate power in constraining transformative power and how movements from below organize to overcome these political obstacles. The analytical attention in these chapters is on: 1) changes in structural/visible power in terms of institutions and legal frameworks; 2) changes in people/networks – new companies, new municipalities, new alliances; and 3) changes in values/beliefs/worldviews.

Section 2, 'From Individual to Institutional Transformations', presents cases from Belgium (resistance strategies against the Brussels mega-prison project), India (women pastoralists and the struggle for transformation in Raika) and Argentina (power transformations led by mining struggles) that address transformations that have taken place at different individual, collective and institutional scales. We look at how transformations scale up, scale out and scale deep, and at the internal conflicts that movements experience during their struggles. The Belgian case study looks inside the movement dynamics and the personal level to understand the role that capabilities/internal relations and interactions in the movement have in bringing about transformations. The Argentinian case study centres on the interaction between movement and state institutions in bringing about transformative change. The Indian case study looks at the role of gender dynamics in pastoral communities struggling to bring about such change.

Section 3, 'Enacting Counter-hegemonic Alternative Politics, Economics and Worldviews', focuses on counter-hegemonic alternatives that movements have carried out to bring about just transformation. These range from alternative forms of democracy and economies to cultural transformations in knowledge systems, discourses and worldviews. Case studies from India (from two regions: Korchi and Kachchh), Bolivia (the Indigenous Territory of Lomerío) and Canada (at a national scale) explore how communities are resisting extractive activities like mining and forestry by defining their own governance space, forms of territorial management and productive activities. Furthermore, this section upends traditional understandings of innovation to argue that reclamation and recreation of traditions and ancestral ways of being and new social institutions provide a more convincing path to transformative change than high-tech solutions. At the same time, it shows that transformation can be internally incoherent, or contradictory, even if there are areas of transformation. It will help explore how the line between types of transformations is blurred, and difficult to separate.

Part III of the book, 'Lessons from Ground-up Transformations', presents two final chapters in which we bring together the learnings from the in-depth case studies, to develop a theory of transformation from the ground up and to draw some final conclusions.

Note

1 Consociational democracies are those in which sectarian groups are allotted 'representations' in public and political offices based on population size, where larger groups are allotted greater representations.

References

Bachrach, P., Baratz, M.S. (1962) Two Faces of Power. *American Political Science Review*, 56(4): 947–52.

Blythe, J., Silver, J., Louisa, J.E., Armitage, D., Bennett, N.J., Moore, M.L., Morrison, T.H., Brown, K. (2018) The Dark Side of Transformation: Latent Risks in Contemporary Sustainability Discourse. *Antipode*, 50(5): 1206–23.

Botes, J. (2003) Conflict Transformation: A Debate over Semantics or a Crucial Shift in the Theory and Practice of Peace and Conflict Studies? *International Journal of Peace Studies*, 8(2): 1–27.

Capasso, C. (2017) Community-based Monitoring to End Oil Contamination in the Peruvian Amazon. *Perspectives*, 26. Coordinated by the UN Environment's Civil Society Unit.

Daga, S. (2014) All the Way to Timbuktu. *Anveshan*, 28 October, https://kalpavriksh.org/wp-content/uploads/2018/12/AllthewaytoTimbuktuProjectAnveshan.pdf.

della Porta, D.D. (2008) Eventful Protest, Global Conflicts. *Distinktion: Journal of Social Theory*, 9(2): 27–56.

Dukes, E.F. (1996) *Resolving Public Conflict: Transforming Community and Governance*. Manchester: Manchester University Press.

Fisher, R., Ury, W. (1981) *Getting to Yes*. Boston: Houghton Mifflin.

Foucault, M. (1971) The Order of Discourse. In Young, R. (ed.) *Untying the Text: A Poststructuralist Reader*. London: Routledge and Kegan Paul.

Fraser, N. (1995) Recognition or Redistribution? A Critical Reading of Iris Young's Justice and the Politics of Difference. *Journal of Political Philosophy*, 3(2): 166–80.

Gadgil, M., Guha, R. (1993) *This Fissured Land: An Ecological History of India*. Berkeley: University of California Press.

Galtung, J. (1990) Cultural Violence. *Journal of Peace Research*, 27(3): 291–305.

Gaventa, J. (1980) *Power and Powerlessness: Quiescence and Rebellion in an Appalachian Valley*. Chicago: University of Illinois Press.

Gibson-Graham, J.K. (2006) *A Postcapitalist Politics*. Minneapolis: University of Minnesota Press.

—— (2008) Diverse Economies: Performative Practices for 'Other Worlds'. *Progress in Human Geography*, 32(5): 613–32.

Giddens, A. (1984) *The Constitution of Society: An Outline of the Theory of Structuration*. Cambridge, MA: Polity Press.

Gifford, R. (2011) The Dragons of Inaction: Psychological Barriers That Limit Climate Change Mitigation and Adaptation. *American Psychologist*, 66(4): 290–302.

Gillard, R., Gouldson, A., Paavola, J., Van Alstine, J. (2016) Transformational Responses to Climate Change: Beyond a Systems Perspective of Social Change in Mitigation and Adaptation. *Wiley Interdisciplinary Reviews: Climate Change*, 7(2): 251–65.

Harvey, D. (2011) *The Enigma of Capital: And the Crises of Capitalism*. New York: Oxford University Press.

Kothari, A. (2016) The Search for Radical Alternatives: Key Elements and Principles. *Counter Currents*, 3 November. www.countercurrents.org/2016/11/03/the-search-for-radical-alternatives-key-elements-and-principles/.

Kurtz, H.E. (2003) Scale Frames and Counter-scale Frames: Constructing the Problem of Environmental Injustice. *Political Geography*, 22(8): 887–916. http://doi.org/10.1016/j.polgeo.2003.09.001.

Lederach, J.P. (1995) *Preparing for Peace: Conflict Transformation Across Cultures*. Syracuse University Press.

Long, N., Van Der Ploeg, J.D. (1989) Demythologizing Planned Intervention: An Actor Perspective. *Sociologia Ruralis*, 29(3/4): 226–49.

Lukes, S. (1974) *Power: A Radical View*. London and New York: Macmillan.

Martinez-Alier, J., Temper, L., Del Bene, D., Scheidel, A. (2016) Is There a Global Environmental Justice Movement? *Journal of Peasant Studies*, 43(3): 731–55. http://doi.org/10.1080/03066150.2016.1141198.

Merlinsky, M.G., Latta, A. (2012) Environmental Collective Action, Justice and Institutional Change in Argentina. In Latta, A., Wittman, H. (eds) *Environment and Citizenship in Latin America: Natures, Subjects and Struggles*. New York: Berghahn.

Moser, S.C. (2016) Can Science on Transformation Transform Science? Lessons from Co-design. *Current Opinion in Environmental Sustainability*, 20: 106–15.

O'Brien, K., Sygna, L. (2013) Responding to Climate Change: The Three Spheres of Transformation. *Proceedings of Transformation in a Changing Climate*, 19–21 June, Oslo, Norway. University of Oslo (pp. 16–23).

Pelling, M. (2011) *Adaptation to Climate Change: From Resilience to Transformation*. London: Routledge.

—— (2012) Resilience and Transformation. In Pelling, M., Manuel-Navarrete, D., Redclift, M. (eds) *Climate Change and the Crisis of Capitalism: A Chance to Reclaim Self, Society and Nature*. Oxford: Routledge.

Pelling, M., O'Brien, K., Matyas, D. (2015) Adaptation and Transformation. *Climatic Change*, 133(1): 113–27.

Robbins, P. (2004) The Hatchet and the Seed. In *Political Ecology: A Critical Introduction*. Malden, MA: Blackwell.

Rodríguez, I. (2020) Latin American Decolonial Environmental Justice. In Brendan, C. (ed.) *Environmental Justice: Key Issues*. Earthscan from Routledge.

Rodríguez, I., Inturias, M. (2018) Conflict Transformation in Indigenous Peoples' Territories: Doing Environmental Justice with a 'Decolonial Turn'. *Development Studies Research*, 5(1): 90–105. http://doi.org/10.1080/21665095.2018.1486220.

Rodríguez, I., Sletto, B., Bilbao, B., Sánchez-Rose, I., Leal, A. (2013) Speaking of Fire: Reflexive Governance in Landscapes of Social Change and Shifting Local Identities. *Journal of Environmental Policy Making and Planning*. http://doi.org/10.1080/1523908X.2013.766579.

Rodríguez, I., Inturias, M., Frank, V., Robledo, J., Sarti, C., Borel, R. (2019) *Conflictividad socioambiental en Latinoamérica: Aportes de la transformación de conflictos socioambientales a la transformación ecológica*. Ciudad de México: Friedrich-Ebert-Stiftung.

Scoones, I. (2016) The Politics of Sustainability and Development. *Annual Review of Environment and Resources*, 41: 293–319.

Scoones, I., Edelman, M., Borras Jr, S.M., Hall, R., Wolford, W., White, B. (2017) Emancipatory Rural Politics: Confronting Authoritarian Populism. *Journal of Peasant Studies*. https://doi.org/10.1080/03066150.2017.1339693.

Staeheli, L. (1994) Empowering Political Struggle: Spaces and Scales of Resistance. *Political Geography*, 13(5): 387–91.

Swim, J.K., Stern, P.C., Doherty, T.J., Clayton, S., Reser, J.P., Weber, E.U., Gifford, R., Howard, G.S. (2011) Psychology's Contributions to Understanding and Addressing Global Climate Change. *American Psychologist*, 66(4): 241–50.

Temper, L., Del Bene, D. (2016) Transforming Knowledge Creation for Environmental and Epistemic Justice. *Current Opinion in Environmental Sustainability*, 20: 41–9.

Temper, L., Del Bene, D., Martinez-Alier, J. (2015) Mapping the Frontiers and Front Lines of Global EJ: The EJAtlas. *Journal of Political Ecology*, 22: 255–78.

Temper, L., Walter, M., Rodriguez, I., Kothari, A., Turhan, E. (2018). A Perspective on Radical Transformations to Sustainability: Resistances, Movements and Alternatives. *Sustainability Science*, 13(3): 747–64.

Thekaekara, M.M. (2015) What an Idea! *Hindu Sunday Magazine*, 28 March. www.vikalpsangam.org/article/what-an-idea/.

Ury, W.L., Brett, J.M., Goldberg, S.B. (1988) *Getting Disputes Resolved: Designing Systems to Cut the Costs of Conflict*. San Francisco: Jossey-Bass.

Visvanathan, S. (1997) *A Carnival for Science: Essays on Science, Technology and Development*. London: Oxford University Press.

Walsh, C. (2005a) *La interculturalidad en la educación*. Lima: Ministerio de Educación DINEBI.

—— (2005b) Interculturalidad, conocimientos y decolonialidad. *Signo y pensamiento. Perspectivas y Convergencia*, 46(24): 31–50.

Walter, M., Urkidi, L. (2017) Community Mining Consultations in Latin America (2002–2012): The Contested Emergence of a Hybrid Institution for Participation. *Geoforum*. https://doi.org/10.1016/j.geoforum.2015.09.007.

Walter, M., Wagner, L. (2021) Mining Struggles in Argentina: The Keys of a Successful Story of Mobilisation. *The Extractive Industries and Society*, 8(4). https://doi.org/10.1016/j.exis.2021.100940.

Walter, M., Latorre Tomás, S., Munda, G., Larrea, C. (2016) A Social Multi-criteria Evaluation Approach to Assess Extractive and Non-extractive Scenarios in Ecuador: Intag Case Study. *Land Use Policy*, 57: 444–58.

Weber, M. (1971) The Three Types of Legitimate Rule. In Etzioni, A. (ed.) *Complex Organizations: A Sociological Reader*. New York: Holt.

PART I

Our Approaches and Methods for Engaging with Transformations

This first part of the book is dedicated to explaining the approaches and the methods we developed in ACKnowl-EJ to engage with transformations and to co-produce knowledge about environmental justice.

Activists and academics meeting at the beginning of the project in Beirut, Lebanon, 2017 (Photo by Iokiñe Rodríguez)

1
Co-production of Knowledge for Environmental Justice: Key Lessons, Challenges and Approaches in the ACKnowl-EJ Project

Lena Weber, Mariana Walter, Leah Temper and Iokiñe Rodríguez

Introduction

> It motivates me to feel that academia can be useful, relevant for communities that are living through situations of injustice. It motivates me to help build justice in the world ... I do not like to do investigations that don't generate change. I am motivated by the idea of transformation ... I wouldn't feel at all comfortable only doing investigations to edify and systematize. I do not think it is ethically correct. (ACKnowl-EJ researcher)

Before we get into our case studies and the research itself, this first chapter describes the methodology underpinning the project; how the activist-researchers in ACKnowl-EJ approached their work and their reflections on that. Academic research has a complicated history. It has often been used as a tool of domination and control, in the service of states, militaries and corporations, not for movements for positive social change (Peake and Kobayashi 2002; Strega and Brown 2005). Also, mainstream academic research often ignores or discredits the knowledge held by communities and movements. This only weakens academic research and makes it less able to adequately address the pressing issues we currently face, like widespread pollution and climate change (Carpenter and Mojab 2017; Saltelli and Funtowicz 2017). This project wanted to do research differently, in collaboration with and in service of movements for positive change. We also wanted to collect and systematize our reflections on these processes, hoping that this will be useful for other like-minded activist-researchers.

Academic-Activist Co-produced Knowledge for Environmental Justice – the ACKnowl-EJ project's full name – describes who we are (a mix of activists, academics, activist-y academics and academically inclined activists), what we fight to transform in our worlds (environmental injustice), and how we aimed to go about doing so in this research project (via an approach called co-production).

Following what some have termed the 'participatory turn', we count ourselves among those researchers working with and being part of social movements that aim to disrupt and upend some of the traditional power relations often embedded in the research and knowledge production paradigms.

As a network of scholars and activists with ties to academic institutions, non-governmental organizations, communities and social movements, ACKnowl-EJ aimed to engage in action and collaborative analysis of the transformative potential of community responses to environmental and social injustices, particularly understood through the lens of extractivism, and alternatives born from resistance. The project was an experiment in co-producing knowledge that could answer the needs of social groups, advocates, citizens and social movements, while empowering communities to push for change. Part of this is also recognizing that our research is inherently biased and has explicit goals. Teams within ACKnowl-EJ were made up of co-researchers from communities and movements around the world, and drew on research conducted in India, Turkey, Bolivia, Canada, Belgium, Lebanon, Venezuela and Argentina. The network was co-coordinated by Kalpavriksh, an environmental action organization based primarily out of Pune, India, and the Institute of Environmental Science and Technology (ICTA) at the Autonomous University of Barcelona.

Research for transformation is akin to science fiction, or magical realism; it is a way to collectively dream about and build realities different from those in which we currently reside. With the complex challenges we face, more typical ways of doing research just aren't up to the task of both understanding *and* transforming the world to a more socially and environmentally just place (Lotz-Sisitka et al. 2016; Saltelli and Functowitz 2017; Temper et al. 2019). Transformations researchers aim to build radically different futures than those that fit into a 'business as usual' mindset – ones actually worthy of our longing (Bell and Pahl 2018; Blythe et al. 2018; Future Earth 2014; McGarry et al. 2021).

A wide variety of terms are used for research that challenges and subverts mainstream academia, including activist scholarship, militant research, participatory action (Chambers 1983; Fals-Borda 1986) and decolonized research (Smith 1999) – and more recently, co-produced knowledge (Temper et al. 2019). The term 'co-production' tends to be associated with science and technology studies (Galopin and Vessuri 2006), a fairly young field that examines knowledge societies and the connections between knowledge, culture and power (Jasanoff et al. 1995). Jasanoff (2004: 2) defines co-production as 'shorthand for the proposition that the ways in which we know and represent the world (both nature and society) are inseparable from the ways in which we choose to live in it.' The concept of co-production helps us highlight power, knowledge and expertise in 'shaping, sustaining, subverting or transforming relations of authority' (ibid.: 4).

Environmental justice is a body of theory and movement born out of Black and Brown communities in the United States experiencing and resisting disproportionate pollution and contamination of their environment compared to their white counterparts (Bullard 1993). Since the 1980s, the term has both spread around the globe and has been pushed (and pulled, sometimes against its will) into academia (Martinez-Alier et al. 2014), coming into contact with countless kindred, though unique, movements born from other communities also resisting disproportionate pollution and contamination, creating a cross-pollination effect.

While the term co-production is widely deployed in transdisciplinary research, the meaning, process and values behind co-production are rarely questioned or explicitly outlined. Moreover, in the ACKnowl-EJ project we aimed to engage in transformative co-production of knowledge for environmental justice (CKEJ), a type of co-production that has yet to be defined and explored. Therefore, we offer this chapter as a reflection and analysis of our research processes, centred on a few key questions:

1. How do we define CKEJ – what are its key characteristics and when and how do we believe co-production processes hold the most transformative potential?
2. What are CKEJ's key values?
3. What obstacles and challenges do we face in realizing our ideal versions of co-production processes and how can we address these?

To do this analysis, ACKnowl-EJ researchers engaged in ongoing reflexive activities to document our research practice, the concept of co-production, our ethics and processes, and our relationships with ourselves, others and knowledge production over the course of the three-and-a-half-year project. We offer the following reflections as an act of transparency and honesty for ourselves and those we have worked with throughout this project, as well as for others engaging in similar processes: 1) a definition of co-production of knowledge for environmental justice (CKEJ), including key aspects and transformative potential; 2) an outline of this type of co-production's key values; and 3) identification of some of CKEJ's key challenges and ways to navigate these. Before turning to these results, the next sections review the literature on co-production and environmental justice as well as our methodology.

Co-production and Environmental Justice

Co-production as a term grew in part from participatory urban and regional planning and public service provision (Ostrom 1990, in Bell and Pahl 2018), often traced to Elinor Ostrom's work from the 1970s (Palmer and Hemström 2020).

It emerged as part of an increased interest in participatory research called the 'participatory turn', greatly motivated by the idea of 'wicked problems': problems that are complex and unsolvable if only using conventional science and planning, like climate change, terrorism and loss of biodiversity (Rittel and Webber 1973, in Palmer and Hemström 2020). Wicked problems emerge at the intersection of how they are framed, the creation of a goal, and an ambition to move towards equity. Sustainability researchers, grappling with how to tackle wicked problems, argued for the value of more collaborative approaches to knowledge production within their own field.

Participatory research was seen as holding potential to address three connected necessities surrounding these complex issues (Felt et al. 2015, in Palmer and Hemström 2020): the need for knowledge democratization, the need for an epistemology enriched by multiple knowledges in order to better address complex current issues, and the need to legitimize academic knowledge production by making science and its institutions more accountable and relevant.

In the 1990s, 'transdisciplinarity' also became popularized as an approach for tackling issues found in post-normal science. Advocates of transdisciplinarity argue for the need to recognize multiple legitimate perspectives. This is a problem-oriented, real-world-based approach that involves non-academic actors and emphasizes practicality (Palmer and Hemström 2020). The involvement of those with first-hand knowledge and experience of the issue at hand is key, something which Corburn (2003) calls the legitimacy of local knowledge. Corburn argues that we should not over-romanticize local knowledge, but instead shift from the idea that science 'speaks truth' to society, to a more 'democratic' perspective that we 'make sense' of things together (Sclove 1995, in Corburn 2003).

One powerful way of 'making sense' together happens in social movements (Cox 2014), with a key characteristic being the tension between what Gramsci calls 'common sense' (stemming from hegemonic relationships) and 'good sense' (stemming from popular practice) (Gramsci 1991, in Cox 2014). Particularly rich learning happens when movements come into dialogue with one another, creating a 'talking between worlds' effect (Conway 2006, in Cox 2014).

Environmental justice movements are a prime example of how diverse movements converge and come into dialogue with one another. In contrast to mainstream environmentalism, environmental justice movements avoid uniformity (Schlosberg 2004) by tackling multiple root causes of the social-ecological crisis (Di Chiro 2008; Schlosberg 2004), building on pre-existing and intersectional anti-racist, feminist and economic justice movements, among others.

This is because environmental justice as a concept initially grew from Black and Latino communities in the United States struggling against multiple and intersecting injustices at the same time. Later, environmental justice also grew in popularity as an analytical framework to understand how diverse factors like race, age, gender and class affect how communities are impacted by environmental injustices. Activists involved in environmental justice movements – now much more globalized and in dialogue with other similar movements – have theorized and coined many of the key concepts used by academic researchers today, including climate justice, ecological debt, leaving oil in the soil, biopiracy, etc. (Martinez-Alier et al. 2014; Temper et al. 2015).

The richness of this type of meaning-making shows how collaborations between activists and academics, through approaches like co-production, can create fertile ground for understanding our realities and building more just futures. However, while there is a rigorous body of research on the challenges of transdisciplinary co-production (Lang et al. 2012; Pohl and Hadorn 2008; Pohl et al. 2017), there is no general agreement on terms/definitions or standards for researchers and funders, and processes and methods have to be tailored to specific problems and contexts (Palmer and Hemström 2020). Further, there is a lack of attention to how co-production specifically oriented towards increased environmental justice works in practice, and how researchers engaged in these processes navigate their own ethics and the challenges and opportunities presented.

Collectively Reflecting On and Documenting Co-production

Within the ACKnowl-EJ project, we engaged in different approaches to co-production. These included work on the EJAtlas, an online map of environmental conflict and justice movements designed by, contributed to and analysed by activist and academic communities from around the world; co-production processes in which ACKnowl-EJ team members formed part of the social movement/community engaging in the research; processes in which ACKnowl-EJ team members collaborated with communities/social movements of which they were not directly a member; and the co-production of knowledge that occurred between ACKnowl-EJ team members in our structured and informal spaces of analysis and reflexivity. This variety of co-production processes relied on using already existing participatory methods, adapting them, or creating new ones for the specific purpose of the research.

Examples of Innovative CKEJ Methods Used by ACKnowl-EJ Researchers

ACKnowl-EJ researchers used numerous methods in their co-production processes. Here we highlight several innovative approaches used as part of broader methodologies. While we present these as examples of methods for reflexivity, analysis, strategizing and evaluation, they are all plural in their potential uses and possible ways of implementation.

Method for reflexivity: the Tarot Activity

The Tarot Activity, developed jointly by Dylan McGarry from the T-Learning project and Lena Weber and Leah Temper from ACKnowl-EJ, aims to operationalize radical reflexivity by guiding researchers through a variety of existing and emergent alternative and critical approaches to research before asking them to use collage materials to reflect on their own identity, positionality and ethics as a researcher and represent this in an artistic manner. The exercise creates space for reflexive exploration into each researcher/activist's unique and plural expressions of their roles and actions, both ideal and actual, which were generatively surfaced.

This can play a key role in CKEJ research design by coaxing out important reflections and allowing for more intentional shaping of the type of role(s) the researcher(s) will engage with, as well as reminding them which ethics and power dynamics they will need and want to attend to.

Readers can find more information on the Tarot Activity and how to replicate it or adapt it in Temper et al. (2019).

Method for empowering: power analysis as part of the Conflict Transformation Framework developed by Grupo Confluencias

Grupo Confluencias is a network of professionals from Latin America working together since 2005 in order to jointly investigate and develop capacities around understandings of power and culture in environmental conflicts and their transformation. The Conflict Transformation Framework (mentioned in the Introduction) aims to strengthen a community's capacities to transform the environmental conflicts they are affected by through strategically targeting three types of hegemonic power (structural power, cultural power and actor-networks). The resulting analysis and actions can help pull up the roots of environmental injustices, by challenging harmful dominant legal, political and economic structures and discourses.

In particular, this lens for understanding power can help CKEJ processes differentiate among types of hegemonic power while also identifying power of agency and strategies to impact upon and transform hegemonic power. Using this method for power analysis is very complementary to other methods used in CKEJ processes aiming to be transformative: in order to change something, and to know how to change it, we first have to understand what we're dealing with and what strengths we have at hand.

Participatory analysis was used in the Lomerío case study in Bolivia to systematize all the strategies of transformative power the Monkoxi Indigenous peoples have used over the last three decades to gain territorial control and property over their lands. The method was adapted from Hunjan and Pettit (2012).

Method for self-evaluation: the Alternatives Transformation Format

The Alternatives Transformation Format[1] is a tool developed over many years through Kalpavriksh's Vikalp Sangam process to help initiatives and organizations to self-assess how holistic and integrated (or conversely, inconsistent and fragmented) their actions and transformations are, and where they may want to make changes. Five overlapping 'petals' guide us through possible indicators of ecological wisdom, integrity and resilience; social well-being and justice; direct and delegated democracy; economic democracy; and cultural diversity and knowledge democracy.

This format, not for use by external actors or for extractive purposes, can also be applied in diverse ways and moments by groups wanting to check in on and/or increase the transformative potential of their project/movement. It is also a useful tool for research communities/teams to reflect on the strengths, weaknesses and possible blind spots of CKEJ projects; activities using the format were held at the final ACKnowl-EJ project meeting and with the Living Aula's alternative research school group for self-analysis and reflection on our own CKEJ processes.

More information on the format can be found here:

https://vikalpsangam.org/wp-content/uploads/migrate/Resources/alternatives_transformation_format_revised_20.2.2017.pdf

Method for strategizing: back-casting and scenario-building

Scenario-building was used in diverse ways over several case studies. In Turkey, the research team built a historical analysis with participatory scenario-planning in order to co-create outputs of greater strategic relevance for members of the Yeni Foça Forum,[2] a specific demand from Forum members involved in the project. One useful output was to make visible historical struggles in the region, which then fed into scenario-planning as a way to dream of alternative futures while recognizing structural limitations. The team used the back-casting technique in order to see how these alternative futures can come into being, instead of just what *could* happen. Back-casting involved looking at a desired outcome and then looking 'backwards' to the steps that need to take place to reach that outcome. This focused the process towards idealized outcomes. Researchers described how this 'two-pronged approach' of both looking backwards and forwards allowed them to explore the various and often-disputed ways of thinking about time in transformation research.

Scenario-building has exciting potential as a CKEJ method due to the innumerable context- and community-specific ways it can be designed and implemented, and the role it can play in movement planning and strategizing to reach transformative goals. Researchers from this team relate it to the type of action-research called for by second-order transformation research, which emphasizes the connection between the researcher and research 'object' while creating space for imagination and experimentation (Fazey et al. 2018). This method is also a good way to facilitate reflexivity, by holding space for collective thinking – even amidst urgent issues and needs.

Method for future visioning: Three Horizons – the patterning of hope

Similarly to back-casting and scenario-building, Three Horizons is a practical framework for thinking about the future – and an exploration of 'future consciousness' and how to develop it, proposed by Sharpe (2013). It is a simple and intuitive framework for thinking about the future. It helps groups explore systemic patterns to identify which of the dominant patterns are no longer fit for purpose, how emerging trends can shape the future, and what visionary action is needed to collectively move us towards a viable future. The future is explored through three lenses: Horizon 1: Continue Business as Usual, Horizon 2: Innovation towards the Vision, and Horizon 3: Vision of a Viable Future.

This method was used in the Lomerío Indigenous Territory of Bolivia to help the Monkoxi Indigenous people think strategically about their desired future. In a two-day workshop with community leaders and representatives from allied organizations, the Monkoxi explored the following guiding question: What transformations are necessary in the way we are managing our territory to consolidate our own government and a good use of our 'big house'? This overarching question was explored in greater detail through the Three Horizons.

Horizon 1 • What difficulties do we have in managing our territory to consolidate our own government and the proper use of our big house? • What is the novelty that we have done that we must maintain?

Horizon 2 • What new things do we need to do to get closer to this vision? • What do we have to stop doing?

Horizon 3 • What is your ideal vision of the future of management of the Lomerío territory in Bolivia to consolidate its own government and a good use of the big house? • What are the seeds that we have sown along the way to achieve this?

This method is a good way to facilitate reflexivity about the positive aspects and achievements in EJ struggles but also the challenges ahead to consolidate a desired vision of the future.

In co-production, reflexivity helps us stay on track for connecting the research process with the real-world problems we are trying to solve (Palmer and Hemström 2020). The methods for ongoing reflexivity applied to knowledge production in the ACKnowl-EJ project are referred to as our 'process documentation'. We define process documentation as *an active reflection on the knowledge practices involved in environmental justice struggles and research and our intention to transcend disciplinary silos and overcome the activist-academic dichotomy, while creating emancipatory theory and a new (critical) research praxis based on an 'ecology of knowledges'* (de Sousa Santos 2014). Our process documentation aimed to create spaces for documenting and analysing the research we did in ACKnowl-EJ, across different levels of the project, from how we worked as a project and framed our research questions, to how we engaged with our case study communities. In particular, we aimed to enquire into the praxis of co-production of academic/activist knowledge for environmental justice.

Through process documentation and radical reflexivity, we aimed to:

- understand the thinking that went into the choices we made, individually and collectively (e.g. which case studies to select),
- be more explicit about the biases we were starting with and introducing into our analyses,
- examine and learn from the hurdles faced and opportunities used,
- reflect on the tools being used and how we came to them, and
- document and reflect on changes made throughout the stages of the case studies in research focus, methodologies, activities implemented, etc.

After jointly elaborating key concepts and a methodology for process documentation, ACKnowl-EJ researchers engaged in a series of reflexive activities and processes throughout the span of the three-year project as a way to document – among other things – our decision-making, ethics, beliefs and positionality surrounding our work. Key moments of reflection and documentation took place in face-to-face project meetings in Lebanon, India and Turkey, as well as from a distance through written prompts and interviews led by the authors (see Figures 1.1, 1.2 and 1.3).

Figure 1.1 ACKnowl-EJ Lebanon meeting, 2017

Figure 1.2 ACKnowl-EJ India meeting, 2018

Figure 1.3 ACKnowl-EJ Turkey meeting, 2019

Results

Our process documentation resulted in a collective definition of CKEJ, the identification of key values, and a deeper reflection on key challenges in CKEJ processes and potential ways to navigate these.

What is co-production of knowledge for environmental justice?

We understand *co-production of knowledge for environmental justice* (CKEJ) as an iterative process of back-and-forth questioning (dialectical and dialogical) with the potential to simultaneously produce facts, values, ideas and plans. It is visionary: it explicitly helps work towards a desired future that is more environmentally just, helping to support alternative-building[3] while aiming to act as an alternative in and of itself.

This type of co-production implies the merging of two or more knowledge systems. It goes beyond learning and sharing to focus on creating. It is more than the sum of its parts. While knowledge is collectively produced all the time, we use this term to signify intentionality from all sides involved in the process. It challenges the binary between researcher and researched, and is dependent on conditions (time, space, etc.) and context (culture, socio-political dynamics, etc.). This is a type of research that takes place with *groups* of people, not just between two individuals from different worldviews, for example.

The type of co-production we aspire towards is one in which relationships are centred throughout the whole process in a way that works to diminish or eliminate harmful, oppressive and exploitative hierarchies common in research processes; different knowledge perspectives are brought into dialogue; and special attention is paid to power dynamics. It aims to be non-extractive. Those most affected by the environmental/social injustice(s) at hand and those most affected by structures of marginalization should be centred throughout with specific efforts made to elevate their perspectives, and strengthen their capacities, knowledges and visions.

Co-production can happen on different scales and in multifaceted ways. Indeed, we can say all knowledge is collectively produced in one way or another across space and time. Therefore, when we speak of 'ideal co-production of knowledge for environmental justice' we are only speaking about *one* type of co-production that we particularly strive for.

This type of co-production works better with long-term, trusting and mutually respectful relationships, and seeks to generate both externally and internally reflexive spaces. It aims to support/facilitate the creation/diffusion of narratives that challenge the dominant worldview in all regards, including narratives of hegemonic development. It also seeks to produce diverse outcomes relevant for multiple audiences, prioritizing usefulness for the community engaged in the process.[4] It respects the community's own rules for negotiating and defining a research agenda.

We believe co-production has more transformative potential – both regarding the situation at hand and our research institutions – when it is immersive, strategic and socially embedded, opens up opportunities for sharing and deepening of connections between groups, and is designed specifically at aiming to challenge diverse/historical harmful norms common in research processes but rooted in broader historical systems of exclusion, discrimination and exploitation.

This style of deeply engaged co-production allows for a peer review process (involving all the actors involved, not just academic peers) to take place *throughout* the knowledge production process – not just regarding final results – which can make the overall process more rigorous and impactful. These peer review processes and tools should be easily accessible, learnable and usable to all the actors involved in the co-production process.

A key aspect of co-production with transformative potential is the collective formulation of the research motivation and question with those most affected by the issue at hand. This ensures the research is in response to a vocalized need regarding transformation, that those closest to the issue are interested in and motivated by the research process, and that the process and results can be useful in working towards the desired transformation.

The process should actively seek to facilitate spaces for mutual learning, validation of discriminated/differential viewpoints and knowledges, development of new research skills for those closest to the issues at hand, sharing of knowledge from other spaces (e.g. organizing tactics that have worked in other struggles with which researchers might be familiar), and broadening/strengthening communities' networks and alliances. It should allow space for evolving ideologies.

Key Takeaway
Research design: This definition of co-production can be used to help researchers and movements/communities design a CKEJ research project. Taken point for point, it can guide those involved through a series of considerations and things to prioritize while trying to foster a process with higher transformative potential. Space must be allowed for adaptation and addition of new, context-specific considerations.

What are the key values of CKEJ?

While some values are mentioned in our overall definition, at our final project meeting we discussed and agreed upon a more explicit list of values that we consider key to this kind of co-production.

These are respectfulness, authenticity, transparency, strategic inclusivity,[5] active reflexivity and mindfulness, resistance to hierarchy when hierarchy manifests in a harmful way,[6] humility, solidarity, and sensitivity towards non-dominant timings, knowledges and needs.

Consent (prior, informed, explicit) is centred, people's well-being and transformative goals are prioritized over institutional requirements, discriminated perspectives are visibilized/strengthened, and a balance is sought between commoning of knowledges and knowledge stewardship.[7]

Solidarity is key. CKEJ should be non-competitive, trust-building, transgressive (boundary-pushing), socially embedded and contextually specific, responsible and reciprocal, diversely justice-oriented, anti-extractive (of knowledge and resources), empowering of personal and collective agency, visionary, empathic, anti-colonial/anti-imperial, anti-assimilation/co-optation, and evidence-based.

Importantly, a final key value is plurality. That is, while we have identified these common values, they may look different depending on the research context and there is not one *correct* way to approach this type of co-production; rather, there is a plurality of variations and possibilities. Plurality is valued in many forms – of knowledges, of cultures, of methods and beyond.

Key Takeaway

Reflexivity: Researchers interested in engaging in CKEJ processes can use this list of values as a discussion starter for themselves, their teams and the movements/communities involved at the start of, during and after a research process. We suggest that they discuss the values they prioritize and how to embody them. For example, what does consent mean in this particular local context, and what is the most appropriate way for it to manifest? What are the needs around the balance between knowledge commoning and knowledge stewardship? How can we be visionary, and how can we conduct our research in an anti-colonial/anti-imperial way in this context?

CKEJ research and political rigour: challenges and opportunities

While a definition of CKEJ and a description of its key values are useful, the reality of CKEJ is much muddier. We identified key challenges of CKEJ throughout our process documentation. In particular, using the relational aspects of CKEJ as a lens, we examined three kinds of relationships: with ourselves, with others and with knowledge. Due to the messy, overlapping nature of these relationships, one useful focus allowing us to understand the interplay between these relationships rests in the tension and complementarities found in our mixed activist-academic identities, and our pursuit of what Borras (2016) calls 'political rigour'.

Borras (2016) argues that political rigour 'means being politically informed and thorough, sensitive and nuanced, and timely and relevant. It means taking a position on political processes that are being researched which in turn runs the risk of compromising the rigour of the academic dimension of the research.' In CKEJ, because of the emphasis on strategic, evidence-based work, 'academic' rigour is just as essential as political rigour. What is important is that the academic rigour is conducted and realized in a politically rigorous way, and while tensions certainly exist, our reflections show that the two can actually strengthen each other. For CKEJ, political rigour is delineated by the specific political needs of the affected community/movement, as well as by the key values and characteristics defined above.

The ACKnowl-EJ project team brought scholar-activists primarily affiliated to academic institutions together with those primarily affiliated to activist institutions, along with others affiliated to both types of institutions or to neither. This unique mixture provided insight into the challenges around these multiple, complex positionalities while attempting to conduct CKEJ research that was both academically and politically rigorous. Here, we highlight three of these challenges and offer insight on how to address them. These are:

1. the difficulties and opportunities around participation and heterogeneity in research teams,
2. navigating the 'quadruple-burden' faced by CKEJ researchers (to the movements and communities they work with, to their research institution, to the project funder and to themselves – their own ethics and health), and
3. the holistic well-being of researchers involved in CKEJ projects.

We now turn to examining each of these separately.

Participation and heterogeneity in research teams

With inclusivity comes heterogeneity and possible conflicts between differing perspectives on the same issue. Heterogeneity was identified as a key strength of CKEJ processes, but attention must be paid to hierarchies and structures of marginalization, including those of gender and sexuality, within the groups engaged in the process. One challenge of these power dynamics noted by ACKnowl-EJ researchers was their observation that some members of a group in one case were more dominant than others, which at times suppressed open conversation. In this case, they decided it would be better to hold separate workshops in order to include more perspectives, as well as including longitudinal individual interviews, as focus groups were obscuring the power relations present.

Those working on the development of the Arab Regional Map grappled with the question of language heterogeneity: what does environmental justice mean to people in the region? To some involved, the term 'environment' implied a place untouched by humans, a place 'over there', while the term 'land' had more concrete connotations. The word 'justice' was also interpreted by some to imply a legal action requiring judicial courts. There were rich discussions and debates around these terms and the implications of directly translating English-language words to the region.

CKEJ faces the challenge of the 'tyranny of participation',[8] with full inclusivity likely extremely difficult if not impossible in many contexts. Processes committed to being as inclusive as possible have to adapt to participants'/team members' lives and other commitments, which might impede progress and draw the timeline out indefinitely. In Bolivia, ACKnowl-EJ researchers reported that the limited number of people able to participate in discussions was a key challenge, as well as the fact that these were mostly leaders. They tried to ensure that the capacity-building process reached out to more participants by rotating locations. While this helped make the process more horizontal, it was still limited to a portion of the population more directly engaged in politics and with CICOL (the Indigenous Union of Lomerío).

One central question raised was if/how research should be interventionist. We take as foundational that all research is biased and all researchers approach research with their own intentions. But should researchers' interventionist intentions and/or biases always be explicit to all parties from the start, keeping in mind that it could affect the study in unanticipated ways? One investigator's anti-war background and strong belief in the destructive impact of capitalism on humans and the natural environment shaped case selection and motivated the decision to create a feature map on militarization and occupation in order to highlight the often-forgotten long-lasting environmental impacts caused by this form of direct violence. The investigators working on the Lomerío case study in Bolivia professed a commitment to principles of social and environmental justice, particularly in Indigenous peoples' territories. They support Indigenous peoples' efforts for self-determination and autonomy, which they say creates a 'clear bias' in their methodology and research agenda. As with other ACKnowl-EJ researchers, they stated a belief that environmental justice research must be relevant and useful for those directly experiencing injustices – thus, their research design tried to go beyond knowledge generation per se to include reflexivity, dialogue and capacity-building processes. We overall see this transparency and self-awareness as positive, but it is important to note that ACKnowl-EJ researchers continued to question the implications of this.

ACKnowl-EJ researchers concluded that this type of co-production is a process of working out how different needs, research questions and perspectives all fit together. It will likely be impossible to meet everyone's needs all the time, so difficult choices must be made, but those closest to the issue should have their needs prioritized. It is important to recognize that the ability to make these decisions is a privilege. The core research team will also likely have conflicting perspectives on different issues. In these tense moments, it is easy to fall (back) into pre-existing dynamics and norms around who has the ultimate 'authority' and power to make decisions and speak on behalf of others, but these are also opportunities to challenge these norms.

Furthermore, at times information might be brought up that some or many involved in the process do not want to be shared publicly. There must be dialogue around what to include and what not to include in research outputs, and for which audiences. We also have to be aware of risks of discrediting movements while still serving as a 'critical comrade' that can unearth and probe uncomfortable questions, and in this way help advance transformative processes (Temper et al. 2019). The process overall must aim to be deliberately participatory. On one hand, if important information is suppressed due to power dynamics, conducting research in a more 'politically rigorous' way that facilitates space for more marginalized perspectives can also strengthen the accuracy and depth of the work, not compromising but rather strengthening its academic rigour. At the same time, not all information has to be included. Academic research regularly limits its scope in order to fit into publishing or institutional requirements, or simply the researchers' specific interests, without undermining the academic rigour of final outputs, and the same can be done in a politically rigorous process. Tensions certainly can arise, but it should not be assumed that there has to be a trade-off.

One potential strategy to help mitigate the challenges mentioned above is to ensure a diverse research team that creates a balance between positionalities. Researchers expressed that it was useful to have multiple positionalities within the same team ('insiders' of the issue at hand as well as 'outsiders'), as it allowed them to ensure a deeper understanding of the issues due to personal experience countered with a more 'neutral' or potentially critical outside perspective. Further, it is important to remain continually attentive to power dynamics within heterogeneous research teams (age, experience, affiliation to a research institution or community group). Teams might seek to work towards a similar perspective through open exchange and dialogue among diverse team members, though remaining vigilant against dominant identities occupying too much space and, in line with the key values of CKEJ, privileging those identities/voices that have been historically (and in general to this day are) marginalized and discriminated against in research processes. Again, we see an example of how conducting research in a more politically rigorous way (ensuring and supporting research teams with diverse positionalities) can help avoid blind spots and foster deeper conversation, debate and understanding.

From 'double burdens' to 'quadruple burdens'

While the double burden and dual loyalties of scholar-activism have been commented on (Borras 2016; Hale 2006), our process documentation brought to our attention the deeper complexity of this issue. In research projects like ACKnowl-EJ there exists a more aptly named 'quadruple burden'.[9] These are responsibilities to/demands from 1) institutions (the NGO/university/other institution that a researcher is affiliated to), 2) project funders (sometimes a project is funded by the same institution, but often multi-institution projects are funded from an external source, for example an EU grant, which has its own requirements and expectations), 3) the movement/community/activist group they are part of/collaborating with, and 4) themselves and their loved ones (their own ethics, needs for mental and physical health, family obligations, etc.).

These burdens manifest in a variety of ways. There are often tensions between demands from research institutions/funders and movement/community priorities, including possible pressure from funders to prioritize fast academic and non-academic outputs (blog entries, for example), which can supersede more time-intensive community processes, or vice versa (slower academic timelines vs urgent movement needs). Researchers from one case study reported feeling like the daily needs and agenda of participants were more 'burning' than any excitement over their research questions, and that they were only marginally able to deliver on participants' expectations for immediate results and answers. ACKnowl-EJ members highlighted that the pressure of the academy to perform often overpowers the activist commitments of researchers, and that the academic demands for nuances and objectivity and rigour can create tension for activism.

Regarding positionality and ethics, one researcher who works as a professor at a university explained:

> going to the field and doing research as [just an] academic is much easier because you go, ask your questions, get your sense of things and write. And this was a different relationship, we tried to not be there just as academics or people who 'know things', [...] in terms of positioning ourselves it was more challenging.

Regarding the tensions between personal ethics, donor expectations and community needs, another researcher lamented:

> we need outputs for the donors, and now with just a few months left I see we won't be able to complete what we set out to because our means and our ethics don't coincide with the timeline that the donors have given us. So which one do we give? And I am adamant that we don't give on the ethics perspective, so then what do we give the donors? How do we ... develop outputs that are realistic and doable and important and yet humble? We maybe demanded too much of ourselves within this time period. If we had demanded less maybe we wouldn't have gotten the grant to begin with. [It is a] double-edged sword.

It is important to note that these tensions can also create richness. For example, when a donor is separate from a host institution, there may be more opportunities for cross-pollination and participation of actors from other institutions. However, donors and host institutions in particular must prioritize flexibility regarding outputs and timelines, for example, allowing outputs to be determined by the movements and communities participating. Increased seed funding to allow for the full research proposal to be developed collaboratively with movements and communities would also be helpful. In particular, ACKnowl-EJ researchers emphasized the necessity of collaboratively coming up with the initial research question. A strong, supportive research and project team can also provide space for difficult conversations while still supporting researchers to do their best work in line with their moral compass and community needs. Again, tension here certainly exists between academic and political rigour, but sacrificing political rigour for institutional requirements is a false solution.

As argued above, a more diverse, inclusive and attentive research process is likely time-intensive, but the result is a deeper and likely more accurate understanding of the issues at hand. Furthermore, researchers working more in line with their own ethics and with time for their personal lives are likely happier and healthier and at a lower risk of burnout (discussed below), which could lead to higher-quality work, both academically and politically.

Holistic well-being of researchers

In large part due to the triple or quadruple burden faced by CKEJ researchers, along with CKEJ's politicized nature, there are important challenges to the holistic well-being of researchers. These include the risk of burnout, physical safety, job security and the emotional toll this type of research can take on those involved.

Within the ACKnowl-EJ project, two proposed case studies had to be abandoned due to safety concerns because of local political contexts. Meanwhile, increasing authoritarianism and fear of state coercion in another case study country restricted the group's ability to grow larger, impeding their ability to transform local relations. However, this same authoritarianism, with its ongoing state of emergency, also turned to environmental movements for 'radical apertures' since already visibly political movements were easier to suppress.

The politicized nature of CKEJ can also cause job insecurity for researchers, depending on their context. In one country, the lead investigator had to leave the university they worked for partway through the project and therefore faced increased bureaucratic challenges, economic uncertainty and a loss of office space.

Our relationship to our own personal values as CKEJ researchers can also be tricky. On the one hand, one researcher reported her personal ethics were strengthened by being part of a group with shared values, though it was difficult to translate this commitment to values into practice in the field. Another researcher expressed that while there is often conflict between the way activist research and academic research is done, the fact that this project explicitly supported a combination of the two allowed for more flexibility: 'responsibilities towards' and 'relationships with' communities were more accepted. However, a third researcher noted that at times it was difficult to always work in coherence with her activist values due to either fatigue or uncertainty.

The emotional weight of CKEJ cannot be underestimated. CKEJ researchers' lives are often either directly implicated in the issues being addressed, or intertwined with those whose lives *are* directly implicated. These researchers witness and grapple with ecological devastation and very real threats against the physical and psychological integrity of those – sometimes including themselves – on the frontlines of highly politicized environmental conflicts. As one ACKnowl-EJ investigator put it:

> This has been the only research project I've done where introspection has been a crucial part of the research. [...] I realised a few times during the meetings over the past few years that actually what I have been going through is extremely emotional

> and there is a lot of baggage I personally carry just because we are learning about this Earth, we're learning about tragedy and trauma and because of the things we are also personally experiencing.

Another researcher spoke to the difficulty of writing about a situation while multiple assassinations were taking place of those directly implicated, saying that, while emotionally difficult, it made it seem all the more urgent to make visible what was taking place. Informal conversations took place around the organization of project meetings due to the possible risks and implications for those participating whose governments might not view project activities kindly, and the possibility of surveillance or criminalization of those attending.

We believe CKEJ researchers are at high risk of burnout, particularly when institutions and donors are less supportive of politically rigorous processes, which in turn risks lessening the 'academic' quality of the work they are conducting. Burnout studies have shown that a mixture of high job demands (role ambiguity, role stress, role conflict, stressful events, pressure and workload) combined with scarce job resources (physical, organizational, psychological or social aspects that help us reach our goals and engage in meaningful work) create burnout. In particular, a lack of fit between someone's 'personality' and preferences (which we frame as our personal ethics in CKEJ) and job demands (demands from institutions, donors and communities/movements) can cause much greater stress (Bakker and Costa 2014).

Burnout limits our ability to process thoughts and to focus on new and global information, and affects our decision-making, but can be remediated in part by increased social support and freedom to craft our own work (ibid.). Our process documentation highlighted how these challenges to the psychological and physical well-being of CKEJ researchers could in part be alleviated through strong extended peer networks. The networks and relationships created through ACKnowl-EJ and within research teams were noted as a key outcome and strength of the project. One researcher stated that 'there were times that the team … when we weren't these researchers working together but during particular times it was friends supporting each other. The research grant gave space for this.' Another shared that within her research team, due to the amount of time they spent together, they were able to build personal bonds and provide each other with emotional support. A third referred to ACKnowl-EJ as 'a sort of family', saying:

> other projects don't really have that, it is always professional relationships, but this project is more personal, you can approach people, express your issues and your angst. I don't know in how many other projects that sort of relationship is there. There is openness, flexibility, trust, warmth, that is great and that is different from my previous research experiences when it involves multiple partners.

CKEJ researchers with strong social support and greater freedom to work in line with their ethics, prioritizing community/movement needs, are more likely to easily reach their goals of both academic and political rigour.

Key Takeaway
For institutions and funders: Reflections on these challenges show that institutions and funders should support heterogenous research teams, give greater flexibility surrounding timelines and outputs, and allow CKEJ researchers to prioritize the needs of the communities and movements they are part of and/or collaborate with. Close attention should be paid to mitigating the quadruple burden faced by researchers, and to facilitating activities and funding that promote the creation and social cohesion of extended peer communities that provide key support to researchers engaging in emotionally, physically and morally complex environmental justice work.

Co-production of Knowledge for Environmental Justice

Defining aspects	Other values/value-imbued characteristics
• Iterative process of questioning • Visionary (towards increased environmental justice) • Aims to support alternative-building; creation-oriented (beyond learning and sharing) • Process itself aims to act as an alternative form of knowledge production • Socially embedded and contextually specific (situationally and methodologically grounded) • Plural • Dialogue between at least two worldviews (held by at least two groups of people) • Relationships are centred, in order to reduce remoteness • Centres those most affected by the environmental injustices and those most affected by structures of marginalization • Works better with long-term, trusting relationships • Generates external and internal spaces for reflexivity • Challenges dominant worldview and narratives, including challenging harmful norms common in research processes • Ground-up design of research question and agenda • Peer reviewed by extended activist/community peer networks • Fosters spaces of mutual learning, skill/strategy sharing and alliance-building • Non-static; evolving	• Authenticity • Anti-hierarchy • Respectfulness • Transparency • Strategic inclusivity • Active reflexivity/mindfulness • Humility • Solidarity • Sensitivity towards non-dominant timings/knowledges/needs • Non-competitive • Centred on consent • Well-being and transformative goals prioritized over institutional requirements • Visibilization/strengthening of discriminated perspectives • Balance between commoning of knowledges and knowledge stewardship • Trust-building • Boundary-pushing/transgressive • Responsibility • Reciprocity • Diversely justice-oriented • Anti-extractivist • Empowering • Empathic • Anti-colonial/anti-imperial • Anti-assimilation/anti-co-optation • Evidence-based • Politically rigorous

Key challenges	Necessities
• Holistic well-being of researchers due to complex and politicized nature of environmental conflicts • Specific risks of burnout, risks to physical safety, risks to job/economic security • Navigating the 'tyranny of participation' • Tension between extended timelines and urgent needs • Decision-making around prioritizing needs • Navigating heterogeneity • Handling sensitive information • Navigating the triple/quadruple burden • Decision-making around outputs • Navigating explicit and implicit bias (recognizing this exists in all research and isn't negative but may be more explicit in CKEJ)	• Institutional flexibility regarding timelines and outputs • Support to develop/maintain strong extended peer networks • Active support for building and maintaining heterogeneous research teams • Measures to mitigate quadruple burden • Active support to avoid and attend to burnout • Attention and support for physical safety of research teams (funds and training, emergency support, flexibility around changing research plans) • Support for challenging and navigating harmful power dynamics and research norms • Mitigation of threats to job security/ economic security of researchers • Awareness of and measures to mitigate risks of criminalization of those involved in research process

Conclusion

While rife with challenges, the visionary and utopian essence of CKEJ is of utmost importance. CKEJ is about both the journey and the destination of environmental justice, but because of its focus on challenging and transforming hegemonic norms through its process, the journey is also *part* of the destination. As it is plural and context-dependent, its visionary aspect must be tied to and born from the ground up, and thus the 'visions' of CKEJ transformation will also be infinitely plural. Each manifestation of CKEJ must work towards its own utopia, even while connected to broader utopian imaginings (Bell and Pahl 2018). One way of understanding this is through the lens of the Alternatives Transformation Format, developed by ACKnowl-EJ co-coordinator Kalpavriksh, which helps us self-assess whether our projects are true alternatives to hegemonic norms, allowing us to see how they fit into broader anti-hegemonic ways of being and doing. CKEJ research teams could use the Alternatives Transformation Format as an aid both in visionary design and reflexivity on their research process and outcomes.

Reflecting on our own ACKnowl-EJ process documentation, we believe we have addressed a key gap in the conversation around co-production of knowledge by explicitly defining one particular type of co-production (CKEJ), identifying many of its central values and key challenges, and identifying some ways to address those challenges and strengthen CKEJ processes. Our research affirms Borras's (2016) assertion that the interplay between different types of scholar-activists, while little understood, can play a key role in social transformation. We have also deepened our understanding of a common issue among environmental justice scholar-activists, which is how to conduct theoretically sound political work and politically sound academic work on environmental justice issues (Weber et al. 2020), and we have shed light on three key challenges in CKEJ researchers' attempts to conduct politically and academically rigorous work.

Our findings include the argument that extended peer networks play an absolutely central role in this type of activist-academic collaboration, not least of all because strong extended peer networks likely improve the psychological and physical well-being of CKEJ researchers, who face unique emotional and intellectual challenges as well as concerns around physical safety, and therefore help create more sustainable and long-term practices. Diverse positionalities within the same research team act as a strength, though heterogeneity also necessitates active work to counter potentially harmful power dynamics. CKEJ researchers in the ACKnowl-EJ project faced a quadruple burden, and donors and host institutions must be aware of these tensions and should be more flexible in their timelines and expectations for outputs.

Affiliations to research institutions and donors offer strategic opportunities, but they also cause tension. In particular, more often than research institutions, obligations to donors created pressure to compromise the ethics of research teams and put the integrity of the CKEJ process at risk. However, flexibility on behalf of the project funders was celebrated, especially the financial and logistical support for building relationships. There are many ways funders could better support more effective CKEJ research – for example, maintaining a legal support fund in case of criminalization of research teams or those they work with, a psychosocial support fund, built-in childcare at conferences and meetings, or funds for an external trainer on anti-oppression. Other methods could include building the mitigation of negative power dynamics in research into the design phase, a sort of 'ombudsperson' available to help address issues as they arise, support funds that can be applied for in case of losing one's job due to conducting research on politicized topics, and the possibility of relocation support if researchers face risks to physical safety or legal threats. These and other measures would provide a more solid foundation for future groundbreaking and boundary-pushing transformations research.

CKEJ is an explicitly value-imbued form of research. It is visionary, creative, relational, reflexive, potentially transformative and collectively shaped, and creates fertile ground for mutual learning. Those collaborating in CKEJ need to be aware that these processes often involve individual transformation, as noted by Palmer and Hemström (2020), but it must go beyond this in order to not become just another extractive research project.

Writers and activists Walidah Imarisha and adrienne maree brown argue that organizing is science fiction, and that we can build new worlds by dreaming the 'impossible' while looking to the past for inspiration (Imarisha and brown 2015). Similarly, we reiterate: CKEJ should be visionary and utopian, while also doggedly reflexive and, above all, plural. As Imarisha paraphrases Arundhati Roy in saying, 'other worlds are not only possible, but are on their way – and we can already hear them breathing.' CKEJ researchers must strive to not just listen to their breathing, but also help pump fresh air into their lungs.

Notes

1 http://www.vikalpsangam.org/about/the-search-for-alternatives-key-aspects-and-principles/.

2 The Yeni Foça Forum is one of the neighbourhood forums (social movements that were created to protect rights to the city) after the Gezi Park protests. The Forum was established in the industrial zone of Aliağa Bay, a key industrial site that has turned into an ecological sacrifice zone since the late 1970s. In 2016 amidst a 'carbon rush', the Forum occupied an unused privatized beach and established a commons while opposing fossil fuel investments.

3 By 'alternative-building' we mean the active creation/assertion of socio-ecological-economic structures that act as an 'alternative' to hegemonic norms, in that they are more just in one or more spheres outlined by the Alternative Transformation Format.

4 We use the term 'community' in a broad sense to refer to a grouping of people affected by and/or heavily engaged in the environmental justice issue at hand. The 'community' could be a town, a forum or organization, or a social movement, for example.

5 Specific strategies are used to work to ensure inclusivity, for example separate meeting spaces just for women in one case study.

6 Including those in positions of power (including the researcher(s)) relinquishing control, letting go and listening, but also recognizing that authority manifests in different ways and is not necessarily always negative.

7 Recognizing that not all knowledge is for everyone, e.g. culturally specific knowledge not appropriate for outsiders.

8 'Tyranny of participation' refers to both the mainstreaming and increasing hegemony of 'participatory research' derived from development studies (Kapoor 2005), and the fact that participatory processes that place a high burden (usually regarding time, if not also resources, energy, etc.) on participants can end up being exploitative and/or harmful. This burden can also exclude potential participants without the time/resources to fully engage (Gaynor 2013). If participatory processes do not engage with power and politics, they risk depoliticizing local processes (Cooke and Kothari 2001).

9 In other CKEJ processes, researchers likely also face at least a triple burden, if not quadruple or more, depending on institutional affiliations, donors, number of communities/movements involved and the positionality of the researcher(s).

References

Bakker, A.B., Costa, P.L. (2014) Chronic Job Burnout and Daily Functioning: A Theoretical Analysis. *Burnout Research*, 1(3): 112–19.

Bell, D.M, Pahl, K. (2018) Co-production: Towards a Utopian Approach. *International Journal of Social Research Methodology*, 21(1): 105–17.

Blythe, J., Silver, J., Evans, L., Armitage, D., Bennett, N.J., Moore, M.-L., Morrison, T.H., Brown, K. (2018) The Dark Side of Transformation: Latent Risks in Contemporary Sustainability Discourse. *Antipode*, 50(5): 1206–23.

Borras, S.M. (2016) *Land Politics, Agrarian Movements, and Scholar-activism*. Paper presented to the International Institute of Social Studies, Erasmus University. http://repub.eur.nl/pub/93021/Jun_Borras_Inaugural_14Apr2016.pdf.

Bullard, R.D. (1993) Race and Environmental Justice in the United States. *Yale Journal of International Law*, 18(1): 319–35.

Chambers, R. (1983) *Rural Development: Putting the Last First*. London: Routledge.

Carpenter S., Mojab, S. (2017) Revolutionary Learning: Marxism, Feminism and Knowledge. London: Pluto Press.

Di Chiro, G. (2008) Living Environmentalisms: Coalition Politics, Social Reproduction, and Environmental Justice. Environmental Politics, 17(2): 276–98.

Cooke, B., Kothari, U. (2001) *Participation: The New Tyranny?* London: Zed Books.

Corburn, J. (2003) Bringing Local Knowledge into Environmental Decision Making. *Journal of Planning Education and Research*, 22(4): 420–33.

Cox, L. (2014) Movements Making Knowledge: A New Wave of Inspiration for Sociology? *Sociology*, 48(5): 954–71.

de Sousa Santos, B. (2008) *Another Knowledge Is Possible: Beyond Northern Epistemologies*. London: Verso.

—— (2014) *Epistemologies of the South: Justice Against Epistemicide*. New York: Routledge.

Di Chiro, G. (2006) Teaching Urban Ecology: Environmental Studies and the Pedagogy of Intersectionality. *Feminist Teacher*, 16(2): 98–109.

Fals-Borda, O. (1986) *El problema de cómo investigar la realidad para transformarla. Capítulo en Una sociología sentipensante para América Latina*, 3rd edn. Bogotá: Tercer Mundo.

Fazey, I., Schäpke, N., Caniglia, G., Patterson, J., Hultman, J., van Mierlo, B., Säwe, F., Wiek, A., Wittmayer, J., Aldunce, P., Al Waer, H., Battacharya, N., Bradbury, H., Carmen, E., Colvin, J., Cvitanovic, C., D'Souza, M., Gopel, M., Goldstein, B., Hämäläinen, T., Harper, G., Henfry, T., Hodgson, A., Howden, M.S., Kerr, A., Klaes, M., Lyon, C., Midgley, G., Moser, S., Mukherjee, N., Müller, K., O'Brien, K., O'Connell, D.A., Olsson, P., Page, G., Reed, M.S., Searle, B., Silvestri, G., Spaiser, V., Strasser, T., Tschakert, P., Uribe-Calvo, N., Waddell, S., Rao-Williams, J., Wise, R., Wolstenholme, R., Woods, M., Wyborn, C. (2018) Ten Essentials for Action-oriented and Second Order Energy Transitions, Transformations and Climate Change Research. *Energy Research & Social Science*, 40: 54–70.

Future Earth (2014) *Strategic Research Agenda 2014: Priorities for a Global Sustainability Research Strategy*. Paris: International Council for Science (ICSU).

Galopin, G., Vessuri, H. (2006) Science for Sustainable Development: Articulating Knowledges. In Guimaraes-Pereira, A., Cabo, M.A., Fuctowicz, S. (eds) *Interface Between Science and Society*. London: British Library.

Gaynor, N. (2013) The Tyranny of Participation Revisited: International Support to Local Governance in Burundi. *Community Development Journal*, 49(2): 295–310.

Hale, C.R. (2006) Activist Research v. Cultural Critique: Indigenous Land Rights and the Contradictions of Politically Engaged Anthropology. *Cultural Anthropology*, 21(1): 96–120.

Hunjan, R. and Pettit, J. (2012) *Power – A Practical Guide for Facilitating Social Change*. UK: Carnegie Trust.

Imarisha, W., Brown, A.B. (2015) *Octavia's Brood: Science Fiction Stories from Social Justice Movements*. Chico, CA: AK Press.

Jasanoff, S. (2004) *The Co-production of Science and the Social Order*. London: Routledge.

Jasanoff, S., Markle, G.E., Peterson, J.C., Pinch, T. (1995) *Handbook of Science and Technology Studies*. Thousand Oaks, CA: Sage Publications.

Kapoor, I. (2005) Participatory Development, Complicity and Desire. *Third World Quarterly*, 26(8): 1203–20.

Lang, D.J., Wiek, A., Bergmann, M., Stauffacher, M., Martens, P., Moll, P., Swilling, M., Thomas, C.J. (2012) Transdisciplinary Research in Sustainability Science: Practice, Principles, and Challenges. *Sustainability Science*, 7: 25–43.

Lotz-Sisitka, H., Belay Ali, M., Mphepo, G., Chaves, M., Macintyre, T., Pesanayi, T., Wals, A., Mukute, M., Kronlid, D., Tuan Tran, D., Joon, D., McGarry, D. (2016) Co-designing Research on Transgressive Learning in Times of Climate Change. *Current Opinion in Environmental Sustainability*, 20: 50–5.

Martinez-Alier, J., Anguelovski, I., Bond, P., Del Bene, D., Demaria, F., Gerber, J., Greyl, L., Haas, W., Healy, H., Marín-Burgos, V., Ojo, G., Porto, M., Rijnhout, L., Rodríguez-Labajos, B., Spangenberg, J., Temper, L., Warlenius, R., Yánez, I. (2014) Between Activism and Science: Grassroots Concepts for Sustainability Coined by Environmental Justice Organizations. *Journal of Politicial Ecology*, 21: 19–60.

McGarry, D., Weber, L., James, A., Kulundu, I., Amit, S., Temper, L., Macintyre, T., Shelton, R., Pereira, T., Chaves, C., Kuany, S., Turhan, E., Cockburn, J., Metelerkamp, L., Bajpai, S., Bengtsson, S., Vermeylen, S., Lotz-Sisitka, H., Khutsoane, T. (2021) The Pluriversity for Stuck Humans: A Queer, Decolonial School Eco-pedagogy. In Russell, J. (ed.) *Queer Ecopedagogies: Explorations in Nature, Sexuality, and Education* (International Explorations in Outdoor and Environmental Education). Cham: Springer.

Palmer, H., Hemström K. (2020) On Participatory Research, Knowledge Integration and Societal Transformation. *Anatomy of a 21st Century Sustainability Project: The Untold Stories*. Gothenburg: Chalmers University of Technology.

Peake L., Kobayashi, A. (2002) Policies and Practices for an Antiracist Geography at the Millennium. The Professional Geographer, 54(1): 50–61.

Pohl, C., Hadorn, G.H. (2008) Methodological Challenges of Transdisciplinary Research. *Natures Sciences Sociétés*, 16: 111–21.

Pohl, M., Krütli, P., Pohl, C. (2017) Ten Reflective Steps for Rendering Research Societally Relevant. *GAIA – Ecological Perspectives for Science and Society*, 26(1): 43–51.

Saltelli, A., Funtowicz, S. (2017) What Is Science's Crisis Really About? *Futures*, 91: 5–11.

Schlosberg, D. (2004) Reconceiving Environmental Justice: Global Movements and Political Theories. *Environmental Politics*, 13(3): 517–40.

Sharpe, B. (2013) *Three Horizons: The Patterning of Hope*. Devon: Triarchy Press.

Smith, L.T. (1999) *Decolonizing Methodologies: Research and Indigenous Peoples*. London: Zed Books.

Strega S., Brown L. (eds) (2005) *Research as Resistance: Critical Indigenous and Antioppressive Approaches*. Toronto: Canadian Scholars Press.

Temper, L., Del Bene, D., Martinez-Alier, J. (2015) Mapping the Frontiers and Front Lines of Global Environmental Justice: The EJAtlas. *Journal of Political Ecology*, 22: 255–78.

Temper, L., McGarry, D., Weber, L. (2019) From Academic to Political Rigour: Insights from the 'Tarot' of Transgressive Research. *Ecological Economics*, 164. https://doi.org/10.1016/j.ecolecon.2019.106379.

Weber, L., Temper, L., Del Bene, D. (2020) Transforming the Map? Examining the Political and Academic Dimensions of the Environmental Justice Atlas. In de Souza, S.P., Rehman, N., Sharma, S. (eds) *Crowdsourcing, Constructing and Collaborating: Methods and Social Impacts of Mapping the World Today*. New Delhi: Bloomsbury.

2

A Conversation on Radical Transformation Frameworks: From Conflicts to Alternatives

Arpita Lulla, Iokiñe Rodríguez, Mirna Inturias and Ashish Kothari

Introduction

This chapter is an interview with Iokiñe Rodríguez, Mirna Inturias and Ashish Kothari conducted by Arpita Lulla at the end of the Academic-Activist Co-produced Knowledge for Environmental Justice project (ACKnowl-EJ), where academics and resistance movements across the world worked together analysing and learning from citizen-led transformations to alternative futures.

One of the challenges of the ACKnowl-EJ project was agreeing on a common way to evaluate radical transformations. Coincidentally, two of the collaborating institutional partners, Grupo Confluencias and Kalpavriksh, had been working separately on this in different continents but did not know about each other's work until they came together for this project (see Box 2.1 for more details about what we refer to as a framework).

The two frameworks, the Conflict Transformation Framework (CTF) and the Alternatives Transformation Format (ATF), emerged in 2014 and 2015 respectively (for the CTF see Rodríguez et al. 2015, for ATF see Vikalp Sangam 2014). Iokiñe and Mirna, from Grupo Confluencias, had been working on the CTF with experiences of resistance movements and of conflict transformation practitioners in Latin America, while the ATF was grounded in grassroots alternatives being put into practice in India, and was developed by Ashish, with inputs from members of Kalpavriksh (the organization where he is based) and Vikalp Sangam ('Alternatives Confluences', an Indian platform for networking between groups and individuals working on alternatives to the currently dominant model of development and governance, in various spheres of life).

In 2016, the teams came together to begin the ACKnowl-EJ project, where the frameworks were expanded and tested by the network. The EJAtlas and five in-depth case studies in Argentina, Bolivia, Venezuela, Beirut and Canada applied the CTF. The ATF was used in India for the Kachchh weavers case study, to analyse the multidimensional transformations that were occurring as a result of recent changes in this livelihood practice.

This interview helps to capture how these two frameworks emerged and were used in the project, how useful they were for the movements and for understanding just transformations, and some of the unexpected surprises that emerged in the process of creating and using the frameworks. For instance, one of the surprises was that although both frameworks were originally designed to be used by different actors in different contexts, there are to a large extent context- and actor-specific. As we will see, both frameworks work better as tools for supporting resistance movements by triggering reflectivity or helping to analyse their own transformative change, rather than as tools for guiding transformative policy intervention from the outside.

While this chapter provides an insight into the making and testing of frameworks, it is also included to ensure that an important part of the story of this project is not lost: how two members of the project, Kalpavriksh and Grupos Confluencias, came together in the and enhanced their perspectives on analysing just transformations to sustainability.

Box 2.1 What Is a Framework?

We are defining framework here as a conceptual structure intended to serve as a support or guide to aid an understanding of the transformation being developed by social movements to push for a more just and sustainable world at the local, national and global scales. The framework is exploratory and may be used in a reflective/participatory way between a group of people, or by individuals in a more externally driven way to systematize and reconstruct transformative change.

Arpita: ***Q1. What was the objective behind creating these two frameworks?***

Iokiñe: In the case of Grupo Confluencias, we have been interested for more than a decade in understanding how environmental conflicts are transformed towards greater justice, but also to help people who are undergoing conflicts, suffering from development activities or conservation practices, to achieve their justice objectives in their struggles. We are a very diverse group. Some work within the academia, others are facilitators or activists in different Latin American countries (Argentina, Bolivia, Venezuela, Chile, Guatemala, Costa Rica, Ecuador, Peru), but we have in common our interest in the transformation of environmental conflicts into situations of greater environmental justice. We have been working together as Grupo Confluencias for more than fifteen years, learning from our different approaches to deal with environmental conflicts. In 2005, we realized that the traditional conflict resolution approach often used by the States, many NGOs and private actors in environmental conflicts has huge deficiencies in terms of engaging with the complexity of the issues underlying conflicts or asymmetries of power relations. In fact, environmental conflicts frequently continue or even escalate despite negotiations and agreements that are put in place to resolve them.

As a group, we wanted to help ourselves and environmental movements and communities understand how conflicts can be, and are, transformed. We also felt there was big naivety about how conflict transformation takes place and that there was a need to capture, explain and give more depth to the understanding of the changes that are necessary to achieve transformative change in conflicts.

Thus, in 2014 we got together, after many years of deliberation and reflection, and decided to develop a framework that could help define and evaluate socio-environmental conflict transformation processes (Rodríguez et al. 2015). The framework is more than anything a tool for analysis, that can help communities unravel the strategies they have used in conflicts to achieve greater justice and develop a deep internal reflection about what has worked and what has not. We felt we needed to create a tool to help systematize the knowledge of communities, because the leaders are always thinking forward, running, and they usually do not have the opportunity to reflect about what works and does not in achieving their objectives. So, it is a tool to systematize knowledge, reflections and questions about what the positive transformations are and what is still hard to achieve in relation to their justice goals. This has been one of the roles of the framework as an instrument of transformation for communities, with primarily a local perspective.

The CTF puts a focus on engaging upfront with power relations in conflicts. Many of the conventional approaches to deal with conflicts tend to put a focus on developing negotiations, consensus and agreements, to deal with what is called the episode or crisis in the conflict. But conflicts have complex causes, which have to do with relational patterns, structural contexts and dominant cultural values that are reproduced over time, and give rise to great power asymmetries and injustices. Examples of this are legal frameworks and policies that exclude certain sectors, or networks between powerful actors that are instrumental for their interests gaining traction in national congresses. This is what is known as the epicentre of conflicts (Maiese and Lederach 2004). If we want to transform conflicts towards greater justice, it is often imperative to produce a change in the relational, structural and cultural factors that give rise to conflicts.

Thus, the framework has two phases (see Figure 2.1). Phase 1 focuses on systematizing the different strategies that have been used by the conflict actors to impact on three spheres of hegemonic power: structural, relational and cultural. Phase 2 identifies the outcomes that have been produced in a set of conflict transformation pillars because of the strategies used. We have identified five key pillars for the long-term transformation of conflicts – recognition and revitalization of cultural diversity, stronger local agency, stronger local institutions and governance, stronger local control of means of production, and greater environmental integrity – and developed a set of key indicators of change that cut across each one of the indicators and spheres of power to evaluate change (see Annex 1 page 68). So, for instance, for the recognition and revitalization of cultural diversity, the framework, among other things, asks, in terms of impacting relational

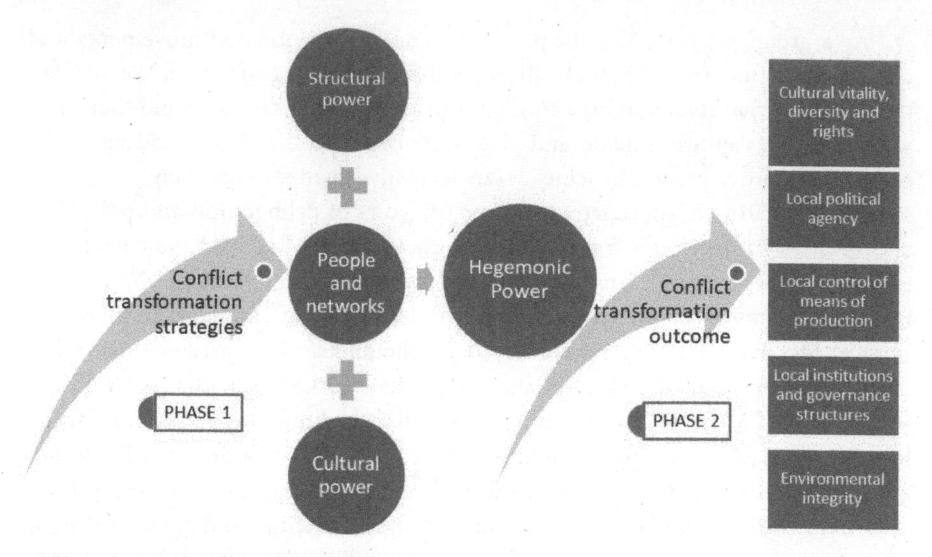

Figure 2.1 The Conflict Transformation Framework

power: have the struggles or strategies used by the communities helped to recover collective memory, create a sense of restorative justice, public narration of the truth, and a revitalization and renegotiation of knowledge and identities? In terms of impacting structural power, its asks: have the institutional practices that generate cultural exclusion in policy-making been made more visible? And in terms of impacting cultural power, it asks: have counter-histories and counter-narratives of the past emerged out of the struggles? Has the violence of stigmatizing discourses and expressions been reduced?

The objective of the tool is not so much to assess impact, but rather to help capture how change is taking place, and whether there are clear signals that progress towards greater environmental justice is being made. In this sense, rather than developing impact indicators what we did was to develop indicators of change.

Since developing the framework, we made some changes based on the conversations we had with our ACKnowl-EJ co-partners and used it in a participatory manner in Lomerío, Bolivia with the Monkoxi Indigenous Nation, to help assess their strategies for gaining control of their territories, against timber extraction. We also used it in a more traditional externally driven way in other case studies (Lebanon, Canada, Venezuela) to analyse conflicts and systematize transformative processes.

Arpita: ***Q2. What about the Alternatives Transformation Format?***

Ashish: I will explain a little bit about how we came up with our process of bringing organizations and movements together. I think what was emerging was that communities, civil society organizations, or academics and activists were asking from all the practical experience on the ground, 'What is the larger

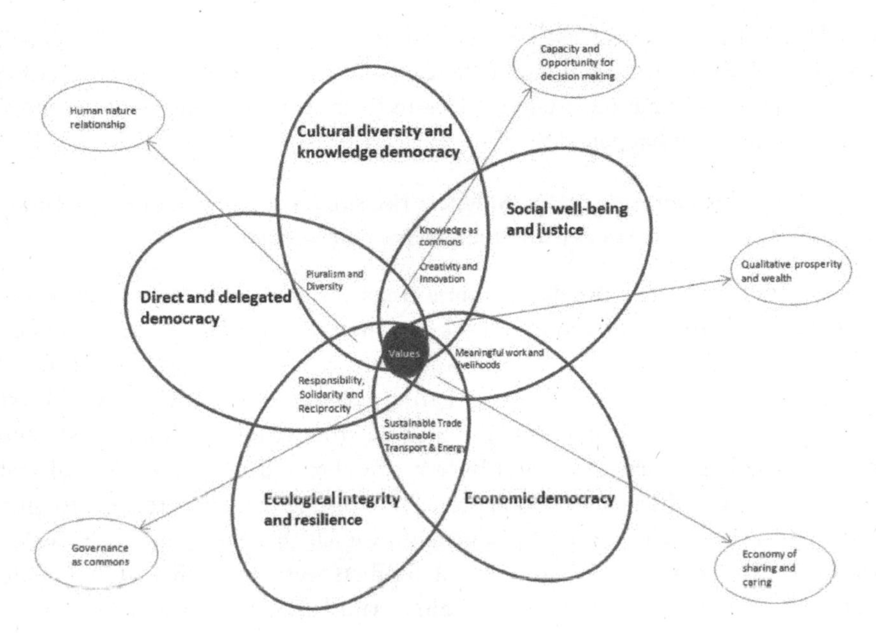

Figure 2.2 The Alternatives Transformation Format

framework of transformation towards a better society? What does it look like and what would be the vision emerging from that?' That process started around 2014 and from that emerged something called the alternative framework[1] (which was looking at the kind of political, social, ecological and cultural transformation. So, somewhat similar to what Iokiñe mentioned about more comprehensive holistic transformations. Then we realized a lot of people were asking 'how do we know whether some things are an alternative?', 'how do we know how comprehensive it is?' It is from that quest that we started working with people who can say 'we have a tool for assessing ourselves'. So, for us, helping movements to assess themselves, or communities or systems, is to say: we are doing really good work on transforming gender relationships or on protecting the forest but are we sure we are working on multiple areas of justice? (Figure 2.2 illustrates the multiple 'spheres' or areas that need to be assessed). Are we sure that we are not actually making some other aspects worse? Typically for instance, in forest protection, the community might be doing very good work, but if certain aspects of traditional decision-making remain strong, the elder men are taking all decisions, women may not be involved, and landless people who are actually gaining their predominant livelihood from the forest may not be part of the decision-making (for an extensive list of indicators within the ecological sphere as an example, see Annex 2 page 72). So out of that, we had the idea 'why don't we have a tool which helps us to assess comprehensively, and how can we integrate it to the transformation?'. So that is how the Alternatives Transformation Format developed. Then we have used it in the last two years with the ACKnowl-EJ project, with the weaving community based in Kachchh, in western India, looking at how the transformation is taking place in their economic, social and political lives. That is how it happened.

Arpita: ***Q3. How can you explain that these two similar frameworks emerged more or less at the same time in two different parts of the world?***

Mirna: There are many points in common between the Latin American reality and the tough reality of India. I think we share a common past, the history of coloniality unites us. Our countries are also rich in biological and cultural diversity and in lived experiences. Latin America and India have diverse ecosystems, species and cultures. If we come from similar colonial pasts and dominant colonial structures, and have a complex and diverse ecological and cultural diversity, then we need an approach to justice that takes this reality into account. In the case of Grupo Confluencias, we felt that the frameworks which we were using to analyse environmental conflicts were not sufficient and could not really encompass such complex reality. And I think this also reflects what happened in India. In our case, we realized we needed to engage with concepts like interculturality[2] to unpack the changes that are needed in existing power structures if we want to achieve global environmental justice. The changes that are needed are hard to achieve as they are embedded in social, economic and political structures, but also in relation to dominant cultural values and discourses. Therefore, in light of this, we needed to think of frameworks that would allow us to make visible the different types of strategies that are needed to think of change at different levels. We needed a conceptual framework that could help to interpret our complex Latin American reality, and that could emerge from the practices of those that are involved in conflicts, trying to bring about change, but that could be translated in theories that would help us propose a transformation of our realities at a regional level.

It is fascinating that we were both trying to develop these frameworks to aid understanding of transformation and create spaces for discussion about these things at the same time in the two continents.

The other interesting coincidence is that Kalpavriksh had been talking about developing a global worldviews dialogue (see Box 2.2) as a need to help people create spaces of reflection but also to share experiences and lessons from alternatives. From our side as well, we have been trying to put together a project to help Indigenous peoples develop dialogue exchanges using our framework, because this is what they have been asking us. Indigenous people we work with insist on the need for more intimate spaces for the sharing of their knowledge, and they don't really have the funding or the means, or even the frameworks to think about their journeys collectively. Framework is an academic term, but it helps to provide a methodological route to initiate a conversation.

Box 2.2 The Global Worldviews Dialogue

What is a worldview?

The world today faces an exponential increase in 'development'-focused projects and extractive activities on one hand, while on the other this 'growth' fetish has resulted in greater centralization of power in the form of patriarchy, domination of the state, markets, corporate, caste and religious institutions. In response to this, several collectives of environmental justice movements have been expanding and diversifying over the last decade, involving both Indigenous and other peoples. The worldviews of groups involved in this resistance and alternative transformation movements around the world are often fundamentally different from the dominant neoliberal or state-centred development narratives. With *worldview* we refer here to a set of ethical principles and beliefs that underlie human behaviour, that set the tone for our relations with the rest of nature and with other humans, that define our spiritualities or ethics, life philosophies. For many peoples this would be equivalent to *cosmology*, but since this term connotes (for many people) a necessarily spiritual or religious frame (in the sense of supernatural phenomena), we use here a broader term that could encompass also non-spiritual ethics.

The Dialogue

As part of the ACKnowl-EJ project, participatory research was conducted to document in detail cases of resistance and alternative transformation with Indigenous people and other local communities (including urban), in Bolivia, Venezuela, India, Turkey, Canada and South Africa. These case studies were the basis for the Worldviews Dialogue. The *Indigenous and Traditional Worldviews Dialogue Process* was conceived as a platform for social learning and knowledge exchange between local Indigenous and other communities and other academics and activists, about alternative worldviews and transformations with a focus on communities in South Asia and Latin America.

Arpita: ***Q4. Is there anything in particular that has surprised you in using the framework? What have been the most revealing or unexpected aspects of the process of putting it into practice?***

Ashish: One thing that has come out sharply when sharing the experiences of the case studies is how transformations in one sphere can have significant impacts in others without the community necessarily even noticing that is happening. Then, as everything may not be happening in a positive direction, some of those impacts might actually be negative. Just to give an example with the weaving community of Kachchh, they have derived significant great economic prosperity, very good social impacts such as women and youth empowerment, but at the same time economic inequality within the community itself seems to be growing, because some are able to access the market more easily or have existing capital to use, while others are getting left behind. So, there are aspects we didn't realize about the community and that we had not spoken about; perhaps many in the community too had not realized it, that came out as part of the study. I think that this process of self-realization or assistance to self-realization has been something quite powerful.

Iokiñe: In my case there have been two surprises. My first surprise with the framework was in the journey of making it before even applying it. Looking back now, we started with Grupo Confluencias fifteen years ago, trying to help people and ourselves to understand and engage with environmental conflicts in a more profound way. We knew that power had to be in there, understanding power, hegemony and changing power relations. And we knew that culture had to be there too because environmental conflicts are in essence conflicts about different meanings of nature and about identity. But we first envisaged it as a framework accessible to all players. Once we started engaging with power, we knew that we had to also contribute to creating justice. And the more we engaged with power, the more it became evident that the framework really could not be useful for everybody, or it shouldn't be directed to everybody. And the more we engaged with issues of coloniality, the more it became clear that the framework had to start from the beginning with strengthening those actors that are in a more vulnerable position in conflicts. So, in a way you can say that our framework (CTF) could have a limitation too, for people that want the framework to work in all contexts. Maybe in urban areas or more modernized societies it does not work so well. Our framework applies very well in Indigenous peoples' contexts. And it works well when you're looking at how the agency and the capacity of vulnerable actors in conflicts has increased or not through their actions. In other words, our framework is particularly useful to help vulnerable actors reflect about their transformation efforts and to assess how effective counter-hegemonic power strategies are. But that has been a surprise because it is by dealing with power that we were forced to rethink transformation from there. Now looking back, it is the whole process and its creation that made us have this realization.

The other thing that surprised us when we applied the framework in Lomerío was that, similarly to what Ashish said for Kachchh, it allowed us to see the great gains and achievements that were reached with the struggles trying to bring out a change, but also the things that are not really going all that well. When we did our workshops in Lomerío using the framework, we asked, what have you gained from all this huge variety of strategies that you have put in practice, but also what have you lost? And to our surprise, they started a very intense brainstorming of ideas of all the things that they still have to work on to make sure that their struggles for autonomy do not get lost on the way because of the huge number of challenges they still have. One thing that came out very clearly from the collective reflection was that although the Monkoxi Nation had spent a lot of energy over the last three decades trying to produce a change in the structural causes of injustices, by gaining territorial property rights, changing the legal and political framework of Bolivia, these strategies were not being paired with strategies to strengthen their cultural identity, and thus the younger generations are increasingly disconnected from their territories. It was surprising to see how it all came out so clearly in these conversations.

Ashish: This shows the usefulness of the frameworks – it enables the community and those working with the community to actually realize things that in our normal day-to-day life we may not realize are happening. The same thing happened with the weaver community in Kachchh, when we worked with them on the ecological impacts of the transformation in living, or on issues of inequality, then discussions began within the community. Hopefully those discussions might take them in a direction where some of the negative trends can also be dealt with.

Arpita: ***Q5. Do you see the frameworks working together and if yes, how?***

Ashish: It has been amazing that this project brought us together because we realized so many things were coincidental. For instance, you have the 'Grupo Confluencias' and we have 'Alternative Confluence'. You are defining a framework of analysis and calling it 'Conflict Transformation Framework', and we call it the 'Alternatives Transformation Format'. So, both of us are looking at transformations in society and in communities, away from the extractivist model of development towards more justice and sustainability. I am sure there are a whole lot of other coincidences that we also have. Then, as we were working over the last three years, we realized how much more complementarity there was between the frameworks. Something we can maybe elaborate more.

I think it might also be useful to kind of dwell a little bit on how the two frameworks speak to each other. The way I think they do is that the Conflict Transformation Framework is very strong on the processes by which power is transformed in the various different dimensions – you have structural aspects or power relational aspects, power narratives, cultural aspects and so on. Whereas the other, our transformation format, actually is not based on process. What it does is say 'at this point in time, where has a community reached?'. And how it will stick or not is the transformation that has been achieved. It can go into process only by asking questions like 'how did you get here?', 'what is your timeline?' It doesn't do that squarely within itself. I'm thinking that if this is the case, then putting the two together in some way, and working with a community using both, could be potentially quite powerful. What do you think?

Iokiñe: Yes, I agree. I think the ATF complements the CTF very well in helping create, and think of, visions of alternatives. Because, transforming conflict towards what? We have been very clear from our perspective that we have to engage with justice, we have to help to reduce violence, we have to contribute to creating intercultural relations in a pluricultural context in Latin America, and we know that we have to challenge the concept of development. But the vision of alternatives to development as the end point or the direction is something that would really enrich our perspective of transformations. It would give a clearer sense of direction of the needed transformations in terms of really engaging with agents of modernity/coloniality and the need for decoloniality more strongly. It is something that is implicit in our framework (Rodríguez and Inturias 2018), but we need to make it more explicit. The other way in which I think they could complement each other very well is to start thinking of more concrete ways of engaging with each one of those spheres. In your case (Kalpavriksh), how do we really test or see or evaluate each one of those different alternative spheres? Because we both develop the indicators in the forms of questions, but it would be really good to have more tangible ways of really evaluating how those changes are taking place. Like you did with the ecological footprint: if in each one of those spheres, we could find ways that are more specific to identify how and why those changes are or are not taking place. I think that is something that we can probably work together on.

Mirna: There is also a point to be made about the intrinsic connection between conflicts and alternatives which we could make more explicit. In different moments during the ACKnowl-EJ project we had interesting conversations about how conflicts and alternatives are connected. You cannot talk of one without the other. For example, we have learned that the words in India like *sangharsha* – struggle/resistance – and *nirman* – creation/alternatives – are not delinked. They have to come together for a process for transformations to take place. Resistance without alternatives can lead to fatigue and alternatives without resistance can lead to stagnation. There is no such thing as reaching an 'ultimate transformed destination'. In fact, transformation is a continuous and dynamic process. The emergence of power structures within which injustices play out is also continuous, and hence alternatives and resistance cannot and must not be delinked. In addition, it is important for social movements to see that continuous thinking, reflexivity, learning and adapting are integral parts of their resistance and alternative movements. We realize now that these frameworks could play a role in such processes of reflection and moving ahead.

Arpita: ***Q6. Have you applied these frameworks beyond this project?***

Ashish: The ATF has been picked up by other institutions and groups (though not nearly as much as we would have liked it to be!). For instance, it is being used by faculty and students of textile design in the National Institute of Design, to assess craft and livelihood initiatives they are involved with. Civil society groups like Kalpavriksh and Blue Ribbon Movement have used it to assess various aspects of their own functioning and structure. In the Vikalp Sangam process, it has had some use in framing discussions on the existence or absence of holism or comprehensiveness of alternative initiatives.

In all these cases, it has helped in creating a sharper recognition of both strengths and weaknesses, in bringing out aspects that may not have been explicit or realized earlier, and through this in generating discussion on steps that could help address weaknesses and gaps.

Iokiñe: We started using CTF in Grupo Confluencias in 2014 as part of capacity-building workshops with Indigenous peoples' leaders in Venezuela to help evaluate conflict in their territories and their strategies for dealing with them (Mirabal 2014). I use it a lot in my teaching at university to explain what conflict transformation is, particularly to help students understand the difference between different types of power in conflicts and to learn from the strategies environmental justice movements use in their struggles. We have also seen in our case that other initiatives have also picked up the CTF framework and adapted it to their particular interests. For instance, Project Transform from the French Institute of Agroecology (INRAE) from the University of Bourgogne, in Dijon, adapted it to study the role of conflict transformations enabling sustainable transformations in the food sector (Skrimizea et al. 2020).

Mirna: In Bolivia, at Nur University we have been carrying out an environmental Socio Conflict Transformation Diploma over the last two years (this year we are running the third edition) in partnership with Grupo Confluencias, the German Development Cooperation Agency (GZI) and the University of East Anglia, and the participants use the CTF to carry out an in-depth analysis of a case study, which they have to study throughout the duration of the course. It has been a useful tool for them.

Arpita: ***Q7. What can be done better with regards to the use of the framework or in the journey of producing the framework, and how?***

Ashish: With the ATF, in hindsight, there are several things that could have been done better or differently. For instance, while initial discussions with weavers in Kachchh helped to narrow down and modify the ATF for manageability and relevance, even earlier discussions while the ATF was being formulated could have helped, though of course this was not possible given the time frame and sequencing of the ACKnowl-EJ project. Having the ATF in local languages and ways of expression would also have been useful, including for better oral communication given lack of literacy in some sections of the population. More visual forms for some of the questions could also help for such sections. A greater integrated use of the ATF along with the CTF before commencing the case studies, especially given the somewhat different focus areas of the two as brought out above, would also have yielded more comprehensive results, with the ATF gaining strength from the *process* orientation of the CTF.

Iokiñe: Yes, I agree, I think that in ACKnowl-EJ we were very respectful of not interfering with each other's frameworks. They were both enriched through our dialogue, but each group wanted to have the freedom to test its frameworks out as it had been originally conceived. And this was logical, because each framework also emerged in relation to, and in dialogue with, their particular contexts. We have done that now, and I think we are ready now to start exploring how perhaps they can be merged into one.

Annexes

Annex 1: Environmental Conflict Transformation Indicators Developed by Grupo Confluencias

	INDICATORS OF CHANGE BY SPHERE OF POWER		
Transformation Pillar	**RELATIONS: People/networks**	**STRUCTURES: Institutions/laws/economic frameworks**	**CULTURE: Worldviews, discourses**
Cultural vitality, recognition and difference	Intracultural level: Have local communities been able to reflect about their identities, cultural and environmental change without external pressure or preconceived agendas? Have the struggles or strategies used helped to recover collective memory, create a sense of restorative justice, public narration of the truth, and a revitalization and renegotiation of knowledge and identities? Is the role of 'knowledgeable' people valued and taken into account more in daily community life (elders: men and women, councils of elders, shamans, chiefs, etc.)? Have there been collective efforts to help build local capacities at different levels, and design social change that can help visualize alternative futures?	Have the institutional practices that generate cultural exclusion in policy-making been made more visible? Do official institutions and legal systems acknowledge the existence and value of Indigenous and local environmental cosmovisions and logics? Are official institutional frameworks more open to an intercultural approach in environmental management – for instance, through new plural institutional arrangements? Are dialogues of knowledge taking place in decision-making processes concerning the environment and its use? Do official institutions and legal systems acknowledge the existence and value of Indigenous and local environmental cosmovisions and logics?	Have counter-histories and counter-narratives of the past emerged out of the struggles? Has the violence of stigmatizing discourses and expressions been reduced? Have counter-narratives and counter-discourses of development, environmental change, well-being, etc. emerged or been created? Have these counter-narratives and counter-discourses been socially accepted to the point of producing systemic change in the dominant conceptions of development, environmental change, well-being, etc.? Has it been possible to awaken the collective awareness of the cause of conflicts through strengthening the dignity, identity and self-esteem of marginalized sectors? Have local identities, their histories and sense of place gained more visibility and recognition in public discourse?

(*Continued*)

	INDICATORS OF CHANGE BY SPHERE OF POWER		
Transformation Pillar	**RELATIONS: People/networks**	**STRUCTURES: Institutions/laws/economic frameworks**	**CULTURE: Worldviews, discourses**
Cultural vitality, recognition and difference (*Continued*)	Intercultural level: Are government officials more open to cultural diversity and realities? Are marginalized or excluded actors and their claims more publicly visible? Have actions been taken to respond to the diversity of cultural, intergeneration and gender perspectives and claims? Are customary norms and regulations linked to local notions of authority, organization and access to natural resources being taken into account and respected in daily interactions among local users and other actors? Have the conditions of dialogue changed to allow for a real exchange between different cultures? Have more respectful, horizontal and equitable intercultural relationships been developed?		Are new public narratives of possible futures emerging? Have the historical causes of cultural exclusion in environmental use and management been made more visible? Are different cosmovisions, ways of valuing nature and knowledge systems being recognized and valued by the different actors? Are tangible and intangible elements of culture recognized in public discourse and practice (views of the territory, local languages, judicial systems, etc.)? Has it been possible to re-signify and re-frame the issues in dispute, giving recognition to previously marginalized value systems? Is there a true dialogue between different knowledge systems and cultures in relation to the use, management and access of the environment? Has an alternative ('other') environmental knowledge emerged as a result of the dialogues of knowledge?

(*Continued*)

Annex 1: (*Continued*)

	INDICATORS OF CHANGE BY SPHERE OF POWER		
Transformation Pillar	**RELATIONS: People/networks**	**STRUCTURES: Institutions/laws/economic frameworks**	**CULTURE: Worldviews, discourses**
Local political agency	How has collective action been strengthened as a result of the struggle? What type of spaces for dialogue, deliberation and decision-making that acknowledge marginalized or excluded actors are created or strengthened? How has the capacity of local actors to monitor environmental impacts been strengthened?	Are new public policies that strengthen participation created? Of what type?	Is there greater respect for the diverse forms of self-organization and representation of social actors? Are customary forms of political participation acknowledged and respected?
Governance (institutional strengthening)	In the case of Indigenous peoples (and/or other local/traditional communities and ethnicities), do they view themselves as part of the nation-state? What level of autonomy do they claim, if any? Have communities engaged in critical dialogue about the effectiveness of their norms, rules and regulations to govern natural resource use and territorial management? Were they strengthened in any way or did proposals emerge pointing at how local governance can be strengthened? Have other actors, such as state actors, engaged in critical dialogue to understand the environmental logic or knowledge systems of local governance practices?	Have state institutional structures changed to respond to local claims/demands for greater control in environmental governance? Do representative bodies at various levels have mechanisms of accountability and transparency? Have there been any opportunities for complementarity between governance systems? Have any policy changes taken place to help strengthen local environmental governance systems? Has local autonomy and control in territorial management been strengthened in any way?	Have governance systems been re-framed in any way to allow for complementarity? Have local environmental governance systems been revitalized, revalued or strengthened in any way?

(*Continued*)

	INDICATORS OF CHANGE BY SPHERE OF POWER		
Transformation Pillar	**RELATIONS: People/networks**	**STRUCTURES: Institutions/laws/economic frameworks**	**CULTURE: Worldviews, discourses**
Local control of means of production and technology	Have local communities increased their control of the means of production over commons as a result of the struggle? Do the different local social actors feel there is greater equity in the access and use of natural resources? Has there been a fair compensation for possible harms that may have been caused by degrading activities?	Have communal productive principles and systems been strengthened? Have processes that seek to increase the equitable access of men and women to the property of land, forests and water, etc., taken place?	Are customary forms of participation in the economic system acknowledged and respected? Have new or alternative value systems that stress the need for equitable distribution of resources emerged out of the struggles?
Environmental integrity	Have new or more solid networks that seek to safeguard the local and/or global environmental integrity emerged out of the struggles? Have ecosystems and landscapes used by marginalized sectors which have been impacted by degrading activities (e.g. oil spills, waste dumps, dams, mining activities) been restored?	Have new institutions, policies or legal frameworks emerged that seek to ensure a greater integration of environmental, social and economic dimensions in territorial planning and use?	Has there been a change in social values where nature is valued as a good on its own (for instance Rights of Nature in Bolivia and Ecuador), or other re-framings of nature?

Annex 2: Alternatives Transformation Format Developed by Kalpavriksh and Vikalp Sangam

Alternatives can be practical activities, policies, processes, technologies or concepts/frameworks, practised or proposed/propagated by any collective or individual. They can be continuations from the past, reasserted in or modified for current times, or brand new; it is important to note that the term does not imply these are always 'marginal' or new, but rather that they stand in contrast to the mainstream or dominant system.

It is proposed that alternatives are built on the following interrelated, interlocking spheres,[3] seen as an integrated whole:

1. **Ecological integrity and resilience,** which includes maintaining the ecoregenerative processes that conserve ecosystems, species, functions, cycles, respect for ecological limits at various levels (local to global), and an ecological ethic in all human endeavour.
2. **Social well-being and justice,** including lives that are fulfilling and satisfactory from physical, social, cultural and spiritual perspectives; where there is equity between communities and individuals in socio-economic and political entitlements, benefits, rights and responsibilities; where there is communal and ethnic harmony; where hierarchies and divisions based on faith, gender, caste, class, ethnicity, ability and other attributes are replaced by non-exploitative, non-oppressive, non-hierarchical and non-discriminatory relations.
3. **Direct and delegated democracy,** where decision-making starts at the smallest unit of human settlement, in which every human has the right, capacity and opportunity to take part, and builds up from this unit to larger levels of governance by delegates that are downwardly accountable to the units of direct democracy; and where decision-making is not simply on a 'one person one vote' basis but rather is consensual, while being respectful and supportive of the needs and rights of those currently marginalized (e.g. some minority groups).
4. **Economic democracy,** in which local communities and individuals (including producers and consumers, wherever possible combined into one as 'prosumers') have control over the means of production, distribution and exchange (including markets); where localization is a key principle, and larger trade and exchange is built on it on the principle of equal exchange; where private property gives way to the commons, removing the distinction between owner and worker.
5. **Cultural diversity and knowledge democracy,**[4] in which pluralism of ways of living, ideas and ideologies is respected, where creativity and innovation are encouraged, and where the generation, transmission and use of knowledge (traditional/modern, including science and technology) are accessible to all.

These five spheres overlap in significant ways, as illustrated in Figure 2.2 page 61 in the main text. They are also based on, and in turn influence, the below set of values that individuals and collectives hold. Each initiative can also be assessed or understood on the basis of whether it displays (or leads to the enhancement of) these (or other related) values and principles. The caveats regarding methodology for assessment or understanding these are the same as given page 74 in Table 2.1, except that given their more abstract or philosophical nature, they are by definition difficult to assess in a quantitative manner (and some may indeed be severely distorted if this were attempted).

- Self-governance / autonomy *(swashasan / swaraj)*
- Cooperation, collectivity, solidarity and 'commons'
- Rights with responsibilities
- Dignity of labour (*shram*)
- Work as livelihood (integrating pleasure, creativity, purpose, meaning)
- Livelihoods as ways of life (*jeevanshali*)
- Respect for subsistence and self-reliance (*swavalamban*)
- Qualitative pursuit of happiness
- Equity / justice / inclusion (gender, caste, class, ethnicity; *sarvodaya*)
- Simplicity / sufficiency / enoughness / living well with less (*aparigraha*)
- Respect for all life forms (*vasudhaiv kutumbakam*)
- Non-violence, peace, harmony (*ahimsa*)
- Reciprocity and interconnectedness
- Pluralism and diversity

This set of values is not like a prescriptive regime; just as it has emerged through the processes of individual and collective reflection and internalization, its further spread, modification and enlargement needs to happen through such processes.

Consistent discussions with Kalpavriksh and the Vikalp Sangam network led to deconstructing each sphere into smaller elements and indicators that can be assessed. Below the sphere 'Ecological integrity and resilience' (from Figure 2.2) has been mentioned as an example, while the main framework document presents these details for all the spheres. This list is not exclusive and for each initiative a relative set of elements and indicators needs to be created with contributions from all stakeholders.

Table 2.1 Spheres, Elements and Indicators of Alternatives Transformation

Alternatives sphere	**Element of circle/sphere** (subject to modifications and additions from local actors)	**Explanation**	Indicators of +ve transformation (subject to modifications and other indicators emerging from local actors)	**Comment**
Ecological	Conservation (taxa and ecosystems)	Sustenance of viable and resilient populations of native taxa, and of integrity and resilience of natural ecosystems	Are the key elements of the ecosystem sustained (if already present), or being restored (if in decline or disappeared) (e.g. a wetland, connections with inflow and outflow)? Is the viability of taxa sustained (if already viable), or being restored (if in decline)?	The term 'native' may be hard to define in practice – some widely acceptable rules of thumb may need to be applied; plus some 'naturalized' elements may also be important
	Diversity	Variability of native (especially endemic) elements as appropriate for ecological conditions	Is the diversity maintained if already healthy, and being restored if in decline?	Diversity is as much a qualitative concept as quantitative, such that *more* diversity is not necessarily better, e.g. if generalist species come into a desert ecosystem due to human introduction of large-scale waterbodies

(*Continued*)

Table 2.1 (*Continued*)

	Sustainability of use	Human use being within renewability limits of species and ecosystems	Is the use of a particular resource maintained within the renewability limits of species and ecosystems, and being re-established or established if not sustainable (e.g. through reduction in overall material and energy use/flows)?	Crucial to connect this to the limits aspect below, to pre-empt view that technology can always make human use sustainable
	Renewable ecological cycles	Sustaining the renewability and maintenance of hydrological, carbon, nitrogen, other cycles	Are the cycles and limits widely understood and respected enough to not be breached, or being re-established where breached?	Connections between 'local' limits and wider (up to global) ones make this complex
	Co-existence/reciprocity between humans and the rest of nature	Living together without unacceptable loss to either, optimizing populations and habitat conditions for both	Is there a common understanding and agreement about what is 'acceptable' loss? Are the processes of co-existence maintained where alive, and being restored where weakened/lost?	This would be a composite of the rows above; linked also to attitudes below
	Environmental factors	Healthy water, air, soil, sound levels	Is the health of environmental elements maintained if already healthy, or being re-established where degraded (e.g. eliminating pollution)?	Needs to be disaggregated

Notes

1 https://vikalpsangam.org/about/the-search-for-alternatives-key-aspects-and-principles/.
2 Interculturality refers to development of horizontal, reciprocal, respectful and symmetrical relationships between different cultures. However, more than the idea of simple interrelation (or communication, as it is often understood in Canada, Europe or the United States), interculturality refers to, and means, the construction of an 'other' process of knowledge production, an 'other' political practice, an 'other' social (and state) power and an 'other' society; an 'other' way to think and act in relation to, and against, modernity and colonialism. An 'other' paradigm that is thought and acted upon, through political praxis (Walsh 2007: 47).
3 The term is used here both as imagery and also in its meaning as areas of activity, interest or society.
4 'Culture' is used here to mean ways of being and acting, including language, rituals, beliefs, norms, ethics, values, worldviews, cosmologies, lifestyles and links with the rest of nature.

References

Maiese, M., Lederach, J.P. (2004) Transformation. In Burgess, H., Burgess, G. (eds) *Beyond Intractability*. University of Colorado Conflict Research Consortium.

Mirabal, G. (2014) Reflexiones de los Pueblos Indígenas de Venezuela sobre la Conflictividad Socio Ambiental y la Construcción de Interculturalidad en nuestros territorios. In Rodríguez, I., Sarti, C., Aguilar, V. (eds) *Transformación de conflictos socio ambientales e interculturalidad. Explorando las interconexiones*. Grupo Confluencias, Grupo de Trabajo de Asuntos Indígenas de la Universidad de los Andes y Organización de los Pueblos Indígenas de Amazonas (ORPIA). Norwich: Swallowtail Print. www.researchgate.net/publication/273694630_Transformacion_de_Conflictos_Socio_Ambientales_e_Interculturalidad_Explorando_las_Interconecciones.

Rodríguez, I., Inturias, M. (2018) Conflict Transformation in Indigenous Peoples' Territories: Doing Environmental Justice with a 'Decolonial Turn'. *Development Studies Research*, 5(1): 90–105. https://doi.org/10.1080/21665095.2018.1486220.

Rodríguez, I., Inturias, M., Robledo, J., Sarti, C., Borel, R., Cabria Melace, A. (2015) Abordando la Justicia Ambiental desde la Transformación de Conflictos: experiencias en América Latina con Pueblos Indígenas. *Revista de Paz y Conflictos*, 2(8): 97–128.

Skrimizea, E., Lecuyer, L., Bunnefeld, N., Butler, J.R.A., Fickel, T., Hodgson, I., Holtkamp, C., Marzano, M., Parra, C., Pereira, L., Petit, S., Pound, D., Rodríguez, I., Ryan, P., Staffler, J., Vanbergen, A.J., Van den Broeck, P., Wittmer, H., Young, J.C. (2020) Sustainable Agriculture: Recognizing the Potential of Conflict as a Positive Driver for Transformative Change. *Advances in Ecological Research*, 63. https://doi.org/10.1016/bs.aecr.2020.08.003.

Vikalp Sangam (2014) *The Search for Alternatives: Key aspects and Principles*. https://vikalpsangam.org/about/the-search-for-alternatives-key-aspects-and-principles/.

Walsh, C. (2007) Interculturalidad y colonialidad del poder. Un pensamiento y posicionamiento 'otro' desde la diferencia colonial. In Castro-Gómez, S., Grosfoguel, R. (eds) *El giro decolonial Reflexiones para una diversidad epistémica más allá del capitalismo global*. Bogotá: Siglo del Hombre Editores; Universidad Central, Instituto de Estudios Sociales Contemporáneos y Pontificia Universidad Javeriana.

PART II

Analysing Transformations from and with Environmental Justice Movements

This part of the book presents a series of case studies analysed to understand how resistance movements across the world are trying to contribute to just transformation to sustainability. We divide the chapters into three blocks that explore key outstanding aspects of the different case studies for understanding the dynamics of transformations from the ground up:

Section 1 analyses the tension between hegemonic and counter-hegemonic power dynamics.

Section 2 discusses different scales on which transformations take place.

Section 3 analyses the role that new forms of democracy, culture and alternatives to development play in winning struggles in the long term.

Yeni Foça Forum, Turkey 2018

Section 1

Double Movements Against State and Market

In this section we present cases from Turkey (transformations and resistance to the fossil fuel rush in Yeni Foça and to the neoliberalization of knowledge production), Venezuela (struggles against extractivism in Canaima National Park) and Lebanon (the 2015 trash crisis) to examine the structures and agency dynamics (double movement) within environmental justice struggles, using power analysis to illustrate the tension between hegemonic and counter-hegemonic power dynamics in transformation struggles. The focus is on the role of the state and corporate power in constraining transformative power and how movements from below organize to overcome these political obstacles. The analytical attention in these chapters is on: 1) changes in structural/visible power in terms of institutions and legal frameworks; 2) changes in people/networks – new companies; new municipalities; new alliances; and 3) changes in values/beliefs/worldviews.

3

'Mirror, Mirror on the Wall': A Reflection on Engaged Just Transformations Research under Turkey's Authoritarian Populist Regime

Begüm Özkaynak, Ethemcan Turhan and Cem İskender Aydın

Introduction

Sustainability transformations call for self-reflection and repositioning researchers' roles in knowledge co-production. While we were investigating the transformations and resistance to fossil fuel rush in Yeni Foça, Turkey, between 2016 and 2019, our lives as engaged and situated researchers, our institutions as leading public research universities, and our country – where we conduct research, co-produce knowledge and put it to use for environmental and social justice – transformed drastically. We have witnessed 'the dark side of transformation' that took its toll on the environmental justice movements we cooperate with, on the public universities we work in and the academics we ally with, and finally, on the political landscape of the country in which we live, work and play (and which more and more frequently – unfortunately – we leave). In this chapter, we want to turn the tables and reflect on transformations based on our experiences and take-homes from Yeni Foça and Boğaziçi University in Turkey, both under intense attack from the authoritarian neoliberalism of the Erdoğan regime. This reflection, hopefully, will provide some food for thought for other researchers in other places also struggling against the clenched fists of populist, conservative authoritarianism unleashed onto their socio-natures while they try to co-produce knowledge with EJ resistance movements.

What is popular in the debates around social transformations is the depiction of and quest for success, where grassroots movements and activists resist some unsustainable socio-economic and ecological practices of state and market, which then drive a fundamental systemic change. A shift for the better in existing structures of meaning, values, identities and patterns of interactions occurs (Feola 2015, Meadowcroft 2009). Yet there are also many other instances where structural conditions and given circumstances push societal resistance to primarily play the role of preventing the worst-case scenario and keep the door open for alternative paths, even if they cannot make existing power structures like the state or market take an immediate step back. Hence, the issue sometimes is not around changing policy for the better but instead fighting a malignant transformation[1] and ensuring that a shift for the worse does not occur at its full pace. This is usually the case in authoritarian regimes where there is a colossal power inequality between the state and the resisting groups: the legal struggle seldom delivers just outcomes, the political battle seems ineffective and the parliament has no real function. Standard rules and procedures do not work, and there is no negotiation logic for revising rules or room for mediation (Cavatorta 2013; Vu 2017). The state is acting more and more boldly and unlawfully, and the movement's braking power is weakening. Then, people resisting usually find themselves asking: What is the purpose of all this? How do we make sense of this situation? Are we even doing the right things?

These were the questions we were asking ourselves when our lives as engaged and situated researchers, in our institutions as renowned public research universities, and in our country – where we conduct research, co-produce knowledge and put it to use for environmental and social justice – transformed drastically. We have witnessed the dark side of transformation[2] (Blythe et al. 2018) in Turkey under the Erdoğan regime. Indeed, the country was presented as a 'model democracy' in the early years of AKP government between 2002 and 2012 (Akyol 2011) following its attempts at political liberalization and at curbing the military influence over the government, but it is nowadays touted as a typical example of 'democratic backsliding', where democracy is reduced to the electoral majority (Tansel 2018).[3] While some claim that there were always signs of the current authoritarian regime even in their early days (Babacan et al. 2021), the so-called 'authoritarian turn' occurred in full swing following the Gezi Park protests in 2013 (Özkaynak et al. 2015) and the coup attempt in July 2016 (Tansel 2018).

As we will depict below, this transformation has had its toll on the environmental justice movements with whom we cooperate, the public universities in which we work, the academics with whom we ally, and finally, the political landscape of the country in which we live, work and play (see also Gambetti 2022). In this process, we realized that in such depressive settings, it is essential to distinguish different layers of transformation in the ultimate quest for a radical change, keep away from simplistic narratives built around victory and defeat, and appreciate the nuances. To sustain social movements, there is also a need to create new ways of thinking about resistance, power and previously unimagined possibilities. Otherwise, when you only focus on the outcome as an indicator of success, it seems like you have been beaten and gained nothing in return. But this is not true, and in this chapter, we explain why.

Today, a growing body of literature recognizes that transformation is a process of ongoing learning about how change happens. In this context, Duncan et al. (2018) highlight the importance of practice-based transformation encounters to inform policy and theory, and argue that transformation is experienced and not delivered. In this chapter, we reflect on our own experiences and the transformation processes in which we are taking part, based on two cases of conflict we were involved with in Turkey – one being in Yeni Foça, an industrial zone and ecological sacrifice setting that has been on the frontline of environmental resistance since the 1980s, and the other in Boğaziçi University, a public university acclaimed for its academic autonomy, critical approach to teaching and research excellence. In both contexts, the state plays a significant top-down interventionist role and structural influences offer an unfriendly setting in many ways. In response resistances play a crucial role as catalysers by pushing the authorities towards a more non-conformist position. Unsurprisingly, the state also brings many new obstacles in the battle for strategic purposes. The critical message to be delivered is that even if these two counter-movements cannot achieve the transformations they desire in the short term, the fact that they have acted as a handbrake to the authoritarian regime and contributed to maintaining hope, as well as conditions for alternative transformations, is an outstanding achievement.

We conducted the Yeni Foça case study as engaged researchers between 2016 and 2019, as part of the ACKnowl-EJ project. Our involvement in the field was facilitated by our earlier acquaintances and engagement in various environmental organizations that have focused on the region in the past decade. Below we report back from our engagement in the field by building on our longitudinal work with a grassroots organization, Yeni Foça Forum, fighting primarily but not solely against fossil fuel infrastructure in the Aliağa region (see Box 3.1).

Box 3.1 Research Methodology for Yeni Foça and Boğaziçi Struggles

Yeni Foça	**Boğaziçi University**
Desktop research: • Analysis of 859 newspaper clippings from two major national newspapers (*Milliyet* and *Cumhuriyet*) to look back (1980–2015) • Review of official reports (investment plans, EIA, court decisions) • Literature review Engaged and situated research: • Ethemcan Turhan has family ties in the region and often visits the site; some family members are actively involved in the grassroots movement Qualitative research: • In-depth interviews and three focus groups (in April 2017, August 2017 and September 2018) with community members • A participatory scenario workshop with Yeni Foça Forum to look forward (2030–50) in August 2017	Desktop research: • Review of books and booklets on Boğaziçi University • Media and news archive • Review of the resistance website as a rich source of material: https://universitybogazici.wordpress.com/ • Literature review Engaged and situated research: • Begüm Özkaynak (since 2005) and Cem İskender Aydın (since 2020) are members of faculty at Boğaziçi University, and actively participated in forums, panels, workshops and commissions

On the other hand, we directly experienced the struggle for academic freedom and autonomy in Boğaziçi University – which has been under attack by the Erdoğan regime from 2016 onwards – as situated actors, faculty members and academics who actively took part in the resistance movement. We believe that, while these social movements' outcomes and impacts might not be apparent initially, both stories in Yeni Foça and at Boğaziçi are worth telling from within. We do not intend to compare the two cases, or to offer their comprehensive accounts, but rather to build on a narrative around some similarities we have seen while fighting against malignant transformations.

Using the Rodríguez and Inturias (2018) Conflict Transformations Framework, we explore the power dynamics between the state and local movements. Our analytical attention is on 1) changes in structural, visible power in terms of institutions and legal frameworks; 2) changes in people and networks, including some new alliances; and 3) changes in culture, values, worldviews and discourses. By exploring how hegemonic power has transformed itself in these cases and how the social movements have responded and transformed themselves in return, our findings hint at the 'unruly politics' of transformation (Scoones 2016) which encompasses multiple ways of knowing and experiencing a place.

We believe that better understanding the dynamics of the struggles in these three forms of power and knowing that transformations usually do not come suddenly or all at once is of great value. We claim that, thanks to their leverage status, grassroots social movements in Yeni Foça and Boğaziçi University provide a window of opportunity for just and sustainable transformations beyond simple success or failure considerations. Our reflections and take-homes will hopefully provide insights into those struggling against the clenched fists of populist, conservative authoritarianism in other places. This investigation will also enhance our understanding that transformations for the better are multi-level, multi-scalar and cascading. An awareness of these matters is essential for reflexive evaluation for social-ecological change and strategy-building for sustainability transformations.

The structure of the chapter is as follows: the second and third parts will consecutively introduce the two cases from Turkey, which both explain the conflicts and oppositional movements in Yeni Foça and Boğaziçi University. Then, the fourth part will show in three subsections – following the power transformation framework with power in institutions, in people and networks, and in worldviews and discourses – that while hegemonic power has transformed itself in these localities, the social movements have responded well in return in all forms and gained significant leverage without a definite short-term visible outcome, but long-term implications. Finally, the chapter concludes by discussing why it is important not to underestimate the different achievements and ruptures and what sets the limits of the possible within a given time frame.

What Has Happened in Yeni Foça?

Driving northbound from the *İzmir–Çanakkale* highway, you arrive at the Yeni Foça district overlooking the Aliağa Bay at night, passing through modern illuminated landscapes. Then, you wake up in the morning in an idyllic seaside town. But just over the hill, there is a coal-fired power plant, a floating LNG (liquefied natural gas) terminal, a whole peninsula owned by Azeri oil giant SOCAR, then another refinery just above it, some petrochemical industries scattered around, and a major shipbreaking site a stone's throw away. Due to a complex history embedded in labour and environmental struggles, this coastal town in western Turkey 50 km north of *İzmir* indeed reflects quite an unsettled character. While overlooking some heavy industrial facilities and energy investments dominated by fossil fuels of all sorts, it is still a summer vacation destination thanks to its lovely sea and agricultural surroundings. The archaeological excavations for the Kyme ruins, dating back more than 2,000 years, which have been continuing on and off since 1985, also sit at the heart of the region's landscape (see Figure 3.1).

Though undoubtedly it has not found its ideal balance or alternative vision, you wonder about the complex story behind this quiet vista and what it would look like if environmental activists had not been defending it since the late 1970s. As is well known, the region has a history of social struggles stretching over the past forty years, featuring the rise and demise of working-class action against large-scale privatizations, with a fierce environmental movement propelled by the local community in tandem with local authorities and other actors throughout the 1990s and after 2000 (Turhan et al. 2019).

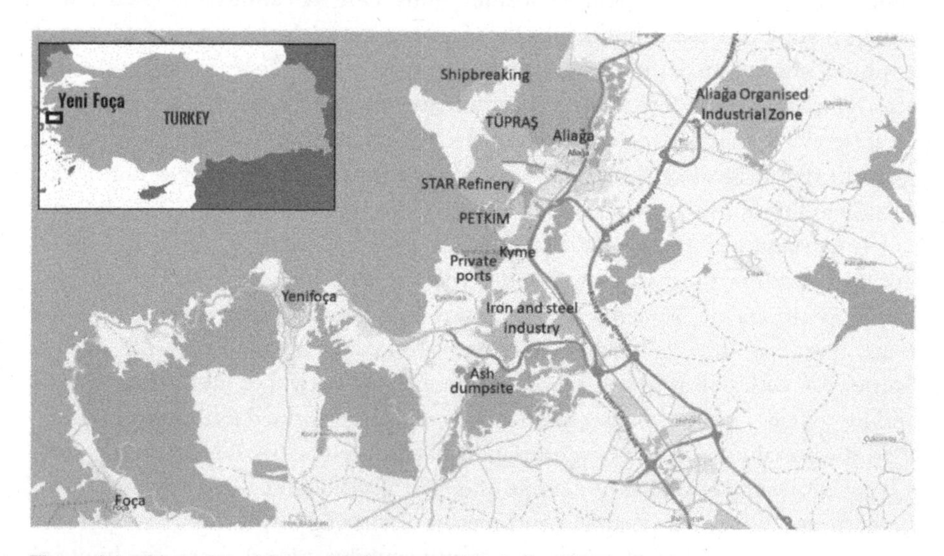

Figure 3.1 Map of Yeni Foça, Aliağa Bay and its industrial environs

Source: Authors' own elaboration.

Indeed, the region's structural transformation followed the global and national trends prevalent in the world since the mid-twentieth century. The area, despite its tourism potential along with its natural and historic sites, was chosen by the state as an industrial development zone back in the 1960s, followed by the establishment of state-led heavy industry infrastructures during the 1980s, namely the establishment of PETKIM (the petrochemical industry) and TÜPRAŞ (the oil refinery). Around these two state-owned facilities, small- and medium-scale industries, such as shipbreaking, iron-steel smelting, cement and energy, were established. Turkey witnessed its economic liberalization in the 1980s – again in line with general trends in the world – led by the centre-right politician Turgut Özal. During this period, private industries were established in the region with substantial formal and informal state support, mainly through direct subsidies and build-operate-transfer schemes (Özkaynak et al. 2020).

Later, during the 1990s, new environmentalism grew in Turkey, on the shoulders of various environmental groups, ranging from the radical and anti-institutional to the scientific and institutional. This was mainly because, in the mid-1990s, the environment was seen as a critical entry point for a civil society where citizens were using the constitutional setting and appealing to the court to cancel project plans (Turhan et al. 2019; Aydın 2005). Legal appeals were, in general, supported by mass mobilization. This was also when the rally against the planned coal-fired thermal power plant started in the region. On 6 May 1990, more than 50,000 people formed a human chain stretching 24 km along the road connecting *İzmir*, the metropolitan city, to its industrial district Aliağa (see Figure 3.2). Following the demonstrations

Figure 3.2 Human chain protest in 1990 against the planned coal-fired thermal power plant (Photo courtesy of *Ümit Otan*)

came the storm of court cases led by lawyers from the *İzmir* Bar Association and an influential MP at the time, Kemal Anadol, who took the case to the Council of State. Anadol would later refer to this 'never-ending fight' as the first instance of organized citizen reaction in the aftermath of the bloody 1980 coup (Anadol 1991: 35). Overall, in the 1990s, the environmental movement in the region at large was lively and used diverse non-violent strategies of resistance (e.g. referendums, court appeals, appeals against EIA, alternative reports, data collection on health impacts, and collaboration between scientists and activists; workers' festivals; signature campaigns; walks between *İzmir, Aliağa and Bergama*; Greek Island–Turkish Aegean coast meetings, etc.). Coupled with the widespread public pressure, the legal process eventually yielded a victory on 8 May 1990, when the Council of State stopped the Aliağa coal thermal power project.

Unfortunately, the story did not end here with the local people living happily ever after in the area. After three major economic crises in 1994, 1999 and 2001, the tectonic shifts in Turkish governments brought Erdoğan's Justice and Development Party (AKP) to power, which embarked on an all-out liberalization of the energy sector in line with IMF plans. One of the most concrete outcomes of this process has been the shift of most energy production from the public sector to the private sector. Forgotten for about a decade, the plans for increased coal-fired power capacity in Aliağa Bay revived in the aftermath of this period. As the country's economic liberalization gathered speed, many previously state-owned enterprises in the region, most notably fuel industries and oil refineries, were privatized between 2000 and 2015 (Öniş 2011).

As expected, local groups started reorganizing and mobilizing against polluting sectors and their new coal ash dumpsites, mainly receiving waste from iron-steel industries in the region and the coal-fired furnaces inside them. The resistance revitalization brought residents together under a grassroots group called FOÇEP (Foça Environment Platform) in 2009. Following the Gezi Park protests in 2013, which culminated in different neighbourhood forums across Turkey (Ergenç and Çelik 2021), local residents also formed the Yeni Foça Forum to go beyond a single-issue movement and develop a more overarching method of local activism against polluting infrastructure. Middle-class, educated, retired citizens who had time and resources to dedicate to local activism started to collaborate with other regional actors (e.g. an environmental platform for the Aegean region, environmental lawyers), national organizations (e.g. Ekoloji Kolektifi Derneği, Ekoloji Birliği) and international NGOs (e.g. Greenpeace, 350.org, Climate Action Network). One significant point in the resistance was during 2015–16 when the global climate activism network 350.org included Aliağa as part of its 'Break Free from Fossil Fuels' campaign (see Figure 3.3), where a mass demonstration took place in Yeni Foça on 15 May 2016 that brought together diverse groups, from local citizen organizations to political parties, national NGOs and mayors (Turhan et al. 2019).

Figure 3.3 Break Free from Fossil Fuels protest on 15 May 2016 (Photo courtesy of Umut Vedat)

Environmental struggles and legal battles in the region are still ongoing, with the case of a discarded Brazilian navy ship loaded with hundreds of tons of asbestos heading to the region for shipbreaking as we write these lines in July 2022. Despite their shortcomings, the mobilizations in the region have been able to impact the decisions of international financial institutions as they withdraw their support from polluting investments, and allegedly influenced the design of a floating LNG terminal due to investors' fear of widespread protests against a land-based one. In addition, a cement factory decided not to invest in the region. In 2019, in a moment of déjà vu, the local court argued against the accuracy of EIA reports on two coal-fired power plant projects in Aliağa, declared them unlawful and annulled the investments again, this time based on a lack of cumulative impact assessment. The Yeni Foça Forum and other local community groups were also heavily involved in following up a major oil spill in the summer of 2019 that resulted from a ship sent to the nearby shipbreaking site in Aliağa – an incident that well illustrates the permanent state of emergency in the region, and the need for environmental defenders continuously in action.

What Has Happened at Boğaziçi University?

On the night of 2 January 2021, the Boğaziçi University community learned from a presidential decree in the official gazette[4] that President Erdoğan had appointed a new rector called Melih Bulu, who is known to have close ties with the ruling party AKP.[5] The controversial appointment was instantly met with intense reaction from the university community (faculty members, students and alumni) and the general public due to concerns about Bulu's questionable merit to run one of the most distinguished public universities in the country. It was quite unusual for Boğaziçi University to host a rector who had not already been working there as an academic and who lacked the necessary credentials (with allegations of plagiarism) and experience for the post (Kirişçi et al. 2022). Bulu's credentials even failed to satisfy the minimum necessary criteria to be appointed to any department at Boğaziçi University as a professor. Therefore, this appointment was considered as a blatant attempt to seize control of the university and another manifestation of the rising authoritarianism in the country. What followed were numerous protests and a long-lived peaceful resistance movement that started eighteen months ago and continues to this day, with several ups and downs along the way.

Before elaborating further on the details of the resistance movement and its transformative character, it is necessary to understand the previous events that led up to this questionable appointment on 2 January 2021, and to recount the historical significance of Boğaziçi University as a critical institution for the democratization and autonomy of all of the universities in Turkey.

Boğaziçi University was transformed into a public university in 1971 from Robert College, the first American college outside the USA. It has since become a symbol of scientific excellence and academic autonomy in the country. However, despite its reputation as an institution with a democratic, bottom-up management approach, it too was significantly impacted by the anti-democratic political developments in the country. Bulu was not even the first rector appointed to Boğaziçi University from the outside. After the 1980 coup, the junta administration of the period established the Council of Higher Education to increase its control over universities. One of the first things the Council did was to appoint a rector from another university to Boğaziçi.

Against the backdrop of the post-coup political re-liberalization at the beginning of the 1990s, Boğaziçi University academics, seeing that this method of rector appointment posed an obstacle to the scientific autonomy of the university, organized an unofficial election among themselves and elected *Üstün Ergüder* as their rector. This was later communicated diplomatically to the government of the era, and the method was accepted as the official method for determining rectors in other public universities. This process was suspended in 2016 (shortly after the coup attempt on 15 July) due to a conflict between the elected rector (who received 82% of the total votes) and President Erdoğan. It was subsequently replaced by the current top-down appointment method, resulting in the appointment of Bulu.

Bulu could only stay in his post for around six months, until 15 July 2021, when he was removed from the office in a fashion similar to his appointment – with a late-night presidential decree published in the official gazette.[6] During his short term, he failed to gain enough support from within the university – only three professors agreed to work with him, one of whom (Naci İnci) became the new appointee-rector after Bulu's dismissal. However, this brief stint did not prevent Bulu from causing severe damage to the university.[7] To increase the number of allies seated in the University Senate and University Executive Board and the number of pro-government academics, two new schools (Law and Communication) were established with an overnight decision and without the consent of the senate or the board. The Faculty of Law (dubbed 'Unlawful' Law Faculty) was immediately appointed a dean from outside the university.[8]

While *İnci* satisfied many more criteria than Bulu as a 'would-be' rector (such as already being a professor at Boğaziçi and having better academic credentials), he was still opposed by 95% of the faculty.[9] However, *İnci* being a Boğaziçi academic did not stop the government from trying to seize complete control of the university; if anything, it aggravated the situation. *İnci* generously used all the powers granted to him by the Council of Higher Education as rector, recruiting new pro-government academics through illegitimate job postings, launching disciplinary proceedings against protesting students and scholars, suing them in court, unlawfully firing some full-time and part-time academics (Gürel 2022) and last but not least, dismissing three elected deans to complete the takeover of the university's executive board and senate (Bianet 2022a; MCO 2022).

In the face of the attack against their institution, the Boğaziçi University community put up strong resistance from the very beginning with the motto 'We do not accept, we do not give up', which characterized the core of all their various acts of resistance (Bianet 2021). The image of the resistance seen by the public on social media is one of daily vigils at noon, where academics gather in front of the rectorate building and silently turn their backs to the appointed rector in a symbolic manner, followed on Fridays by weekly bulletins[10] about the status of the resistance (see Figure 3.4). Academics also wielded other democratic resistance means, such as using the university senate and executive board (before these seats were taken over with the dismissal of the elected deans) to enforce the university bylaws and rules and prevent the appointed rector from evading them through illegitimate tactics. On top of this, academics have taken to court most of the presidential decrees concerning the university (such as the establishment of new schools and institutes) and unlawful decisions by the rectors (such as illegitimate recruitments and the arbitrary dismissal of academics) (Kirişçi et al. 2022).

There are evident similarities between the Boğaziçi resistance and other well-known mobilizations in Turkey, such as the *Gezi Park* protests (Kirişçi et al. 2022; Özkaynak et al. 2015), in that, as Gökarıksel (2022) also argues, both are part of a counter- or dissident body politic, which strives to rethink and rejuvenate rising neoliberal authoritarianism in Turkey. Indeed, Boğaziçi University is one of the few institutions in the country that have so far remained somewhat (although only partially) excluded from the role assigned to the universities by global neoliberalism – acting as another cog in the capitalist machinery (Gambetti

Figure 3.4 Daily vigil at noon by Boğaziçi University academics on 22 July 2022 (Photo courtesy of Can Candan)

and Gökarıksel 2022). Finally, as also voiced very clearly in the demands of the resisting academics, this is not only a matter of saving Boğaziçi University from the authoritarian takeover, but a struggle for free, autonomous and democratic universities all around the country (Gürel 2022).

How Transformation(s) Take(s) Place (or Why They Fail When They Do)

This section considers the elements of power and power relations encountered in the two cases – the anti-coal movement in Yeni Foça and the resistance at Boğaziçi University – with actors involved in the dynamics of change and transformation from the ground up. Overall, we aim to provide a combination of the theoretical framework and the practical knowledge we acquired in the field as academic activists. Both cases presented here are illustrative of the social-political-economic crisis in Turkey and the oppositional movement that has emerged against it. This is in line with Bourdieu (Burawoy 2018), arguing that what is usually taken for granted becomes questioned in crisis contexts to a greater extent. Accordingly, people are more inclined than usual towards symbolic and political mobilization strategies. And there is always a role for activists and researchers in such conjunctures for sustainability-aligned transformations.

On the one hand, in the Yeni Foça case, the economic crises that Turkey experienced throughout the 1990s and at the beginning of the 2000s provided the grounds for the state to discredit the pre-existing policy regime, and hence facilitated the neoliberal turn in the country, which then opened the way to large-scale industrial investments in the Yeni Foça region. The plans for increased coal-fired power capacity and LNG terminal construction were also implemented after the 2001 crisis, which led to devastating environmental impacts on the area, triggering an existing clash and socio-environmental resistance. On the other hand, at the root of Boğaziçi University's problematic situation lies a political crisis for President Erdoğan, which started early on with the Gezi Park protests of 2013 (Özkaynak et al. 2015) and continues with his increasingly authoritarian regime (Tansel 2018). Then, as Erdoğan's AKP lost Turkey's major cities in mayoral elections in 2019, and hence the control over resources of metropolitan municipalities, the hegemonic system needed new sources to exploit to maintain the power bloc. For years, the incumbent Turkish government was systematically altering the country's higher education institutions with a long-term political agenda, and Boğaziçi University seemed like the last stronghold of secular education built on academic autonomy and scientific excellence to be conquered, at a time when this was much needed politically (Kadıoğlu 2021). Indeed, these two economic and political crisis settings provided entry points, corresponding openings for change and opportunities for opposition movements.

We explain below in three subsections how the opposition movements in these two distinct contexts experienced and reacted to power shifts under the same authoritarian regime. For both cases, we look at how hegemonic power is confronted, contested and to some extent impacted, first and foremost in a visible manner and structurally through institutions and legal frameworks; then behind the scenes by people's organizations and networks; and finally in an invisible form through discursive practices, worldviews and values embedded in knowledge systems and vocabularies. We also discuss the opposing groups' tactical shifts and strategies in due process. We mainly observe that while the easiest method for the hegemonic system is structural control, the real strength lies in people, networks and their discourses – in the ability to speak to society. This is a point of crucial importance for any counter-hegemonic alternative.

Changes in the outset – in institutions and legal frameworks

To elaborate on how counter-hegemonic challenges drive transformations in Turkey, it is useful to first assess the institutional and legal settings that govern these 'terrains of resistance' (Routledge 1994). Drawing on our fieldwork and individual experiences, we can assert that both conflicts result from a top-down imposition of regulations and policies (new industrial developments and energy production on the one side, and a wholly irrelevant rector appointment and sudden establishment of two faculties on the other), and the lack of meaningful participation and representation of stakeholders in decisions that have a severe and negative impact on their everyday life. Overall, both Yeni Foça and Boğaziçi University are sites of contestation born from crisis settings in which injustices over access to information and recognition of rights and a shared legitimate grievance about government choices loom large.

We observe that in both cases the Turkish state – now fully transformed under twenty years of Erdoğan's rule – reacted to collective action and resistance at these sites by changing the rules of the game (i.e. the institutional and legal frameworks) whenever necessary and to some extent possible. The direct use of legislative and executive powers in authoritarian regimes is no surprise but well established as an easy and effective instrument of domination in the short run (Akçay 2021). For instance, in the case of Yeni Foça, while the opposition was growing against the Aliağa coal-power plant in 1989, the Council of Ministers issued a governmental decree officially announcing the establishment of a joint venture company (70% Japanese, 30% Turkish capital) for the investment, which would become the country's first plant running on imported coal owned mainly by foreign investors. The critical legal trick was using the free trade zones law, which was meant to facilitate land allocation for export-oriented purposes. So many things could be done with the legal apparatuses of the state in one night! This was also the case for the governance setting at Boğaziçi University when the rector elections for Turkey's state universities were suddenly lifted in 2016, with a late-night emergency decree when President Erdoğan did not want to appoint the rector-elect. A new regulation was passed in a day, giving Erdoğan the authority to appoint whomever he wishes as rector at Boğaziçi University, and across all universities in Turkey (Coşkun and Kölemen 2020).

Legislative changes in both circumstances exacerbated the conflicts and further intensified political struggles. And, as the crises deepened, procedural violence over these territories became a common form of domination at the expense of consensus-based rule. In this sense, the move in the political climate away from democratic governance principles at Yeni Foça and Boğaziçi was remarkable but no surprise under the Erdoğan regime. For example, when in Yeni Foça, the activists, as environmental defenders of their touristic and historical sites, problematized the EIA processes – seen as a bureaucratic stage to complete by the investors – of the coal-power facility planned near historical ruins, they were initially able to paralyse the project proposed in 2017. Yet instead of seeking independent consultation and monitoring, the state changed the EIA regulations and designed them to overcome the court's verdict and bypass the laws with an EIA exempt judgement for energy, mining and construction projects (Erensü 2018). This eventually led the court to stop investments after new legal appeals, based on the lack of cumulative EIA arguments. It is also telling that, despite environmental regulations, critical environmental data supposed to be produced by government agencies (such as air pollution measurements) in an industrial zone are never made publicly available.

Examples of procedural violence by the appointed and illegitimate administration at Boğaziçi included – against all established customary practices and democratic procedures – suspending elections for the director of the graduate institutes, expelling those elected from the official meetings and inviting some who had no right to be there, disregarding the quorum in the University Executive Board meeting, allowing double voting in the Senate to achieve the majority, and finally shutting down microphones of elected members in online platforms against legitimate complaints about irregularities in these meetings. The ultimate aim was to pave the way for politically motivated appointments to the university's decision-making bodies like the University Executive Council and the Senate and hire new faculty in a top-down manner by changing standards and procedures for academic promotions and posts.

On various occasions, the state bureaucracy in both contexts concentrated its strategic moves on direct economic and structural exploitation and harassment, through criminal investigations and arrests within and outside formal decision-making spaces. At Yeni Foça, environmental activists were labelled as enemies of the state and anti-developmentalist. At Boğaziçi, security cameras were installed all around the campus; the number of private security guards increased dramatically; over 500 disciplinary investigations and over 150 lawsuits were initiated on baseless pretexts against students and members of academic and administrative staff, and penalties were applied for some, including termination of employment, dismissal from the campus and random arrests; student clubs (e.g. LGBTQ+) and activities (e.g. photography and film exhibitions) were banned. The rectorate also made top-down decisions to close or move the offices of several research centres, and stopped the activities of Boğaziçi University Press under the pretext of space and budget limitations.

Note also that in both contexts, it was important not to view the state as one single monolithic body or the only part of the problem, but also as the solution. Against increased pressure, both social movements used every available safe way to mobilize and seize opportunities for legitimate action. As such, the constituents of these struggles appealed in judicial activism and engaged in legal battles on every occasion by making applications to the Council of the State and opening lawsuits in various administrative courts for the cancellation of decisions (in the case of Yeni Foça, regarding biased EIA reports and in the case of Boğaziçi, regarding the appointment of the rector and opening of new faculties with a presidential pen stroke). As resistances often suffer from lack of access to information, another key strategy was to mobilize the Right to Information Act as much as possible. Even if lawlessness is the order of the day, using all legal avenues in tandem becomes important, not only as an effective way of publicizing the struggle and making violations heard but also because such legal openings are crucial for maintaining hope in bad times, sustaining objections under formal public records and pushing open new windows of opportunity in the quest for justice.

Changes behind the scenes – in people and networks

The presence of grassroots resistance and activism in Yeni Foça and Boğaziçi under authoritarian constraints not only altered the initial state of affairs in these localities but also relations and connections of practice in an unprecedented manner at personal, local and collective levels. In both communities, issue-specific activism brought a diverse set of actors together, helped them to get to know and understand one another, and allowed them to network and collaborate despite their differences. More importantly, these crisis settings enhanced the sense of belonging and collective agency by strengthening the community culture beyond imagination. The people of Yeni Foça and Boğaziçi University were reminded why they care about localities and identities and how they relate to local and institutional histories.

Indeed, the rise and consolidation of these movements as respected and known counter-hegemonic actors at regional levels, their continuity over long periods and their impacts beyond their borders are quite remarkable. One significant achievement for Yeni Foça was forming a coalition of different opposition movements and groups – villagers, local authorities, environmentalists, opposition parties in parliament, professional associations and labour unions – around an environmental justice claim. The anti-coal struggle in Aliağa marks a significant point in the history of environmental movements in Turkey in building politically conscious environmental resistance towards the emblematic Bergama gold mine case and beyond (Gönenç 2022). Moreover, the environmental lawsuits were the first of their kind in Turkey, which led to a pro bono lawyer group, which later facilitated the formation of nationally coordinated lawyers of environmental and ecologist movements. In a similar vein, the state interference in Boğaziçi University's democratic governance structures led to a genuine struggle for academic freedoms nationwide where Boğaziçi academics, students, personnel and alumni were spontaneously united around a common goal: 'defending and maintaining with greater determination than ever the values that make this university *a universitas*' (Bianet 2022b).

Local power did not of course emerge automatically in either context, but due to a devotion to organizational structures and participatory processes that were able to carry some tension while strategies were built in small increments. Therefore, in both cases, it was essential to keep the culture of participatory democracy within the movement along the way. Individuals' powerlessness against structural influences was broken within political activism spaces such as forums, working groups and commissions, and decentralized local networks. While deciding on arenas of contestation to be mobilized, care was taken to respect the common wisdom that discussions bring. In many instances, alternatives and practices with no clear legitimate or consensual grounds were discarded. Retirees, in both contexts, played a crucial role in ensuring the integrity of the movements as they brought their experience of constructive and self-reflexive criticism into the discursive processes. From time to time, different activists and researchers became spokespersons of their resistance in the media. The visibility and effectiveness of both oppositions were also enhanced by direct relations with national and international organizations. As Silva et al. (2018) point out, connections to political parties served as bridging mechanisms that allowed both movements to directly influence the drafting of policy agendas at election times. Overall, solidarity was forged within these localities and with national and international partners, which were crucial for transforming despair and grief into a valuable and productive rage.

Needless to say, for activists surviving a long-term struggle while keeping up with daily responsibilities and a multitude of tasks is tremendously challenging. Given the burning issues that the community members deal with and act on every day, it was also crucial for them to pause working on their immediate problems and collectively reflect on their roles, capabilities and desired futures from time to time. As engaged researchers, our involvement in the Yeni Foça and Boğaziçi struggles is also meant to open up such spaces for the community to reflect on actions and challenges and better understand the opposition's role in these dynamic, ongoing conflictual processes. Yeni Foça Forum members, for instance, as our knowledge co-producers, have explicitly asked us to produce outputs with strategic relevance, including an overview of the historical struggles in the region to enhance their visibility at national and international levels.

On the other front, the dominant and pro-government circles, apart from antagonizing resisting groups, also dedicated time and energy to appropriating critical positions and controlling political decisions, lobbying, and image-remaking as a counter-strategy (Özen 2022). At Yeni Foça, for instance, new companies and sectors were introduced. SOCAR tried once to sideline potential local opposition by taking the chiefs (*muhtar*) of the nearby villages on a fully paid trip to Germany to show how similar 'clean' power plants operated. At Boğaziçi, the hidden actor on the government side seemed to be the Boğaziçi University Reunion Association (BURA). This conservative organization was established in 2003, with President Erdoğan joining its general assembly in 2018. While the most experienced academics and managers were dismissed or demoted to second- and third-degree positions in unlawful manners, the ex-chairman of BURA was appointed as the general secretariat – a critical senior managerial role on campus, now left to a person who gained his PhD just before the appointment and was inexperienced in administrative matters. One can only wonder at the possibility of such a coincidence! Among many illegitimate acts, the most disgraceful and scandalous was perhaps the appointment of three new deans from outside the university after the dismissal of the three elected deans, as if no one else in the university would be eligible for these posts. It is therefore not surprising that the positions of these three appointees hang now on a tightrope.

Changes in invisible power – in values and discourses

What impact did the proliferation of action against top-down policies and decisions have in these regions, if any? These social movements, above all, wanted to act as a brake on the hegemonic powers' malignant transformation through oppositional politics. Taken together, the evidence suggests they have already achieved this, as the state could not carry out its agenda as smoothly as it would have liked. In both contexts, political mobilization, in many ways, interrupted some state actions, led to delays and partly redirected them along the way (see again Turhan et al. 2019 for Yeni Foça, and Gambetti 2022 for Boğaziçi). But above all, thanks to the crucial role played by these movements, and their vocal and legitimate stance, today, the majority of the public acknowledges that what happened in Yeni Foça and Boğaziçi is disgraceful and illegitimate. In the case of Boğaziçi, for instance, according to a nationwide survey taken in January 2021, during the early stages of resistance, 73% of respondents among those who were knowledgeable about the resistance stated that 'the university faculty members should be able to choose their rectors' (Duvar 2021). Another survey conducted in March 2022 delivered similar results – 83% of the respondents said the faculty members should be consulted during the process of rector appointment (Medyascope 2022) – confirming the legitimacy of the demands of the resisting academics and the public support for the Boğaziçi community.

From the beginning, there was a clash of worldviews between these localities and the government, and domination was manifested invisibly but purposively through populist discursive practices and control of ideas. In both cases, activists felt the responsibility to discredit false rhetoric or ungrounded accusations and tried to disseminate all available factual evidence when necessary, without fear. The fact that the Aliağa region has been declared an ecological sacrifice zone on several occasions by governments, for instance, made the opposition feel weak. This was on top of energy scarcity, and energy discourses were often used at the national level to legitimize top-down decisions, the former being imminent in the country from the early 1980s and the latter as part of the Erdoğan regime's national development programme (Özkaynak et al. 2020; Turhan et al. 2019). The social movement in Yeni Foça ultimately found the strength to go against these discourses, believed in social action, and acknowledged the importance of being more explicit about alternative knowledge and futures. In this context, the scenario workshops we conducted as part of the ACKnowl-EJ project were helpful, as people felt they gained a better awareness of their agency and alternative visions for the future. Of course, local activists also knew they needed to sustain their discourses on solid grounds and scientific knowledge. So, whenever they could, they consulted academics for peer-reviewed, high-quality scientific knowledge documenting the impacts of the industrial activities in the region and ordered reports through professional chambers of engineers and health professionals.

After years of experience, the social movement in Yeni Foça produced a discursive and material transformation for the area by claiming to 'defend life' beyond the polluting fossil fuel projects. The movement's decisive and openly political stand against polluting investments and active engagement with all other actors (including local authorities, national authorities and other grassroots groups in the region) have given them leverage to amplify their messages. The principal statement of the Yeni Foça Forum is telling in this context:

> With every passing day, knowing that our environment is under a systematic assault and considering the environment as the basic right to life, we as the witnesses of this assault are coming together to form Yeni Foça Forum to build environmental awareness, strengthen it and widen the solidarity, reclaim our historical and cultural values and pass them on to the next generations, defend life with all its diversity and colors.

Of course, while investors still have power over locals and as always, themes like jobs and national energy act as influential narratives for the investors to maintain locals' buy-in, establishing a positive framing for the struggle that would motivate people to become more active in the political arena – next to their everyday environmentalism – was a vital move for agenda setting (Kelz 2019).

In a similar vein, for Boğaziçi's struggle, being transparent and explicit regarding moral principles, values and a campus culture embraced by an old, respected university and its constituents, as well as a dedication to excellence in education and research, was extremely important. This was especially the case at a time when post-truth politics was at its peak in Turkey as protests were designated as elitism and terrorism by President Erdoğan himself; the interior minister deemed opposing the will of the president 'fascistic', the new vice-rector responsible for research announced on his Twitter that 'Boğaziçi is, at last, doing science' and the rectorate's rhetoric at large implied that an authoritarian, hierarchical structure is natural in university settings, and even necessary for efficiency (Gambetti 2022). Against such organized lying, fundamental principles[11] on academic functioning and governance embraced by Boğaziçi and approved by the Senate in 2012 were publicized (e.g. Boğaziçi Ayakta, 2021, 2022; Kolluoglu and Akarun 2023; Çolakoğlu and Demirci 2022; Freely 2013) published on Boğaziçi's legacy and history were disseminated; public speeches, video artworks and short classes were organized by students, alumni and academics on campus and broadcasted regularly through internet platforms; a university governance proposal[12] was drafted and disseminated nationwide; and panels and conferences were organized on the future of higher education and academic freedoms, again in collaboration with academics and alumni.

These counter-acts not only served to disqualify untruths and set the agenda for Turkey's academic landscape but also synergized people outside the university. Undoubtedly thanks to these events, there have been references to Boğaziçi's spirit on several occasions inspiring everybody in the country to fight against injustices.[13] Knowing Boğaziçi is fighting, others are fighting (e.g. the Middle East Technical University (METU)). Therefore, as previously mentioned, here we mainly assert that while the easiest method for the hegemonic system is structural control, the real strength lies in people, networks and their discourses – in the ability to speak to society. This is crucial for any counter-hegemonic struggle. The fact that the authoritarian government, despite all efforts around antagonism and populism, cannot construct a legitimate discourse to produce consent while the resistance has a natural one and can shout truth to power is what gives people hope.

What Sets the Limits of the Possible? The Way Forward

In this chapter, we have provided a combination of theoretical knowledge and the practical insights we acquired in two oppositional movement contexts of which we were a part: Yeni Foça and Boğaziçi University. In doing so, we discussed different layers of power transformation within a double movement setting (Ford 2003). In particular, we depicted how the contradictory forces, the ruling and subordinate groups, try to improve their relative positions in the three spheres of power elaborated by Rodríguez et al. (2015) and Rodríguez and Inturias (2018). We also demonstrated how oppositional movements fighting on multiple fronts, though seemingly beaten, challenge the hegemonic system and complicate our understanding of the change process. Such awareness is fundamental since, as we have already pointed out, people are primarily inclined to disregard the dynamic and multiple ways in which transformations take place and often instead prioritize a narrow set of outcomes designated as successes or failures. Yet the issue is sometimes around fighting a malignant transformation first and ensuring that a shift for the worse does not occur at its full pace. Therefore, it is important not to expect too much at once and to be conscious and patient with the necessary process for transformative change, particularly in the case of an authoritarian breakdown.

Moreover, we sense that people often get hung up on power transformations they observe at the outset, and miss or overlook the importance of the hidden and invisible power spheres. Presumably, such omissions make people feel powerless in the battleground and keep them away from acting from time to time. However, it is crucial to remember that from an institutional perspective, changes at the outset, without public support and coherent and legitimate discourse, can neither produce expected results nor remain stable and sustainable in the long run. Therefore, it is crucial to keep believing in collective agency, take responsibility as needed at the forefront by reacting to institutional pressures, and never stop proactively imagining and building alternative futures. Insights from these cases also demonstrate that for social movements under an authoritarian turn, one key area to keep an eye on is the increasing threat to democratic knowledge co-production and the free press. Surely, a solid understanding of complex forms of knowledge politics and anti-politics will undoubtedly be essential for those struggling for a more just, egalitarian and plural society.

Outside our analytical gaze, our individual life experiences were shaped and influenced by the events in Yeni Foça and Boğaziçi. As researchers with family ties to Yeni Foça (in Ethemcan's case) or as academics in both senior and junior positions (in Begüm and Cem's case), we became the subjects of ongoing changes in Yeni Foça and Boğaziçi and, therefore, naturally had to embrace the changes, join the struggles and maintain the pickets we witnessed. Consequently, we paid the utmost attention to catering to these ongoing struggles' needs through our research engagement. We collectively distilled from these experiences that the second-worst option is still better than the worst-case scenario; hence, there is always something worth fighting for. This is in line with Out of the Woods Collective's (2020: 34) formulation of hope against hope, which does not translate as an expectation of change or undue optimism but rather as a way to build different futures with solidarity and struggle.

While we caution against overly optimistic approaches to policy changes without durable shifts in power structures (Silva et al. 2018), our central contention is that those social movements acting as a handbrake on malign transformations are valuable. Moreover, even in bad times there are reasons to be hopeful, since resistance movements consciously or unconsciously impact multiple power spheres, creating favourable conditions for positive and sustainability-aligned change in the medium and long run. The societal experience itself is very precious, and presumably is the one that makes a difference in the long term. It is true that sometimes, social movements abandon some of their radical ideas, put them to bed temporarily, and make concessions – but still they know that with a potential structural change or any other new window of opportunity, the revolutionary ideas and built-in values are there and ready to be used. Therefore, no matter what, it is essential to keep asking and discussing how alternative transformative futures can come into being, and confronting whatever sets the limits of the possible.

Notes

1 At this juncture, we argue that it is important to embrace the inherent uncertainty in transformations and answer the questions put forth by Scoones and Stirling (2020) clearly before branding any transformation as benign or malign: 'What methods, processes and mobilizations can tilt the balance towards more positive outcomes? How can alternatives be prefigured to reinforce this new politics? Who is centred in transformatory spaces, and who is to the side? And what solidarities, ethics and styles of reflexivity are required for this new politics of uncertainty?' (ibid., pp. 20–1).

2 Following Blythe et al. (2018), we refer to the dark side of transformation as 'the risks associated with discourse and practice that constructs transformation as apolitical, inevitable, or universally beneficial, has the potential to produce significant material and discursive consequences'. Our working definition of malignant transformation can, therefore, be understood as thorough structural changes that produce worse outcomes than the initial condition.

3 The AKP (and Erdoğan) initially rose to power in 2002 following the previous governments' critical political and economic failures. It delivered the promise of a liberal welfare state with major structural economic and political reforms (Turhan et al. 2019) and gained major electoral support.
4 See Official Gazette no: 31352, www.resmigazete.gov.tr/eskiler/2021/01/20210102-7.pdf.
5 Bulu was a parliamentary candidate for Erdoğan's AKP in 2015 (AKP: Adalet ve Kalkınma Partisi – Justice and Development Party).
6 See Official Gazette no: 31542, www.resmigazete.gov.tr/eskiler/2021/07/20210715M1-2.pdf.
7 For a complete list of damages compiled by Boğaziçi academics, see https://universitybogazici.wordpress.com/hasarlar-talepler-damages-demands/.
8 The appointed dean resigned on 26 July 2022 (one day before the faculty presentation meeting to the candidate students) citing health problems, and the Faculty of Law had to begin education in the autumn 2022 term with no full professors on its academic staff list.
9 Boğaziçi academics organized a 'vote of no confidence' on 31 July 2021; 82% of 746 eligible/registered voters participated in the voting.
10 See all bulletins here: https://universitybogazici.wordpress.com/bulten/.
11 The key motivation behind this statement regarding the structure and functioning of public universities was to contribute to the discussion of the university system and the rules in relation to higher education in Turkey. See the English version: https://universitybogazici.files.wordpress.com/2021/07/bogazici-university-academic-principles_senate-2012-resolution.pdf.
12 See the report in Turkish: https://universitybogazici.files.wordpress.com/2021/07/uyykrapor13temmuz2021.pdf.
13 For a list of awards as a reflection of inspirations see: https://universitybogazici.wordpress.com/oduller-awards/.

References

Akçay, Ü. (2021) Authoritarian Consolidation Dynamics in Turkey. *Contemporary Politics*, 27(1): 79–104.

Akyol, M. (2011) Turkey's Maturing Foreign Policy. *Foreign Affairs*, 7 July, https://foreignaffairs.org/articles/turkey/2011-07-07/turkeys-maturing-foreign-policy. Accessed 3 October 2022.

Anadol, K. (1991) *Termik Santrallere Hayır*. Ankara: V Yayınları.

Aydın, Z. (2005) The State, Civil Society, and Environmentalism. In Adaman, F., Arsel, M. (eds) *Environmentalism in Turkey: Between Democracy and Development?* London: Routledge.

Babacan, E., Kutun, M., Pınar, E., Yılmaz, Z. (2021) Introduction: Debating Regime Transformation in Turkey: Myths, critiques and challenges. In *Regime Change in Turkey*. London: Routledge.

Bianet (2021) Boğaziçi University Faculty: 'Academic administrators can be appointed only after being elected'. 3 January, https://bianet.org/english/education/236977-academic-administrators-can-be-appointed-only-after-being-elected. Accessed 20 April 2022.

—— Boğaziçi Deans Dismissed due to 'Disciplinary Offenses,' Says Rector. 21 January, https://m.bianet.org/bianet/education/256559-bogazici-deans-dismissed-due-to-disciplinary-offenses-says-rector. Accessed 20 April 2022.

—— (2022b) Call for International Solidarity with Boğaziçi Resistance. 5 January, https://m.bianet.org/english/education/255753-call-for-international-solidarity-with-bogazici-resistance. Accessed 22 July 2022.

Blythe, J., Silver, J., Louisa, J.E., Armitage, D., Bennett, N.J., Moore, M.L., Morrison, T.H., Brown, K. (2018) The Dark Side of Transformation: Latent Risks in Contemporary Sustainability Discourse. *Antipode*, 50(5): 1206–23.

Boğaziçi Ayakta (2021) Boğaziçi Ayakta: Kabul Etmiyoruz, Vazgeçmiyoruz, Booklet I.

—— (2022) Boğaziçi Ayakta: Bir Devlet Üniversitesinin Bahçesi, Booklet II.

Bourdieu, P. (2000) *Pascalian Meditations*. Stanford, CA: Stanford University Press.

Burawoy, M. (2018) Making Sense of Bourdieu: From Demolition to Recuperation and Critique. *Catalyst*, 2(1): 51–87.

Cavatorta, F. (2013) *Civil Society Activism under Authoritarian Rule: A Comparative Perspective*. London: Routledge.

Çolakoğlu, N., Demirci, A. (2022) *Boğaziçi'nde Yanan Meşale*. Remzi Kitabevi.

Coşkun, G.B., Kölemen, A. (2020) Illiberal Democracy or Electoral Autocracy: The Case of Turkey. In Vormann, B., Weinman, M.D. (eds) *The Emergence of Illiberalism: Understanding a Global Phenomenon*. London: Routledge.

Duncan, R., Robson-Williams, M., Nicholas, G., Turner, J.A., Smith, R., Diprose, D. (2018) Transformation Is 'Experienced, Not Delivered': Insights from Grounding the Discourse in Practice to Inform Policy and Theory. *Sustainability*, 10(9): 3177.

Duvar (2021) Turks Want Politically Independent Universities, Disapprove of Appointment of Rectors with Political Connections: Survey. *Duvar English*, 4 February, www.duvarenglish.com/turks-want-politically-independent-universities-disapprove-of-appointment-of-rectors-with-political-connections-survey-shows-news-56139. Accessed 20 April 2022.

Erensü, S. (2018) The Contradictions of Turkey's Rush to Energy. *Middle East Report*, 288: 32–5.

Ergenç, C., Çelik, Ö. (2021) Urban Neighbourhood Forums in Ankara as a Commoning Practice. *Antipode*, 53(4): 1038–61.

Feola, G. (2015) Societal Transformation in Response to Global Environmental Change: A Review of Emerging Concepts. *Ambio*, 44: 376–90.

Ford, L.H. (2003) Challenging Global Environmental Governance: Social Movement Agency and Global Civil Society. *Global Environmental Politics*, 3(2): 120–34.

Freely, J. (2013) *A Bridge of Culture: Robert College – Boğaziçi University (How an American College in Istanbul Became a Turkish University)*. Boğaziçi University Press.

Gambetti, Z. (2022) The Struggle for Academic Freedom in an Age of Post-truth. *South Atlantic Quarterly*, 121(1): 178–87.

Gambetti, Z., Gökarıksel, S. (2022) Introduction: Universities as New Battlegrounds. *South Atlantic Quarterly*, 121(1): 174–7.

Gökarıksel, S. (2022) University Embodied: The Struggle for Autonomy and Democracy. *South Atlantic Quarterly*, 121(1): 188–98.

Gönenç, D. (2022). Litigation as a Strategy for Environmental Movements Questioned: An Examination of Bergama and Artvin-Cerattepe Struggles. *Journal of Balkan and Near Eastern Studies*, 24(2): 303–22.

Gürel, A. (2022) Boğaziçi a Year On: A Damage Report. *Bianet*, 8 February. https://bianet.org/5/27/257321-bogazici-a-year-on-a-damage-report. Accessed 20 April 2022.

Kadıoğlu, A. (2021) Autocratic Legalism in New Turkey. *Social Research: An International Quarterly*, 88(2): 445–71.

Kelz, R. (2019) Thinking About Future/Democracy: Towards a Political Theory of Futurity. *Sustainability Science*, 14(4): 905–13.

Kirişçi, K., Eder, M., Arslanalp, M. (2022) Resistance to Erdoğan's Encroachment at Turkey's Top University, One Year On. *Brookings*, 21 January, www.brookings.edu/blog/order-from-chaos/2022/01/21/resistance-to-erdogans-encroachment-at-turkeys-top-university-one-year-on/. Accessed 20 April 2022.

Kolluoglu, B., Akarun, L. (2023) Standing up for the university. Nature Human Behaviour 7: 668–9.

Meadowcroft, J. (2009) What About the Politics? Sustainable Development, Transition Management, and Long-term Energy Transitions. *Policy Sciences*, 42: 323–40.

Medyascope (2022) KONDA Barometresi: Boğaziçi olaylarında kamuoyu demokratik seçimden yana. 7 March, https://medyascope.tv/2022/03/07/konda-barometresi-bogazici-olaylarinda-kamuoyu-demokratik-secimden-yana/. Accessed 21 April 2022.

MCO (2022) MCO Public Statement on Academic Freedom in Turkey. *Observatory of Magna Charta Universitatum*, www.magna-charta.org/news/mco-public-statement-on-academic-freedom-in-turkey. Accessed 20 April 2022.

Out of the Woods Collective (2020) *Hope Against Hope: Writings on Ecological Crisis*. New York: Common Notion.

Öniş, Z. (2011) Power, Interests and Coalitions: The Political Economy of Mass Privatisation in Turkey. *Third World Quarterly*, 32(4): 707–24.

Özen, H. (2022) Interpellating 'the People' Against Environmentalists: The Authoritarian Populist Response to Environmental Mobilizations in Turkey. *Political Geography*, 97: 102695.

Özkaynak, B., Aydın, C.İ., Ertör-Akyazı, P., Ertör. I. (2015) The Gezi Park Resistance from an Environmental Justice and Social Metabolism Perspective. *Capitalism, Nature, Socialism*, 26(1): 99–114.

Özkaynak, B., Turhan, E., Aydın, C.İ. (2020) The Politics of Energy in Turkey: Running Engines on Geopolitical, Discursive, and Coercive Power. In Tezcür, G.M. (ed.) *The Oxford Handbook of Turkish Politics*. Oxford: Oxford University Press.

Rodríguez, I., Inturias, M. (2018) Conflict Transformation in Indigenous Peoples' Territories: Doing Environmental Justice with a 'Decolonial Turn'. *Development Studies Research*, 5(1): 90–105, https://doi.org/10.1080/21665095.2018.1486220.

Routledge, P. (1994) Backstreets, Barricades, and Blackouts: Urban Terrains of Resistance in Nepal. *Environment and Planning D: Society and Space*, 12(5): 559–78.

Scoones, I. (2016) The Politics of Sustainability and Development. *Annual Review of Environment and Resources*, 41(1): 293–319.

Scoones, I., Stirling, A. (2020) *The Politics of Uncertainty: Challenges of Transformation*. London: Routledge.

Silva, E., Akchurin, M., Bebbington, A.J. (2018) Policy Effects of Resistance Against Mega-Projects in Latin America: An Introduction. *European Review of Latin American and Caribbean Studies*, 106: 27–47.

Tansel, C.B. (2018) Authoritarian Neoliberalism and Democratic Backsliding in Turkey: Beyond the Narratives of Progress. *South European Society and Politics*, 23(2): 197–217.

Turhan, E., Özkaynak, B., Aydın, C.İ. (2019) Coal, Ash, and Other Tales: The Making and Remaking of the Anti-Coal Movement in Aliağa, Turkey. In Turhan, E., İnal, O. (eds) *Transforming Socio-Natures in Turkey: Landscapes, State and Environmental Movements*. London: Routledge.

Vu, N.A. (2017) Grassroots Environmental Activism in an Authoritarian Context: The Trees Movement in Vietnam. *Voluntas*, 28: 1180–208.

4
Games of Power in Conflicts over Extractivism in Canaima National Park, Venezuela

Iokiñe Rodríguez and Vladimir Aguilar

Introduction

The 1997 conflict between the Venezuelan state and the Pemon, an Indigenous people inhabiting the Canaima National Park in the state of Bolívar, over the construction of a high voltage power line to Brazil is among the most emblematic and well-known conflicts of the last two decades in Venezuela. The Pemon were noted for their tenacity, unity and perseverance in the struggle, and were able to force important structural transformations because of their resistance to the project. Two decades later, however, the Pemon have been demobilized, fragmented and in some places displaced from their traditional homelands. This chapter uses the Conflict Transformation Framework explained in Chapter 2 to help understand why; unfortunately, the Pemon are not the first to experience circumstances like these.[1]

Like many recent environmental conflicts in Indigenous peoples' territories in Latin America, this was not merely a struggle over an infrastructure project (Rodríguez and Inturias 2018). The core of this situation was a conflict between two different models of the nation-state: a modernist vision of development and homogeneity, versus a pluralist view demanding rights for special types of citizens. This clash persisted after the power line conflict, even as the country transformed to a pluricultural nation-state. Understanding the tension over different forms of citizenship as part of the push for pluricultural politics in Indigenous peoples' territories is key for the study of just and radical transformations for alternative futures, particularly for understanding why they often go wrong.

As this case study will show, for five consecutive years from 1997 to 2001, the Pemon fought assiduously against the power line project because they saw it as a threat to their cultural and environmental integrity. One argument against the project was that the power line was the beginning of a long-term plan to open up the south of the state of Bolívar to mining. The Pemon systematically demanded territorial land rights as part of their struggle, and succeeded in temporarily suspending construction thorough a variety of resistance and political action strategies. Most significantly, in 1999, with Hugo Chavez newly in power, the Pemon played a key role in forcing a constitutional change to include a chapter on Indigenous rights as part of the new pluricultural Venezuelan nation-state. For the first time in contemporary Venezuelan history, Indigenous peoples were granted ownership rights over their traditionally occupied ancestral lands (Constitución de la República Bolivariana de Venezuela 1999: Art. 119). The constitutional reform was vital for reaching an agreement where the Pemon accepted the completion of the project, under conditions including, within a week of signing, initiating a process of demarcation and titling of Indigenous peoples' 'habitats'. This process has yet to be put in place, and Pemon territory is under increasing pressure from mining, carried out partly by the same people who were against the power line project in 1997.

Strikingly, in the long term the Chavista government used the transformative potential of the conflict and the wider transformative socialist political project to demobilize the Pemon Indigenous rights movement, and to advance an extractivist agenda for the region that in the end gave no option to the Pemon except becoming miners. Furthermore, public policies reproduced the same homogenizing and modernizing intention towards Indigenous peoples as those of previous governments. This case therefore has much to say about how transformation itself can be used as a strategy for domination, and how plurinational politics can become a vehicle for maintaining the coloniality of power through the workings of a nation-state.[2]

This case also uncovers the complex strategic manoeuvring and anti-hegemonic power strategies used in some Pemon factions to defend their lives, livelihoods, territory and right to self-determination. Contradictory as it may seem, we must understand the Pemon's recent shift to mining in the Canaima National Park (CNP) as one action within their wide repertory of resistance to state policies over the last two decades. This only becomes clear when we carry out a longitudinal analysis of the interactions between the state and the Pemon in the definition of the development agenda for CNP and southern Bolívar State. The power line conflict offers a useful entry point for such an analysis.

We will analyse and describe the power dynamics between the Pemon and the state during this conflict and in the twenty years that followed, using a retrospective analysis to reveal the power games present during pluricultural transformation politics. This involves analysing the interplay between the strategies used by the Pemon to impact hegemonic power and give rise to an alternative and plural model of Venezuelan society, and those used by the government to reinforce its power and advance an extractivist agenda in the south. Analysing these power dynamics in socio-environmental conflicts is key to understanding why radical transformations often go wrong.

In the next section, we explain the key methods and concepts guiding the study and analysis. The third section then provides background information for the case study, and in the fourth we present the main findings in relation to transformation politics, with a focus on power relations. In the fifth section we analyse the transformations that have taken place because of the conflicts, and we close with a final discussion on key lessons of the study.

Key Methods and Concepts

This case study is based on data we have collected in three different stages. The first stage was between 1995 and 1997 when, while working in the region as part of unrelated conflict resolution projects, we were able to witness the gestation and eruption of the power line conflict. The second stage was between 1999 and 2002 when we studied in detail how the conflict unfolded as part of our respective PhD fieldwork (Aguilar 2004; Rodríguez 2004), which used action-research methods to help strengthen the Pemon conflict transformation strategies. The later stage has been in the last ten years, in which we have continued supporting some Pemon factions in their cultural reaffirmation strategies through further action-research working on a Pemon Life Plan, the publication of a Pemon self-authored book about the history of one village (Kumarakapay), and capacity-building on Indigenous rights. This allowed us to see how tensions developed over the expansion of mining.

During these periods, we collected qualitative data through a variety of methods: participant observation, interviews, notes from conversations, recordings and transcriptions from meetings and workshops, and a collection of secondary sources. These include newspaper and academic articles, manifestos, reports, laws, official government projects, minutes from meetings, and transcriptions from radio and television programmes. Most recently, we have collected similar information through the internet, the press and personal communications.

For our analysis we used the Conflict Transformation Framework developed by Grupo Confluencias (Rodríguez and Inturias 2018). This framework focuses on power analysis, differentiating between different forms of hegemonic power (structural, actor-networks and cultural), and the counter-hegemonic strategies used by movements to bring about just transformations. It also provides five pillars of conflict transformation to help evaluate transformations that have resulted from resistance movements' different strategies: cultural vitality, political agency, control of local means of production, local institutions and governance structures, and environmental integrity. To discern the extent to which resistance strategies helped advance just transformations for the Pemon, we therefore seek to answer the following questions:

- Was the Pemon culture revitalized and did their rights gain greater recognition in society, allowing dialogues to take place between different knowledge systems and worldviews?
- Was the Pemon's political agency strengthened during or as a result of their struggles?

- Did the Pemon gain greater local control of the means of production, to thereby influence the distribution of environmental damages and benefits?
- Were the Pemon's institutions and governance structures strengthened?
- Did the quality and integrity of the environment improve?

Background to the Conflict

Geographical location

CNP is located in south-eastern Venezuela, near the borders of neighbouring Brazil and Guyana (see Figure 4.1). It protects the north-western portion of the Guyana Shield, an ancient geological formation shared with Brazil, Colombia and the Guyanas (Guyana, Suriname and French Guiana). The park was created in 1962 with an initial area of 10,000km^2 which was extended to 30,000km^2 in 1975 to protect its watershed. The best-known landscape components of CNP are the 'tepuyes', sacred ancient mountains in the form of a plateau, receiving their name from the Pemon word *tüpü* (which literally means 'sprouting stone' (Juvencio Gomez, pers. comm.). CNP's vegetation is markedly divided, between a forest-savannah mosaic in the eastern sector known as the Gran Savana, and an evergreen forest in the western zone. In 1994, the park was registered on the list of UNESCO Natural World Heritage Sites in recognition of its extraordinary landscapes and geological and biological value (see Figure 4.2).

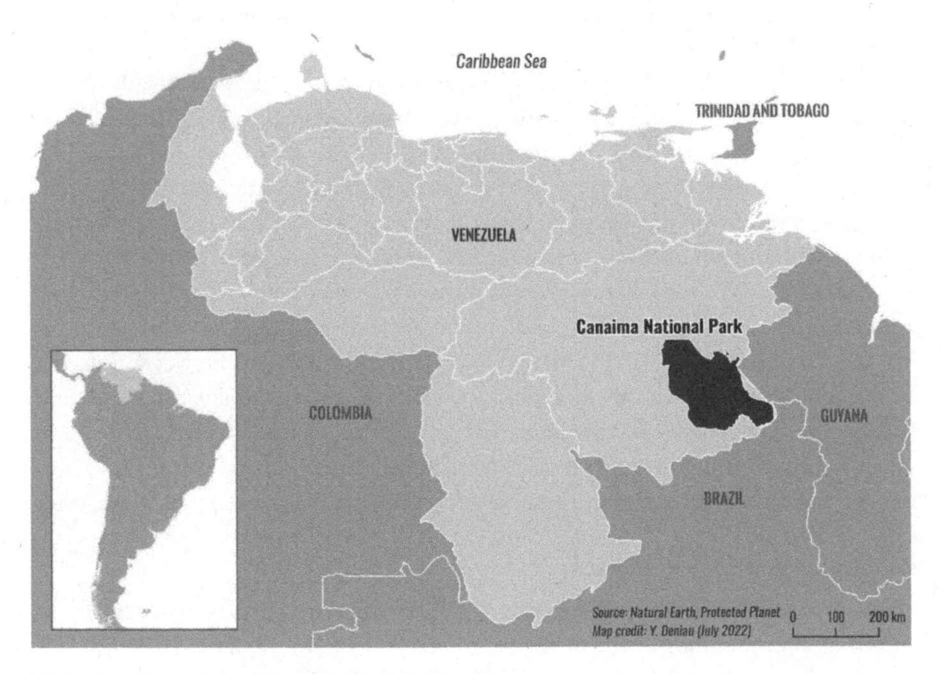

Figure 4.1 Location of Canaima National Park

Figure 4.2 Roraima *tüpü* (Photo by Roman Rangel)

CNP is the site of considerable tension between competing interests, largely because the protected area was established without any prior consultation on territory occupied ancestrally by the Pemon people, the main inhabitants of this vast area. With an estimated population of 25,000, most of the CNP Pemon live in settlements of 100 to 1,000 inhabitants, although some still maintain the traditional system of scattered, nuclear family settlements (see Figure 4.3). Their lifestyle is based largely on traditional activities: agriculture, fishing, hunting and gathering, although there is increasing work in tourism and associated activities (e.g. handicrafts), as well as public administration posts (e.g. teachers, nurses, community police and municipal staff) and more recently, mining (Rodríguez 2014).

Although CNP is the main focus of the analysis that follows, the regional and national scope of the power line conflict requires reference to actors and processes beyond the boundaries of the national park itself. Tensions between different Pemon factions both within and outside CNP, which eventually fractured into various groups opposing and supporting the power line, are particularly important. The faction that remained more unified against the power line was the one that lives in CNP, with the community of Kumarakapay taking an important part of the lead.

Factors contributing to the conflict

On 29 January 1997, in a closed meeting, the Venezuelan and Brazilian governments signed a Memorandum of Intent (MOI) to connect the Macagua II Hydroelectric Dam (located on the outskirts of the city of Puerto Ordaz, Venezuela) to the Brazilian

Figure 4.3 Kumarakapay village, 1999 (Photo by Iokiñe Rodríguez)

city of Boa Vista by an electric power line, with the aim of supporting industrial development in northern Brazil. Three months later, the electricity companies Edelca in Venezuela and Electronorte in Brazil signed a contract to construct a 680 km high-voltage power line, of which 480 km were to be built in Venezuela and the remaining 200 km in Brazil. Venezuela promoted the power line as a symbol of regional economic integration (Toro Hardy 1997), but this joint undertaking between Venezuela and Brazil was not celebrated locally.

The areas slated to be affected by the power line in Venezuela included lands traditionally occupied by Pemon Indigenous peoples that overlap with three protected areas: the Imataca Forest Reserve, the Southern Bolívar State Protective Zone, and Canaima National Park. To many Pemon inhabiting these areas, the power line represented only the most recent in a long history of threats to their land tenure, livelihoods and forms of local governance. Shortly after the MOI for the power line was made public, conflict broke out.

Several factors played a role in the way the conflict emerged and developed:

- prior governmental expansion of mining in southern Bolívar in response to falling oil prices,
- secret planning and illegal implementation of the power line,
- an emerging Indigenous rights movement in Venezuela, and
- the historical approach to dealing with 'the Indigenous issues' in Venezuela.

Up until the late 1980s, the south of Bolívar State had remained free from large-scale development and extractive activities thanks to stable oil prices and the high percentage (76%) of the state's land designated as protected areas (MARNR 2000: 137).[3] However, the oil price crash of the 1980s, combined with the ensuing macroeconomic crisis, led the Venezuelan government to diversify the economy, including opening southern Bolívar to extractive activities. In the late 1980s, southern Bolívar experienced a gold- and diamond-mining boom, particularly in small-scale operations. By 1991, taxes on foreign investment were substantially reduced, and by 1994, the government had granted 367 concessions and contracts in southern Bolívar State to national and international mining companies (Colchester and Watson 1995).

Apart from exporting electricity to Brazil, power line supporters aimed to aid the expansion of extractive activities in southern Venezuela, particularly gold- and diamond-mining.

In the early 1990s, the Ministry of Energy and Mining agreed to supply Placer Dome (one of the largest multinational gold-mining corporations) with high-voltage electricity as part of a negotiated mining concession in Las Claritas (INFORECO 1997, in Audubon 1998). Just one month before the power line project was signed, President Caldera issued a presidential decree which lacked the backing of the National Congress (Decree 1850), converting 38% of the Imataca Forest Reserve (1.4 million hectares) from forest management to mining. The power line would provide energy to expand the mining industry in this area.

Uncontrolled growth in mining, however, had already created an unfavourable context for the power line project. First, the expansion of small-scale mining had caused acute environmental degradation in some areas (Barreto and Perez Puelles 1991). Second, the majority of the mining activity was illegal: many of the concessions granted lacked the required permits from the Ministry of the Environment and some were even illegally awarded in and around CNP (Miranda et al. 1998). The mining boom also attracted many immigrant miners working without the necessary permits. Thirdly, because most mining occurred in undermarketed Indigenous peoples' territories, invading miners displaced Indigenous people on a large scale (Colchester and Watson 1995). Thus, while the state viewed the power line as crucial to strengthening the nation's economic and political future, many Pemon saw it as a pillar of a neo-colonial process of displacement from their homelands.

The mining boom was compounded by the second factor leading to the 1997 conflict: the autocratic and illegal way in which Edelca planned and built the power line. The project was planned and implemented without prior local consultation or the approval of the national congress. Furthermore, President Caldera pressured the National Parks Institute (INPARQUES) and the Ministry of the Environment (MARNR) into overlooking legal restrictions. As a result, the legal frameworks of the National Park (República de Venezuela 1991) and Natural World Heritage Site (Convention Concerning the Protection of the World

Cultural and Natural Heritage – The World Heritage Convention of 1972) were ignored. All this contributed to Pemon distrust of the legal system, particularly in protecting the well-being of Indigenous people.

The Pemon disillusionment with the legal system leads to discussion of the third factor setting the stage for this conflict: an emerging Indigenous rights movement. In the early 1990s, some Pemon leaders, the Indigenous Federation of Bolívar State (FIB) and the Venezuelan Council for Indigenous People (CONIVE), joined a network of Latin American Indigenous organizations openly challenging the relationship between the state and Indigenous peoples, and claiming greater autonomy in their homelands. By 1993, some Venezuelan Indigenous leaders were publicly demanding differentiated rights in Venezuelan society (Bello 1999).

Almost immediately after Edelca disclosed plans to construct the power line, the Pemon began to demand recognition of rights to help secure their material and cultural survival. This included territorial rights, the right to define and control their own development processes, the right to retain their customary decision-making institutions, and the right to participate in regional and national development planning. However, Caldera's government was reluctant to recognize Pemon claims. This has to do with the fourth important factor leading to the conflict: the government's historical approach to Indigenous issues in Venezuela.

Along with many other Latin American nation-states, Venezuela's legal system had from its inception been based on the notions of the unity of the state and the equality of all citizens. This has direct consequences for how Indigenous issues have been approached in the last century.

Since the birth of the Republic, the aim of national Indigenous policies has been to assimilate Indigenous peoples and 'make' them into Venezuelan citizens. Except for the 1811 and the 1999 National Constitutions, the remaining twenty-four constitutions of Venezuela have made little mention of Indigenous people and their rights. Ironically, there was greater acceptance of Indigenous peoples' cultural differences and rights for legal territorial protection during colonial times than throughout the entire Republican era (Arvelo de Jiménez et al. 1977). The 'Leyes de Indias' (Laws of the Indies), enacted by the Spanish government in the seventeenth century, sought specifically to protect Indigenous people. Through this law the Crown designated a large proportion of Indigenous lands as '*resguardos*', giving Indigenous peoples the right to own, use and administer their communal territories.

In contrast, throughout the Republican era, the Indigenous issue has been regulated through a series of decrees, resolutions and regulations, destined to 'protect', 'civilize', 'reduce' or 'assimilate' Indigenous people (Bello 1999: 42). Although since the 1970s some policies have reflected a greater sensitivity to 'cultural pluralism' (Arvelo de Jiménez et al. 1977),[4] the predominant aim of assimilation has been the replacement of Indigenous cultures with a single national '*campesino*' ideal.

Throughout most of Venezuela's democracy, the law has treated Indigenous peoples as a special type of rural peasant. This is clearly illustrated in Article 77 of the 1961 Constitution (valid until 1999), which equated the obligations of the state towards Indigenous peoples to those it had towards the broader mestizo peasant population of mixed Spanish-Indigenous origin. By suggesting that Indigenous peoples are merely a category of peasant, the legal system overlooked major differences in land use and ownership, educational systems, social and economic organization, and cosmogony. Treating Indigenous peoples as peasants is part of a broader goal of assimilation, replacing the main cultural features of Indigenous peoples with those conventionally accepted by the dominant society. In this historical context, it is clear why Pemon claims for differentiated rights during the power line conflict were seen as a threat to the well-established notion of unity of the state and as an obstruction to modern nation-state-building objectives.

The power line conflict thus became a contest between two clashing models of the nation-state: one driven by a modernist vision of development, seeking to create a homogeneous and unitary society; and another demanding rights for special types of citizens, seeking pluralism. Within this contest between models of society, three particular areas of tension emerged: visions of development, land rights, and decision-making systems. These same tensions persisted after the conflict within the new pluricultural model of a national state, because the modernist vision of development and the intentions of assimilating Indigenous peoples to a unitary model of society have not essentially changed over the last two decades.

What follows is only a short account of the power relations in the struggle against the power line and extractivism in this area. A more detailed analysis can be found in Rodríguez and Aguilar (2021).

Power Relations in the Struggle Against Extractivism in CNP: From the Power Line to the Arrival of Mining

This case study developed in two parts: the first (the power line conflict) took place between 1997 and 2001, when the biggest efforts were made to transform the relationship between the Pemon and the state. The second (the conflict against mining) demonstrates how transformation was co-opted by the state to forcefully advance its extractivist and modernizing agenda for the south of the country.

Power relations during the power line conflict

From the outset, this conflict was marked by the difficulty of developing a productive dialogue about the clash between the state's nation-building objectives and the Pemon's demands for cultural recognition. A full year passed between the first public announcement of the power line and the commencement of formal consultation with the Pemon, consultations which began a month after the Ministry of the Environment and INPARQUES had granted permits to Edelca to start construction. The Pemon therefore had no real opportunity to contribute to decisions about the project.

The tension between nation-building and the defence of cultural difference was first apparent in different notions of landownership. The Pemon's demand for information and consultation on the project was based on their ancestral ties to the area, which they considered gave them the right and responsibility to look after their homelands (INNA PEMONTON 1997). State representatives dismissed their claim, since under national law the property upon which the power line was to be built belonged to the state and general public.

This same cultural tension also became apparent in the respective actors' different ways of defining decision-making. The Pemon's demand to be consulted about the project was rooted in a sense of ownership, but also in a tradition of broadly based deliberative decision-making processes. To the Pemon, building the power line demanded exhaustive discussion at the community level, followed by a consensual decision. For the state, on the other hand, the representative democratic system would follow the tradition of an efficient and productive exchange of ideas between formal leaders, followed by a majority decision.

Finally, the tension became apparent in the groups' clashing visions of development. While the Pemon feared the cultural and environmental impacts of the development model behind the power line, government representatives defended it as an obvious necessity:

> *Governor of Bolivar State*: 'We are talking of a project that costs the state $500 million and which is going to reactivate the mines in Las Claritas, generate 3600 jobs and take electricity to Ikabaru and some Indigenous communities ... We are trying to bring development alternatives to this area. But development only comes after public services have been established, and one of the most important public services is electricity ... Sooner or later you will have to build electricity generators in your own communities.' (Jorge Carvajal)
>
> *Pemon Chief*: 'I would like to comment on what you are saying. We still have not seen a city in which people live happily and in a harmonious way with nature, where there is no poverty. Until we see it, how can we guarantee that we will be happy here in the Gran Sabana with that type of development?' (Silviano Castro, inhabitant of San Rafael de Kamoiran)
>
> (Extract from power line meeting, Santa Elena de Uairén, 5 June 1997, Rodríguez 2004)

While the Pemon persistently fought the power line by forcing a dialogue about cultural identity and rights, the state sought to perpetuate the values of a unified and homogeneous nation-state. This clash between different models of a nation-state persisted even during the few opportunities for negotiation that arose. Because the source of the conflict was recreated in the very dialogue intended to solve it, negotiations were systematically ineffective, and the conflict escalated in cycles until the very end.

Most of the strategies used by the Pemon concentrated on impacting *structural power* through a variety of direct action and political mobilization strategies.

Figure 4.4 Pemon mobilization against the power line in front of the National Congress, Caracas 1999 (Photo by Iokiñe Rodríguez)

These included writing public manifestos, producing press releases, giving press conferences, blockading roads, carrying out marches, filing lawsuits, lobbying in the national congress (see Figure 4.4), and at times taking violent action such as expelling power line surveyors and workers from their lands and pulling down power line pylons. The Pemon pulled down a total of twelve electric pylons over the course of the conflict (see Figure 4.5).

The Pemon were also effective in strategically *building alliances* with national and international Indigenous rights and environmental NGOs to give visibility to the conflict and gain public support in the media and political lobbies. At the international level, these included COICA (the Federation of Indigenous Organizations of the Amazon Basin), KLIMA BÜNDNIS (Climate Alliance), Survival International and Amazon Watch. Their national allies included the Audubon Society, Amigransa and AVVA Gran Sabana.

Though the initial aim of these strategies was simply to be properly informed and consulted about the project and the customary decision-making procedures, the government's lack of will to engage with underlying problems exacerbated the conflict.

The government tried to maintain hegemonic power during the conflict by ignoring and attempting to divide the Pemon in their struggle, using a variety of strategies. These included creating alliances with Pemon sectors from outside CNP that were more aligned with the state's development views and extractive interests, offering compensations, using repression and intimidation strategies, and creating changes in the legal framework reinforcing the government's assimilating intentions, including partial amendments to the land reform law. The strategies persisted

Figure 4.5 One of twelve power line pylons pulled down by the Pemon throughout the power line conflict (Photo by Iokiñe Rodríguez)

even after Hugo Chavez's arrival to power when opportunities for a settlement arose thorough the constitutional reform process.

As the conflict escalated, Pemon strategies focused on demanding territorial property rights. Because the monocultural nation-state's legal framework made this impossible, the Pemon began to focus on producing a change in the state's model. The constitutional reform process set up in 1998 by Chavez's new presidency was favourable to these changes. The Pemon, along with other Indigenous peoples, lobbied for a new constitution to acknowledge:

- Venezuela as a multicultural and multilingual country;
- Indigenous peoples' ownership rights over their ancestral territories (land rights arising from being the oldest surviving pre-colonial social group in an area);
- respect for Indigenous peoples' autonomy and forms of social and political organization;
- Indigenous peoples' rights to the preservation of their natural resources;
- Indigenous peoples' traditional forms of biodiversity use; and
- Indigenous peoples' territories as autonomous territorial entities.

In 1999, therefore, for the first time in Venezuela's Republican history a chapter on the differentiated rights of Indigenous people was included in a new national constitution. The most salient Indigenous rights included: the right to maintain their culture, customs, traditions, languages, religions, and social, economic and political organization; rights over their ancestral and traditionally occupied lands;

the right to be informed or consulted on any exploitative activities planned in their lands; and the right of political participation.

This new legal framework opened the way to negotiating a power line agreement that stated:

- Within a week of signing the agreement, the process of demarcation and titling of Indigenous peoples' 'habitats' would be initiated.
- The government would ratify the Indigenous and Tribal Peoples Convention No. 169 of the International Labour Organization (ILO), which ratifies the aspirations of Indigenous peoples to exercise control over their own institutions, ways of life and economic development and to maintain and develop their identities, languages and religions, within the framework of the states in which they live.
- The Executive would ensure that the Indigenous peoples are involved in monitoring the cultural and environmental impacts of power line construction.
- Indigenous peoples and the state would collaborate to manage protected areas.
- The government would contribute to the formation of a Sustainable Development Fund for Indigenous people.

When this negotiation proposal was made public, the Pemon from CNP rejected it. In line with their position throughout the conflict, the group argued that the power line was not negotiable. Furthermore, they alleged that most of the conditions were rights already granted to them in the constitution, and that the state was obliged to respect them.

An alliance between the President of the Indigenous Federation of Bolívar State (FIEB) (Jose Luis González) and President Chavez's administration during the constitutional reform process was instrumental in negotiating a solution to the conflict. The FIEB's defence of the proposal was seen by a faction from CNP as a betrayal of their struggle, and as disrespectful of their decision-making procedures.

All the same, the FIEB organized a series of meetings with Pemon chiefs from both inside and outside CNP to decide whether or not to accept the proposal. Consensus could not be reached, because most Pemon from CNP continued to consider the power line non-negotiable. In the end, a decision was made by vote, with the majority deciding to support the power line agreement (Flores et al. 2000). On 21 July 2000 the proposal was signed, and the power line was inaugurated on 13 August 2001 (see Figure 4.6).

The power line agreement had a high cost for the CNP Pemon. It ended up fragmenting the Pemon organization and leadership structure, and severely crushing the morale of those who had strongly and consistently opposed the project. One Pemon leader, who eventually became the first Pemon mayor of Gran Sabana, refers to this episode in their lives as 'the trauma of the power line'. Furthermore, with the

Figure 4.6 The functioning power line, outside the north-west boundary of Canaima National Park (Photo by Iokiñe Rodríguez)

exception of the ratification on Convention 169 of the ILO, none of the remaining points in the proposal have been accomplished to date.

Another compromising aspect of the Pemon resistance strategies during the power line conflict, in addition to the internal tensions between different Pemon factions, was the lack of attention paid to impacting cultural power through revitalizing their knowledge and culture and collectively defining alternatives to development. There were some exceptions, however, such as the II Pemon Congress in 1998 and the efforts by the village of Kumarakapay in the midst of conflict to define a Pemon Life Plan in order to reconstruct their identity and define their own views of development (see Box 4.1 for details).

This process of internal strengthening of cultural identity among the Pemon of Kumarakapay played an important part in balancing power relations in the park, particularly in the post-conflict phase (Rodríguez 2016). In addition to the value of this experience for the people of Kumarakapay, the process of cultural reaffirmation was key in reasserting Pemon cultural identity at a wider level, particularly regarding the need to advance a Pemon Life Plan more globally in CNP. As explained in Box 4.1, after the 1999 constitutional reforms, the Life Plan came to be seen as a platform for intercultural dialogue with external actors about the current and future well-being of the Pemon, as well as for articulating different institutional agendas.

Box 4.1 The Kumarakapay Pemon Life Plan

The village of Kumarakapay in 1999 pioneered the development of a Life Plan for themselves, contributing to a new counter-narrative of development in CNP during the peak stage of the power line conflict. The Kumarakapay chief, Juvencio Gomez, saw clearly that internal discussions were urgently needed among the Pemon about the type of development they wanted for the area.

The Kumarakapay Life Plan involved a year-long participatory research process, led with the village's consent by Gomez and a group of thirty elders and young Pemon to reconstruct their local histories, evaluate their present situation and imagine a desired future. Through this process, important aspects of their past fading from the community's oral memory were discussed and made visible. The elders issued a request to the younger generations to put this history into writing in order to make it known to younger Pemon and wider Venezuelan society; this request was eventually realized in 2010 (see below). The Kumarakapay Life Plan's vision for desired elements of a future society was summarized in the following points (Roraimökok 2010):

- A Pemon society with awareness of who we are, and with a sense of identity and of belonging.
- Knowledgeable about our history, culture, tradition and language.
- Owners of our land – territory, knowledge, culture and destiny.
- A society educated with ancestral and modern knowledge.
- A society that values its wise people (parents and grandparents).
- A respectful, hard-working, obedient, kind, courteous, cheerful, generous, harmonious, understanding society where there is love.
- A productive, autonomous society.
- A society that defends its rights and is ready to confront pressures from Venezuelan society.

Power games in the conflict against mining

Despite the CNP Pemon's strong disillusionment with the outcome of the power line conflict, after the constitutional reforms there was a new air of cooperation in state–Pemon relations: the Pemon expected that the new model of nation-state would eventually start producing new opportunities for them to participate in policy-making and thereby benefit from public policies. Tensions mounted again, however, when President Chavez announced new national and regional mining policies for the country, including the nationalization of gold-mining and opening up of the central area of the national territory south of the Orinoco River to large-scale mining through the Mining Arc project.

The Mining Arc (Arco Minero) project, spanning over 110 square kilometres, is a new massive mining project seeking to open up 12% of the national territory to large-scale mining of gold, coltan, diamonds, copper, iron and bauxite. The Pemon are only one among several Indigenous peoples to be affected by the project.

The Mining Arc directly overlaps with the ancestral territories of the Mapoyo, Inga, Kariña, Arawak and Akawako Indigenous peoples, and its area of influence includes the homelands of the Yekuana, Sanemá, Pemon, Waike, Sapé, and Eñepá y Hoti o Jodi in Bolívar State; the Yabarana, Hoti and Wotjuja in Amazonas; and the Warao in the Amacuro Delta. The Mining Arc followed the same pattern of total disregard to prior and informed consent procedures as that seen in the power line dispute, violating once again Chapter VIII of the National Constitution, the new national Indigenous peoples' legal framework and the international conventions on Indigenous peoples rights of which Venezuela is signatory.

In the readjustment phase after the power line conflict, the Pemon from CNP substantially decreased their confrontations with the state, largely because they were divided and demoralized as a result of the struggle. However, they did not abandon their intention to gain territorial rights, and continued developing strategies to impact *structural power* (the new legal framework) to assert their ownership of their lands, first by carrying out self-demarcation processes and then by introducing their land rights claim in 2005.

In the development of these strategies, a new board of directors of the FIEB was set up and the Pemon from CNP were strategically building *alliances* with a new set of national and international organizations from NGOs and the academic sector, attaining funding, technical advice and expertise on Indigenous self-demarcation processes in preparation for the official land demarcation and titling process, which would start in 2005.

In this period, contrary to the power line conflict, the Pemon also took important steps to impact *cultural power* by positioning a counter-narrative of development into the public policy arena, based on Kumarakapay's 1999 Life Plan. The Pemon Life Plan was highlighted as the articulating concept for any new public policies to be developed in the area.

After that point, the Life Plan agenda was incorporated into the political discourse of most Pemon leaders and started forming an important part of the well-being agendas of some Pemon villages (Pizarro 2006). The Pemon Life Plan was viewed as a critical space for self-analysis of their current situation, their changes and their cultural values, in order to help them reflect about who they are as a people and who they want to be in the future. By developing a clear vision of their identity, needs and desires, a Life Plan sought to allow them to negotiate more strategically with state and other external institutions: 'our own Life Plan will not only strengthen us as a people, but also facilitate the necessary interactions with the institutions with which the Pemon interact, helping such institutions structure their initiatives and activities with the communities' (Juvencio Gomez, in World Bank 2006).

The idea of developing an overarching Life Plan for the Pemon in CNP was adopted soon after by INPARQUES as part of a new approach to manage the national park. In response to the need for a collaborative strategy to manage CNP (which had been one of the commitments of the power line agreement), in 2006 a $6 million project was formulated for World Bank GEF funding to develop better working relationships between the National Parks Institute and the Pemon ('Canaima Project' in 2004–6). At that time, one of the Pemon's conditions for taking part in the participatory management of CNP was that the project would be implemented in coordination with communities' Life Plans. A series of workshops defined a preliminary version of the Pemon Life Plan, emphasizing the following components: 1) Indigenous Territory and Habitat; 2) Education and Culture; 3) Organization-building; 4) Health and Culture; 5) Social Infrastructure; and 6) Production and Economic Alternatives (Pizarro 2006). However, although the World Bank approved this project, because of changes in the Venezuelan government's political priorities and the decision to cut off ties of cooperation with the World Bank, it was never implemented. The Life Plan idea was aborted from public policy formulation and never incorporated by the government into the plans that followed.

In the years that followed, furthermore, park management became severely constrained by diminishing funding, insufficient personnel, and lack of both inter-institutional coordination and will from the national government to support the national system of protected areas (Bevilacqua et al. 2009; Novo and Díaz 2007). Environmental policies were overshadowed by social policies geared towards major development, which jeopardized the Pemon Life Plan by reinforcing rather than reducing their historical relationship of dependence on the government.

Similarly, the process of territorial demarcation and titling ended up being slow, complicated and even futile for most Indigenous peoples (Caballero Arias 2007; Mansutti 2006), including the Pemon. A combination of factors contributed to this: 1) lack of political will, 2) processes of disarticulation of Indigenous peoples, and most importantly, 3) a biased geopolitical view still prevailing in some sectors of the government, like the military, who continued to believe in the idea of 'nation' as a unifying space at risk of fragmentation if territorial rights were granted (Caballero Arias 2016).

Despite the promises of the new constitutional reforms, the Chavista government thus failed to respond to the Pemon territorial rights claim. After the Chavista government announced the nationalization of mining and plans for large-scale mining developments in Bolívar, confrontational strategies began to re-emerge among the Pemon.

The year 2006 marked a tipping point in these renewed state–Pemon relations. This was the moment at which the government began intensifying its hegemonic view of development in the area through a variety of strategies intended to advance the extractivist agenda in the south, including a new large-scale development project in CNP (a transcontinental pipeline that was eventually abandoned, and the construction of a satellite substation, finished in 2008), the implementation of new social welfare programmes, and the creation of new community organization and political structures. Along with these new development projects and social policies came renewed development discourses offering the Pemon access to new information technologies, education, productive projects and economic resources via communal councils. These were part of the new (and old) strategies to assimilate Indigenous peoples to the 'new' national state.

Direct and structural violence were also increasingly used to impose new mining regulations through militarized Indigenous territories, with complex alliances forming between informal miners (including some Indigenous sectors), transnational corporations and the Colombian guerrilla (FARC and ELN), which is now occupying an important part of the state of Bolívar. In addition, over the last decade armed mining syndicates and gangs have spread over the whole of the south of the country, and also seek control in mining areas.

In this new phase, the Pemon put a new plan into motion: to control mining in their lands. The reasons for this shift towards mining, which largely reinforced the hegemonic view of development, are explained in the following extract:

> You all know that the Indigenous leaders always declare themselves as pro-environmentalists, that we defended Mother Nature, the land, the rivers. But precisely because of the discourse we had, we believed in President Chávez when he proposed all that he was going to do for us. But in the face of the situation that is now being presented, that gold being nationalized to be delivered to transnational companies, the same ones that the government calls 'savage capitalist,' so what is the coherence of the discourse? What do we do, keep guarding our lands so that others come to exploit them? The lands that they will destroy! Where do we stand in this? They talk about mining royalties, but since when have the communities benefited from royalties? In the face of this situation, we are proposing that we ourselves begin to exploit these resources, but looking for the least harmful form of exploitation. That we ourselves regulate mining. We are in the middle of that process now. Let's see how we do it, why we do it, or if we continue to exploit only to take it to the bar, which is what mining has always been used for. We are talking about the communities themselves presenting a work plan, a life plan for those who want to exploit mining, to define what they want to achieve with that. Already several communities have responded. Many of the communities that were not mining are now in the mines. I think there are 100 small communities, in the Gran Sabana, that are now living from mining. (Interview with Alexis Romero, quoted in El Toro 2013).

Thus, from a Life Plan that had placed restoring cultural identity at the centre of their well-being, the Pemon now began to build a Life Plan to help them define what type of mining they should carry out. To accomplish this, they have used a variety of strategies to impact on *structural power* and *actor-networks*. The first has included the use of direct violence (kidnapping national guards), forcing agreements with the government for exclusive rights over mining in their communal areas, and creating Indigenous mining enterprises (many communities in CNP are currently engaging in mining activities). The second has involved creating new organizational structures for decision-making and territorial control (Indigenous guards).

The future of this competition over mining in southern Bolívar State is difficult to predict, and will depend on how the various actors use their respective power resources. The Mining Arc project does not offer any security to Indigenous peoples to control mining activities on their lands. The risks of dispossession have, if anything, increased with the recent arrival of new players in mining areas. Most of the mining areas of Bolívar State, both north and south of CNP as well as within it, have become war zones linked to smuggling and illegal gold-mining (Ebus 2019a, 2019b).

Furthermore, although many communities had agreed to practise mining as a form of territorial control, no agreement has been made as to how to deal with this complex network of external actors. Similarly to the power line conflict, more recently the Pemon have begun to face divisions over different approaches to territorial control. Whereas in the western sector of the CNP the communities have reached agreements with military sectors to exercise mining and prevent the entry of armed gangs and the Colombian guerrillas into their territories, the eastern sector chose to make use of their customary practices to form a territorial guard, similar to Indigenous people in the Colombian Cauca, who protect their territories against entry by the Colombian guerrilla.

Tensions between the Pemon groups in the east and in the west of CNP, and their different approaches to territorial control, have made it difficult for them to collaborate in facing the expansion of the mining company in the CNP. The bloc working with the support of the military sector has in recent times been consolidating, while the other is weakened after a military attack in the Kumarakapay community in February 2019, the result of a protest against government blockades of humanitarian aid entering southern Venezuela. This attack ended up dismembering the territorial guard, and led to the displacement to Brazil of an important part of this community.

The political, legal and institutional reforms emerging from the new 'pluricultural' nation state proved ineffective in advancing Indigenous rights. The dominant discourse on Indigenous–state interactions over the last two decades has been the modernizing imaginaries and intentions of key government actors, which has influenced the Pemon's view of development and demobilized the Indigenous rights movement. This has been instrumental in the government's advancement of its extractivist agenda for southern Venezuela.

Transformations for an Alternative Future?

Despite this bleak outlook, we can conclude that the Pemon's resistance strategies produced mixed outcomes, by using the five conflict transformation pillars discussed above (cultural revitalization, political agency, governance, means of production and environmental integrity).

Cultural revitalization

Some of the Pemon strategies, like political mobilization, production of public statements and manifestos, and the use of public media and litigation effectively led to important structural changes in the cultural sphere. For instance, they revealed institutional practices and political legal frameworks that have systematically excluded Indigenous peoples from policy-making, and they influenced major constitutional changes to acknowledge Indigenous rights. This was followed by important legal innovations for the protection of Indigenous rights, such as the 2001 Law for Demarcation and Guarantee of Indigenous Peoples Habitat and Land, and the 2005 Organic Law of Indigenous Communities and Peoples (LOPCI). Internationally, in 2001 Venezuela ratified ILO Convention 169 on Indigenous and Tribal Peoples,[5] and in 2007 the United Nations Declaration of Indigenous Peoples. This new legal framework, however, has not yet been put into practice. The Pemon continue to wait for their territorial property rights, and no new plural institutional decision-making arrangements have emerged after the constitution of the pluricultural nation-state. In fact, with the creation of communal councils in 2006 (a new country-wide local decision-making structure) the government has superimposed an external form of social organization onto the existing customary decision-making institutions, often creating tension and conflicts of interest at the community level (Sanchez and Vesuri 2017).

Interculturally, however, even though the Pemon are still waiting for recognition of their property rights, conflict helped to instil a sense of restorative justice in the Pemon. Being recognized in the national legal framework as citizens with differentiated rights, even if only nominally, was important in terms of strengthening the Pemon's dignity and cultural identity. Important intracultural transformations, such as recovery of collective memory and revitalization of knowledge and identities, also occurred as a result of Pemon strategies related to development of Life Plans and the self-demarcation process.

Although the Life Plan concept became increasingly anchored in the Pemon's discourse as an alternative to conventional approaches to development, only the Kumarakapay village has made progress in this regard, out of thirty individual communities. The Life Plan concept cannot, therefore, be said to have had a widespread impact in reshaping the nature of Pemon–state relationships. If anything, during the last decade the Pemon seem to have been increasingly persuaded to adapt to the dominant development model, to the point that they are now following the dominant extractive development pathway. Their intention to control mining in their territory is clear, but to what degree it will differ from other forms of mining is still to be seen.

Local political agency

The Pemon's political agency was undoubtedly strengthened during the power line conflict. The Pemon showed tenacity in opposing the power line and in bringing about the political and legal changes necessary for the recognition of their cultural and territorial rights. They gained visibility as political actors, opening opportunities for them to participate in the National Constitutional Assembly, in the National Assembly and in the regional legislative assembly. Rather than advancing the Indigenous rights agenda, however, the Pemon's participation in decision-making bodies ended up incorporating them into the state's agenda and interests, and demobilizing the leadership.

Governance (institutional strengthening)

The Pemon's claim to political self-determination and autonomy was an important part of their struggle against the power line. Apart from land rights issues, the government's continual disregard for the Pemon's customary decision-making procedures and institutions was also contentious. Constitutional reforms nominally changed this: Article 119 of the 1999 constitution acknowledges the right of Indigenous people to maintain their 'forms of social, political and economic organization'. In this sense, perhaps, the strategies used by the Pemon effectively impacted structural power to strengthen their own governance systems. Yet the Pemon governance structures have not been strengthened as a result of the conflicts. For instance, no progress has been made since the constitutional reforms to formally grant autonomy and political rights to Indigenous peoples. Indeed, in contrast to other Latin American countries like Honduras, Mexico and Bolivia, where formally recognized Indigenous autonomous regimes are in place, in Venezuela the subject has yet to be addressed.

Intraculturally, the strategies used in the power line conflict seem to have offered little opportunity for Pemon communities to engage in a critical dialogue about the effectiveness of their norms, rules and regulations over natural resource use and territorial management. This seems to be changing, though, as a result of more recent conflicts over the expansion of mining that are forcing the Pemon to realign themselves and their organizations to protect their territory.

Local means of production

The Pemon of CNP have historically been effective in ensuring their control over means of production. Over the past three decades, for example, tourism services inside the national park have been provided almost exclusively by the Pemon, particularly in the eastern sector. Ensuring such control over their livelihood activities was an important part of the logic behind their struggle against the power line. Although in the years after the power line conflict there were no direct threats to the local means of production, this began to change when the government announced its policies to expand mining. The Pemon's conversion to mining in the national park is an expression of their determination to keep control over their means of production. Whether Pemon mining is carried out in an equitable way, and whether it integrates communal productive principles and systems, is yet to be determined.

Environmental integrity

Ensuring the environmental integrity of the Pemon territory and CNP is the pillar that has seen the least transformation as a result of the conflicts. Although protecting the environmental integrity of the Pemon territory was an important part of the reason for opposing the power line, this issue seems to have lost importance after the end of the conflict. No new or improved networks safeguarding CNP's environmental integrity emerged from the struggle, nor have either the state or the Pemon made express efforts to restore areas degraded by human activities like the power line or mining. Similarly, no new institutions, policies or legal frameworks have emerged seeking to ensure a greater integration of environmental, social and economic dimensions in territorial planning and use.

The 2000 power line agreement offered an important opportunity to strengthen the institutional and policy frameworks for a more just and intercultural approach for protected area management, as can be seen in two points from the agreement:

- The Executive will ensure that the Indigenous peoples are involved in monitoring the cultural and environmental impacts caused during the construction works of the power line.
- The management of protected areas will be carried out in a collaborative way between Indigenous peoples and the state.

The World Bank GEF project might have contributed to this aim, because it sought to develop a co-management approach for the national park. Yet, as mentioned above, the project was aborted after having been approved for funding, because of the Venezuelan government's political priorities and the decision to cease cooperation with the World Bank. In the years that followed, park management became severely constrained by dwindling funding, insufficient personnel, and lack of inter-institutional coordination and will from the national government to support the national system of protected areas (Bevilacqua et al. 2009; Novo and Diaz 2007). Indeed, with their conversion to mining the Pemon are likely to contribute considerably in the future to environmental degradation in their homelands.

Finally, in contrast to other Latin American countries like Bolivia and Ecuador, where rights of nature have been integrated into new cultural framings as a result of Indigenous struggles and a shift to a plurinational nation-state, in Venezuela's case, the Pemon's struggles had no impact on the emergence of new national cultural framings of nature.

Key Lessons

Some key lessons can be drawn from this analysis, relating to the role of the Pemon resistance and mobilization in the struggle for just transformations in CNP during the power line conflict and the two decades that followed.

The power line conflict played an important part in constitutional reforms and in the emergence of a new national legal framework for the protection of Indigenous rights in Venezuela. In this respect, we can say that the conflict was instrumental in transforming the political system to allow a greater recognition of Indigenous peoples' rights.

However, this new legal framework has not led to any significant changes in the relationship between the state and Indigenous peoples in the country. If anything, the state has showed a systematic effort to demobilize the Indigenous rights movement even after these significant political and legal changes, by exercising different forms of direct, structural and cultural violence on the Pemon. Although the Pemon have been effective in confronting some forms of structural and direct violence, they have been ineffective in confronting cultural violence, which largely explains why the extractivist development model ended up encroaching over the whole of the south of Bolívar State.

Although the Pemon effectively challenged the development model discursively through opposition to mining expansion and large-scale development, they did not invest the same effort inwardly to develop and create alternative forms of development. The Kumarakapay Life plan is an exception to this, though it is largely confined to one village. Until 2010, tourism was the main source of income in CNP, but this has changed over the last decade with Venezuela's economic and political crisis. This partly explains why mining is now so widespread in the national park.

The expansion of mining in the area must, however, be seen as mainly the result of an increasingly aggressive advance by the Venezuelan government of extractivism and of modern visions of development on Indigenous territories, which currently relies on a dangerous mixture of chaos, crime and disorder in the use of natural resources. As a result, there has been a recent resurgence of Pemon mobilization in favour of the protection of their ways of life, territories and livelihoods, though playing to some extent the game of the state. The long-term effectiveness of the Pemon's control over mining in their territories in this new context of aggressive and anarchic advance of extractive activity will largely depend on their strategies and on how state actors and their allies continue to articulate against them. In turn, the evolution of political conditions in the country will also have a significant influence on all of the above.

More specifically, we can derive the following lessons from this case study, which are also applicable to transformation politics more broadly.

About power relations

Although Indigenous peoples' struggle against extractivism urgently requires fighting for differentiated rights in the modern and liberal nation-state, this is insufficient for safeguarding their cultural and physical integrity and survival. During and after the power line conflict, the Pemon invested most of their efforts into reshaping the Indigenous–state relationship by producing changes in the legal and institutional frameworks of the state, paying little attention to strengthening their own decision-making procedures and frameworks. This was and continues to be the great weakness of the Pemon mobilizations against the power line and extractivism. The decentralized decision-making structure of Pemon society poses challenges for identifying the right decision-making procedures to mediate Pemon–state relationship in large-scale conflicts. The Pemon have no tradition of making decisions at the scale demanded by this conflict, which led to difficulties in reaching a unified and consensual view in the power line negotiations. This explains the fragmentation that occurred in the movement. Deliberative and consensus-based decision-making seems to work best at the community level, when settlements remain relatively small. In bigger settlements (such as Kumarakapay, with approximately 700 inhabitants), there is already a noticeable trend towards voting on certain decisions because of difficulties in arriving at decisions through consensus. In complex decisions involving the entire Pemon population, such as those involved in this conflict, these difficulties become nearly insurmountable. The state took advantage of these internal weaknesses to fragment and weaken Indigenous peoples in their struggles and to advance its extractivist agenda, even within the framework of a pluricultural nation-state. Moving beyond rights-based conflict transformation strategies to ones that help strengthen and reshape the internal decision-making procedures of the Pemon in large-scale conflicts would be important for advancing their interests.

Intracultural revitalization is key in fighting against extractivism. Despite the fact that the Life Plan concept has not widely been taken up to develop a counter-narrative of development in CNP, it continues to be pertinent to the transformation of conflicts with the state. The Pemon, like many other Indigenous peoples of the world, are increasingly experiencing a disconnection from nature and the local environment because of rapid processes of cultural change and assimilation policies (Pilgrim and Pretty 2013. This gives rise to intra-community and intergenerational tensions, and conflicts over development projects and the use of the environment, potentially limiting the possibility of articulating clear views against external development agendas. Development counter-narratives cannot emerge or be sustained over time unless long-term and continual endogenous processes of cultural revitalization are put in place to help strengthen Indigenous peoples' own knowledge and value systems and cultural identities.

In developing counter-hegemonic strategies against extractivism, Indigenous movements must develop alliances with a variety of actors that can help increase their mobilization capacities and work at the community level to develop proposals and visions of the future. In the last decade, the Pemon have suffered significantly from the loss of support networks that helped them strategize their actions against the state's extractivist agenda. The government has in part contributed to this by expelling international NGOs and international funds from the country and by reducing funding for conservation.

About transformations

Unless coupled with economic and cultural transformation, political transformations will not transcend and develop into new long-term intercultural relations. Transformation can in fact become a new form of domination, or of renewed hegemonic power. In this sense, pluricultural nation-states and extractivist development models can be a dangerous mix if political transformations are not accompanied by processes that effectively allow Indigenous communities to develop alternative solidarity economies and plural democratic systems at the local level.

There are limits to transformation 'as we know it' under a breakdown of democratic institutions and rule of law. As the latter part of this case study shows, the prevailing context of chaos, violence and institutional breakdown in southern Bolívar State makes it very difficult to envisage how transformations towards sustainability and the construction of interculturality can make any progress. For just transformations to help Indigenous movements develop alternatives to extractivism, they must address the increasing interconnection between extractivism and armed conflict in Latin America.

A cultural revitalization agenda for and by the Pemon people is urgently needed. The process of organizational and community fragmentation imposed on the Pemon by the Venezuelan state has been consubstantial with the fracture of their territories, and the persecution and criminalization of the Indigenous struggle. Venezuela has a twenty-year history of political polarization to which the Indigenous people are not oblivious. Indigenous people in general, and the Pemon in particular, need to push for the revitalization of a threatened cultural identity. Some strategies for the restitution of the right to cultural identity include:

- Activation of a special Pemon Indigenous jurisdiction for the restitution of the right to cultural identity.
- A new ordering of the ancestral and traditional territory of the Pemon Indigenous people of Venezuela, through a co-governance management approach in national parks.
- The pending task of intra- and intercultural dialogues among the Pemon: from 'inward' to 'outward' dialogues.

Notes

1 According to the EJAtlas (https://ejatlas.org), only 18% of the 3,506 environmental conflicts that have so far been recorded at a global scale have concluded in successes for environmental justice, where court cases were won, communities strengthened, access to the commons reclaimed or development projects scrapped.

2 The *coloniality of power* is exercised through two primary mechanisms: first is the codification of racial difference between Europeans and non-Europeans, aimed at making the latter appear naturally inferior. This manifests in normative rules such as definitions of development or progress. The second is the use of modern Western institutional

forms of power (like the nation-state) in non-Western societies to organize and control labour, resources and products (Quijano 2000). Hence, although coloniality is intrinsically linked to global capitalism, it cannot be reduced to economics alone, because it also involves other invisible cultural mechanisms of domination.

3 This includes 22% in the form of national parks and national monuments, 24% in forest reserves and 30% in protection zones.

4 One such policy was the establishment in 1979 of a Programme for Bilingual Intercultural Education, which sought to adapt the education system to the cultural reality of Indigenous peoples. However, due to a lack of continuity and funds, the programme was gradually suspended (Arvelo Jiménez 1993; Colchester and Watson 1995).

5 The only one of five points from the 2000 power line agreement that the government has honoured so far.

References

Aguilar Castro, V. (2004) *Les enjeux dans les relations internationales actuelles sur la question autochtone. Resistance et dissidence au Venezuela.* IUHEID, Université de Genève.

Arvelo de Jiménez, N., Coppens, W., Lizarralde, R., Heinen, D. (1977) Indian Policy. In Martz, J.D., Myers D.J. (eds) *Venezuela: The Democratic Experience.* New York: Praeger Publishers.

Audubon (1998) *Análisis de impactos ambientales, sociopolíticos y económicos del tendido eléctrico a Brasil.* Caracas: Sociedad Conservacionista AUBUDON de Venezuela.

Barreto, A., Perez Puelles, S. (1991) *Estudio integral de la actividad minera en la Cuenca Alta del Río Caroní. Segunda Jornada de Profesionales de EDELCA.* Puerto Ordaz, Venezuela: EDELCA.

Bello, L.J. (1999) *Los derechos de los pueblos indígenas en Venezuela.* Copenhagen: Grupo Internacional de Trabajo sobre Asuntos Indigenas (IWGIA) & Forest People Programme (FPP).

Bevilacqua, M., Medina, D.A., Cardenas, L. (2009) Manejo de Recursos Naturales en el Parque Nacional Canaima: desafíos institucionales para la conservación. In Senaris, J.C., Lew, D., Lasso, C. (eds) *Biodiversidad del Parque Nacional Canaima: Bases Tecnicas para la Conservacion de la Guayana Venezolana.* Caracas: La Salle Natural Science Foundation and The Nature Conservancy.

Caballero Arias, H. (2007) La demarcación de tierras indígenas en Venezuela. *Revista Venezolana de Economía y Ciencias Sociales,* 13(3): 189–208.

—— (2016) Entre los marcos jurídicos y las cartografías indígenas. *Revue d'ethnoécologie* [En ligne], 9, https://doi.org/10.4000/ethnoecologie.2633.

Colchester, M., Watson, F. (1995) *Venezuela: violaciones de los derechos indígenas. Informe para la OIT sobre la observación del Convenio de la OIT.* Chadlington, UK: World Rainforest Movement.

Ebus, B. (2019a) Venezuela's Mining Arc: A Legal Veneer for Armed Groups to Plunder. *The Guardian,* 8 June, www.theguardian.com/world/2019/jun/08/venezuela-gold-mines-rival-armed-groups-gangs.

—— (2019b) *Venezuela, el paraíso para los contrabanditas.* Human Rights Foundation/Dutch Fund for Journalism Projects. https://smugglersparadise.infoamazonia.org/about.

El Toro, P. (2013) Interview with Alexis Romero, Pemon Resistance: 'Ahora se nacionaliza el oro para entregarlo a transnacionales'. *El Libertario*. http://periodicoellibertario.blogspot.com/2013/05/entrevista-con-alexis-romero-pemon-en.html.

Flores, M., Castro, D., Ortíz, G., Monagas, J.M., Castro, E. Castro, S. (2000) *Ampáro Constitucional a favor de la suspensión del tendido eléctrico*. Caracas: Tribunal Supremo de Justicia, Sala Constitucional.

INNA PEMONTON (1997) *Declaración de los pueblos indígenas de 'La Gran Sabana'*. Unpublished, 28 June, Caracas.

Mansutti, A. (2006) La demarcación de territorios indígenas en Venezuela: Algunas condiciones de funcionamiento y el rol de los antropólogos. *Antropologíca*, 105–6: 13–39.

MARNR (2000) *Primer informe de Venezuela sobre Diversidad Biológica*. Caracas: Oficina Nacional de Diversidad Biológica, Ministerio del Ambiene y de los Recursos Naturales.

Miranda, M., Blanco-Uribe, A.Q., Hernández, L., Ochoa, J.G., Yerena, E. (1998) *All That Glitters Is Not Gold: Balancing Conservation and Development in Venezuela's Frontier Forests*. Washington, DC: World Resources Institute.

Novo, I., Díaz, D. (2007) *Final Report on the Evaluation of the Canaima National Park, Venezuela, as a Natural Heritage of Humankind Site. Project on Improving Our Heritage*. Caracas: Vitalis.

Pilgrim, S., Pretty, J. (eds) 2013. *Nature and Culture: Rebuilding Lost Connections*. Earthscan.

Pizarro, I. (2006) El Plan de Vida del Pueblo Pemon. In Medina J., Vladimir, A. (eds) *Conservación de la Biodiversidad en los Territorios Indígenas Pemon de Venezuela: una Construcción de Futuro*. Caracas: The Nature Conservancy.

Quijano, A. (2000) Coloniality of Power and Eurocentrism in Latin America. *International Sociology*, 15: 215–32.

República de Venezuela (1991) *Decreto 1.640. Plan de Ordenamiento y Reglamento de Uso del Sector Oriental del Parque Nacional Canaima*. Caracas: República de Venezuela.

Rodríguez, I. (2004) *The Transformative Role of Conflicts: Beyond Conflict Management in National Parks. A Case Study of Canaima National Park, Venezuela*. Doctoral thesis. Institute for Development Studies, University of Sussex, UK.

—— (2014) Canaima National Park and World Heritage Site: Spirit of Evil? In Disko, S., Tungendhat, H. (eds) *World Heritage Sites and Indigenous Peoples' Rights*. Copenhagen: IWGIA, Forest Peoples Programme and Gundjeihmi Aboriginal Corporation.

—— (2016) Historical Reconstruction and Cultural Identity Building as a Local Pathway to 'Living Well' amongst the Pemon of Venezuela. In White, S., Blackmore, C. (eds) *Cultures of Wellbeing: Method, Place and Policy*. Hampshire: Palgrave Macmillan.

Rodríguez, I., Inturias, M. (2018) Conflict Transformation in Indigenous Peoples' Territories: Doing Environmental Justice with a 'Decolonial Turn'. *Development Studies Research*, 5(1): 90–105. https://doi.org/10.1080/21665095.2018.1486220.

Rodríguez, I., Aguilar, V. (2021) *Juegos de poder en la conquista del Sur: dominación, resistencias y transformación en la lucha contra el extractivismo en el Parque Nacional Canaima, Venezuela*. Venezuela: Fundacion Buría. https://ueaeprints.uea.ac.uk/id/eprint/81034/1/Published_Version.pdf.

Roraimökok, D. (2010) *La Historia de los Pemon de Kumarakapay*, ed. Rodríguez, I., Gómez, J., Fernández, Y. Caracas: Ediciones IVIC.

Sanchez, I., Vesuri, H. (2017) Derecho a la autogestión del pueblo Pemón y la construcción del Estado Comunal en Venezuela. ¿Un contrasentido? In En Ledesma, M. (ed.) *Justicia e Interculturalidad Análisis y pensamiento plural en América y Europa*. Lima: Tribunal Supremo Constitucional del Peru.

Toro Hardy, A. (1997) La interconexión electrica con Brasil. *El Globo*, 3 February.

Walsh, K. (2005) Interculturalidad, Conocimientos y De-colonalidad. *Revist: Signo y Pensamiento*, 24(46).

World Bank (2006) Annex 20. Project brief on a proposed grant from the Global Environment Facility Trust fund in the amount of USD 6 million to the government of Venezuela for a Venezuela-expanding partnerships for the National Parks System Project.

5

Lebanon and the 'Trash Revolution': Constraints, Challenges and Opportunities to Transformation: 2015 Onwards

Rania Masri

Introduction

The 2015 national trash protests in Lebanon attracted international attention. Many during this time, particularly the tens of thousands of protestors and their supporters, felt that this so-called 'Trash Revolution' could bring about not merely wise waste management policy, but socio-political change. This chapter examines the ways in which we can assess the transformative potential of those protests, beyond environmental and waste management public policy changes. A critical question arises, one often raised by activists and observers: were these 2015 protests successful? While the 2015 protests have inspired many published analyses and interpretations, I am interested in understanding certain aspects of the protests' impact. Did the 2015 protests impact public policies related to waste management? Were individuals and their networks, and public discourses and narratives, transformed? What are the constraints and challenges of citizen-led transformative change in Lebanon? What were the organizational challenges of the 2015 protests? How would one reflect on the state of the protests, one year after the 2019 uprising? This analysis is written based on studies and reflections as of December 2020.

Key Concepts That Guide the Case Study

Environmentally sustainable policies are intricately tied to sustainable economic and political polices, and so the pathway for transformation for sustainability is tied to economic and political changes, and not just to compartmentalized environmental policies. Transformation implies 'radical, systemic shifts in deeply held values and beliefs, patterns of social behavior, and multi-level governance and management regimes' (Westley et al. 2011). 'Transformative power is the capacity of actors to develop new structures and institutions, be it a new legal structure, physical infrastructure, economic paradigm or religious ideology' (Avelino 2017).

For multilevel transformation to occur, changes at three different levels are needed, forming the conflict transformation framework that I will use for this analysis: the legal and economic institutional framework; the networks within which people organize; and the discourse narratives and the ways we see the world (Rodríguez and Inturias 2018). I do not claim to be a neutral observer, if such a person indeed exists; in this analysis I am an academic-activist. I have not merely studied the protests from afar, but have actively organized discussions about them. Furthermore, I participated and organized within the protests, since 2015. In 2016, I joined a nascent oppositional political party, which I represented as a political candidate in the 2018 parliamentary elections. I also was on the ground organizing in the 2019 uprising. My experience thus further grounds the discussion provided in this chapter.

Background of Waste (Mis)management and the 2015 Waste Crisis

For decades, the Lebanese political authorities have pursued policies that deliberately weaken public services and encourage privatization. Waste management was not immune to these policies. Although Lebanese law grants responsibility for waste management to municipalities, in Beirut and Mount Lebanon (the most populated and most municipal-solid-waste-producing areas in Lebanon), waste management was privatized in 1994. In violation of laws specifying municipalities' responsibilities for waste management, a private company, Sukleen, was awarded the contract without transparency or bidding, and made enormous profits from it.[1] Their US$500,000/day contract included collection, sorting, transportation, treatment and dumping of waste, but 70% of the waste they collected was simply dumped into landfills without treatment or sorting, even though 80% of the trash collected was recyclable or compostable (Sweepnet 2014). Landfills were set up, each with an official plan to be opened only for a few years, but, each time, the landfill was kept open until significant public pressure deemed it necessary to close, and then another landfill was opened, and the process continued (see Figure 5.0).

The Bourj Hammoud landfill is located in a densely populated residential and commercial area inhabited by a majority Armenian and working-class population. The landfill was born from the chaos of the civil war when various governmental agencies and other groups began dumping garbage there haphazardly in 1977. By 1991, after the end of the civil war, approximately 3,000 tons of waste were being thrown in the landfill daily. Twenty years after its establishment, the Bourj Hammoud landfill closed due to protests from nearby communities.[2] Rather than responding to the solid waste management problem by seeking a practical and sustainable solution, the government simply created another landfill as an interim measure, without scientific study, without adequate treatment or sorting, and without the support of local residents.

The Naameh landfill south of Beirut was thus opened in 1997. As with other landfills, the Naameh landfill was intended to be 'provisionary', an 'emergency plan', to be open for only six years (1997–2003). Successive governments, however, continued to postpone its closure. Though the landfill was slated to receive *only*

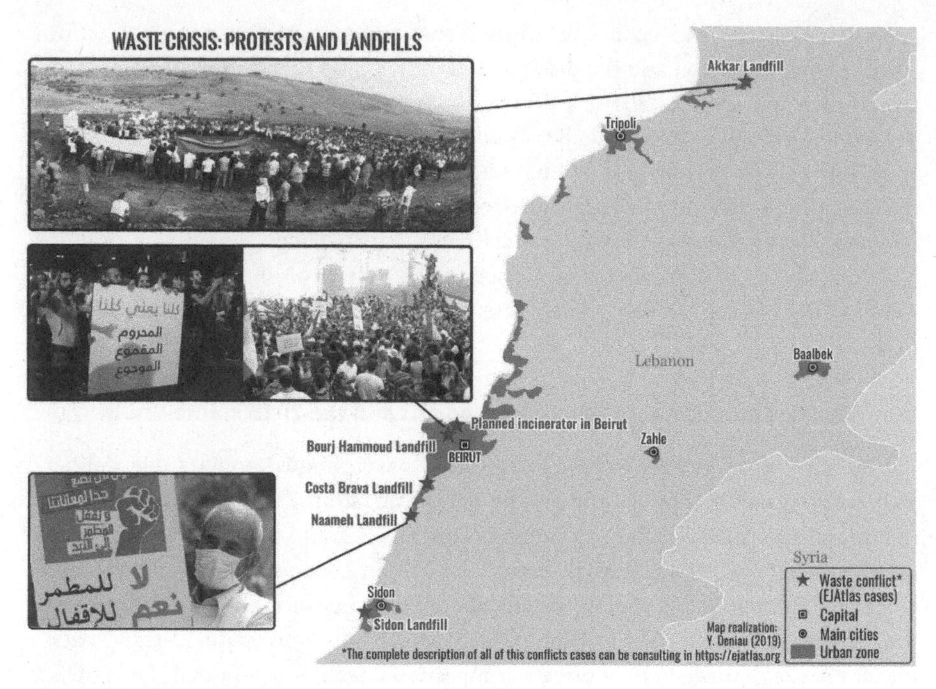

Figure 5.0 Protests and landfills

two million tons of waste, more than fifteen million tons had accumulated by 2015. It had been receiving 2,850 tons of waste each day, five times its intended capacity, and was producing 20,000 cubic metres of methane per hour owing to the high organic content of the waste and the lack of treatment.[3]

The government's contract with Sukleen continued through this time. Indeed, when the contract was brought to the table for negotiation in 2010, Prime Minister Saad Hariri said, 'either the contracts are extended or you will drown in garbage' (Zbeeb 2012). So, the contracts were extended, again and again. In January 2015, when the government took yet another procedural decision to close the Naameh landfill *in six months*, it was not surprising, six months later, to discover that no preparations had been made for an alternative. Local protests to close the landfill, which had been ongoing throughout 2014, continued. In 2015, police did not break up the protests, and larger numbers of people rallied to close the landfill. Importantly, the political support for the protests leading to the landfill's closure was more a result of internal disagreements among politicians over profit distribution in the contracts, rather than being a grassroots success as it might superficially appear.

The Naameh landfill was closed in July 2015, and trash collection halted. Hundreds of thousands of tons of garbage filled the streets of Beirut and Mount Lebanon. Within eight months, more than 700,000 tons of garbage covered the streets, underpasses, valleys and rivers in Beirut and Mount Lebanon. By February

2016, the situation in Beirut had attracted international attention with its 'river of stinking garbage bags snak[ing] its way through the suburbs, [an] overflowing landfill, stretching for hundreds of meters through Jdeideh in the city's suburb' (Hume and Tawfeeq 2016).

A public health emergency also began to emerge. Unmanaged waste threatened the water supply, and municipalities and individuals burning their garbage filled the air with stinking smoke containing high levels of dangerous pollutants. Studies by the American University of Beirut's Air Quality Associated Research Unit warned of unprecedented carcinogens in the ambient air resulting from the trash-burning trend (Safa 2015).

The Rise and Fall of the 'Protest Movement'

Protests against the waste management crisis began in local communities and suburbs, like Naameh, that had been suffering from their proximity to the landfills, and then spread to the capital. The protests expanded as garbage continued to accumulate in the main streets of Beirut; thousands of people protested regularly in the streets during the first few weeks of the crisis. It was the first instance of large-scale non-sectarian political mobilization in Lebanon since 2012, when the Union Coordination Committee of public employees and schoolteachers had gone on strike to demand the implementation of their legal wage increase. Indeed, protests in Beirut gave the movement its national dimension both because of Beirut's central economic and political position, and because of media focus. From Beirut, the protests later extended to many regions such as Nabatieh and Sidon in the south, Tripoli and Akkar in the north, and the Bekaa in central Lebanon.

Nationwide, there was an outcry for proper waste management, and, in some instances, a clear discourse linking the garbage crisis with the greater crisis of governance.

The protests became known as 'the movement' (*El-Harak*). At one point, on 29 August 2015, 100,000 people protested in the streets of Beirut. However, after those first few weeks of protests, participation dwindled, perhaps because of fatigue, disillusionment, instances of targeted fear,[4] or increasing repression, police violence and arrest campaigns (Kerbage 2017). Most of those who remained were long-time activists. With the decreased size of the protests, police violence against activists increased, further dissuading others from joining, and thus effectively immobilizing the movement.

Before assessing the impacts of these 2015 protests, it is important to understand their context: what were the national constraints and what were the organizational challenges to transformation?

Constraints to Transformation

Understanding the national constraints to transformation, i.e. real institutional and policy changes, is important to then address organizational challenges.

Is the political system and its accompanying national narrative conducive to change, or are they a barrier and an impediment to change? If the latter, how should discourses, narratives and strategies be developed to support positive change?

Lebanon's political system is a consociational democracy in which sectarian groups are allotted 'representations' in public and political offices based on population size,[5] where larger groups are allotted greater representations. This form of government is also being touted[6] as suitable for Iraq, Syria and Libya. Some political theorists argue that this system is the foundation for peaceful 'co-existence' and 'stable' democracy, while others label it an inept form of conflict management which dangerously separates individuals into groups rather than encouraging a shared civic identity (Horowitz 1985). This political sectarianism is also accused of creating dysfunctional political institutions (Salloukh et al. 2015), entrenching antagonistic 'ethnic' divisions (Haddad 2009) and encouraging corruption and clientelism (Leenders 2012).

This sectarian system has, by design, created a small, politically connected elite which appropriates the bulk of economic surplus[7] and redistributes it through communal clientelism, while maintaining their own individual wealth. Through the alliance of former warlords and a new 'contractor bourgeoisie' (Baumann 2013), vulnerability has been deliberately manufactured for Lebanese citizens to facilitate their subjugation, by curtailment of their rights for social equity, public education, affordable housing and public healthcare (Saghieh 2015). As explained by Baumann (2016):

> this political economy of sectarianism is built on two premises. The first premise holds that the majority of Lebanese remains dependent on patronage resources controlled by the politicians, for instance jobs, education or healthcare. The second premise holds that the politicians must themselves maintain control over the distribution of resources. Income and wealth distribution must be unequal and public services must be distributed through clientelism rather than impersonal rules.

Thus, social movements demanding equality, or even a re-orientation of resources, threaten the very political and economic system. It is not surprising that sectarian elites engage in strategies to coerce movements, as was most exemplified in the repression of independent trade unions in the 1990s (Baumann 2016), or to tame, co-opt and infiltrate non-sectarian movements, especially through clientelist behaviour, so that these movements become de facto auxiliaries of the sectarian political parties themselves (Kingston 2013; Nagle 2018). Not only are non-sectarian groups rendered invisible (Clark and Salloukh 2013; Finlay 2010; Nagle 2013), but the very 'politicization of sectarianism *requires* the depoliticization of alternative social visions' (Baumann 2016).

One critical question surrounding the sectarian political system: just what are these sectarian groups or alleged 'ethnic groups', and how do they impact the national narrative? Individuals are not free to join and leave, thus violating the Lebanese constitutional right of freedom of belief and association. Sects are entities based on the affiliation of children to their father's sect. Citizens are seen only as members of the sect into which they were born. Consequently, society is erased, and this sectarian system becomes the antithesis of community management, which is precisely the function of the state (Nahas 2019a). Sects also weave a history for themselves to create their own self-justification, and to present themselves as eternal political entities. This sectarian discourse becomes an instrument of unaccountable elites to maintain political power and enrich themselves (Ofeish 1999).

An additional harsh barrier to change is created: the narrative that 'it has always been like this', a refrain that is often repeated throughout the country, a refrain that implies that the only alternative is merely different faces for a 'power-sharing' of sectarian leaders, rather than the development of a civil state. A civil state means that a citizen's relationship with the state becomes direct, rather than via a family, a tribe, a political party or, in the case of Lebanon, a sect. A civil state would, *ideally*, also treat all its citizens as equals, and not as components of sects, and would develop its foreign policy based on the objectives of what is best for the country instead of personal political interests that are then presented as the interests of the sect.

Lebanon, in addition to these internal challenges, also faces external pressure to maintain itself in a 'stable form', as perceived by leading international actors. In addition to external foreign interference from stronger states, within Lebanon, leaders prefer to maintain 'the resilience of the sectarian system' (Mouawad 2017) rather than support state-building:

> The stability of Lebanon is the resilience of its elites, which in turn hides the seeds of conflict and instability that might erupt at any point. Resilience is a discourse, often adopted by international actors and appropriated by national elites, that conceals a set of practices by several actors, ranging from the international community to the ruling political elite, that hollow out state public institutions. (Moawad 2017)

Consequently, a social movement that is successful in transformation, or even successful in pushing the country closer *towards* a needed transformation to alternative futures, needs to both build a strategy protecting itself from depoliticization and taming, and develop a discourse strong enough to break through the dominant sectarian narrative, perhaps by presenting an alternative political system. How successful was the 2015 movement in achieving these objectives? Would it even be reasonable to demand of a movement to achieve such lofty goals?

Organizational Challenges

The protests in Beirut during the trash crisis were first organized by a small group of media-savvy individuals. They came up with the slogan 'YouStink',[8] comparing the garbage to the politicians themselves. As the dynamics of the protest changed, so did the organizers. A network of openly political individuals and representatives from secular opposition political parties was formed, under the slogan 'We Want Accountability'.[9] These two groups became the main organizers of the significant Beirut-based protests. Various smaller groups also formed, many of which had splintered off each other. Although all agreed during internal meetings that this was a political crisis and not merely an environmental one, the public discourse of the organizers reflected a different analysis of the situation: some argued and campaigned for political accountability and transformation, where others adopted a discourse of technical fixes.

As a natural consequence of their own organizational composition, We Want Accountability argued for political clarity. Their discourse centred around a 'corrupt class of political elites' and they, along with other left-leaning groups, 'found it impossible not to place the crisis in its larger political and structural context' (Khneisser 2019). Their activism and discourse were not singularly focused on garbage, but constantly sought to target various points of political accountability. Thus, they adopted a discourse grounded in political and economic justice.

Other key organizers, including YouStink, focused singularly on garbage in a largely technocratic discourse, and sidelined the political and structural coronaries of the conflict and the socio-economic grievances underlining the majority of the protestors' demands. There were three key incidents that highlighted this perspective and its impact on organizing and strategy. During the first week, protests dynamically increased in size and in composition. In one particularly large protest, YouStink's Facebook page posted a video of hundreds of young men entering the protest en masse, referring to them as 'hooligans' who were purposefully inciting violence. A YouStink organizer then called for people to leave the protest to facilitate the arrest of the 'infiltrators' by the police. The use of provocative language and calls for police intervention received condemnation by leftist organizers. 'The protest exposed the classist colors of the YouStink organizers who betrayed the street [by] calling on state security to clean the protest of agent provocateurs while thousands of protesters stood their ground defying state repression', wrote two organizers (Nayel and Moghnieh 2015). Noted sociologist and activist Nahla Chahhal stated in a public panel that

> the use of the term infiltrators has a social understanding and not a security understanding. It seems to me that it goes beyond what is usually meant by security, i.e. sending police or relevant security forces to any protest due to harm caused by individuals. There are always infiltrators, in various forms, including those who have the right to participate in a public protest because they, after all, came to express their different views. (Chahhal 2015)

A second critical point occurred when a significant segment of the organizers, particularly those who had organized solely under an environmental banner, regarded positively the replacement of the Minister of Environment Fadi Jreissati with seasoned, and notoriously problematic, sectarian politician Akram Chehayeb.

Failing to contextualize the crisis politically, they agreed to meet with Chehayeb and form a committee to develop waste management plans. No plan was developed; rather, as expected by the We Want Accountability organizers and others, their credibility was damaged and protest groups experienced further splintering as a result. Meanwhile, other environmental groups started to encourage individual citizens to pick up trash themselves.

A third incident was the development of a task force on 'the waste management problem' by a group of professors at the American University of Beirut. They recommended interventions in effective waste management techniques, and directly avoided engagement on the wider politics of the trash problem. They thus presented the problem as being solely scientific, rather than political. These three activities were more suitable for campaign organizers rather than social movement organizers (Khneisser 2019).

Meanwhile, the media demanded a unified face to the movement. Regular meetings were held between various groups, including YouStink, We Want Accountability and a number of well-known oppositional figures. However, no strategic unified plan was produced. Instead, organizers found themselves organizing point by point, in a constantly reactive mode. As one lead organizer for We Want Accountability said, the political authorities 'were playing chess while we were forced to play backgammon'.[10] For example, state police would typically arrest tens of people after a large protest. Despite this regularity, no plan of action was built in advance. For example: Should protests continue as if nothing had happened? Should protests be moved to the police station to call for the release of the detainees, or should they be held at the Ministry of Interior to apply political pressure to release the captured activists? In addition, because some organizers wanted to discuss politics openly while others sought to influence political decisions without adopting a political discourse, a diluted political message arose. Criticism of those in political authority was worrisome, and organizers decided to criticize all politicians equally. Thus, the slogan 'All of them means all of them' was born. In discourses about the waste mismanagement, conversation was often directed towards the issue of trash alone, and if the privatization of public services was mentioned, it would not be linked to the larger political economy or to the political sectarian system.

Assessing the Impacts: Transformational?

The 2015 protests did not occur in a vacuum. They built on the energy, networks and experiences of previous protests, including the 2014 campaign against domestic violence, the 2013 civil marriage campaign, the 2012 'right to know' campaign for the disappeared, the 2011 'people want the fall of the sectarian regime' campaign, and, most importantly, the vibrant and quite large Union Coordination Committee labour rights protests of 2011 and 2012. The dynamic energy raised during those electric months in 2015 had grown from the amassing energies of the many previous national protests within the past ten years.

The challenge in achieving conflict transformation is to generate strategies that impact three areas in which power is concentrated: 1) in the institutions, the legal framework and the economic frameworks; 2) on people and their networks; and 3) in discourses, narratives and ways of seeing the world. 'The final outcome of the struggles in terms of achieving the desired transformation depends on knowing how and when to impact on each one of the types of hegemonic power' (Temper et al. 2018). How successful were the 2015 protests in impacting those areas?

Policy changes because of the protests?

Was a new waste management system developed? No. Was there, indeed, a policy change of any kind? No. The government responded to the crisis with the same policy that had been in place for decades: more landfills and talk of incinerators (Boutros 2015). The Lebanese parliament chose to 'resolve' the problem using the same techniques for waste management that had already failed, which included private profiteering by maintaining an exorbitantly expensive privatized waste management service. The parliament approved a plan to establish two new landfills [Costa Brava (Meraaby 2017) on the shoreline south of Beirut, and another landfill in Bourj Hammoud, on top of the old one already present], to reopen the Naameh landfill for another four years, and to build a landfill in Akkar (in north Lebanon). No Environmental Impact Assessment studies were done, due to the 'urgent nature' of the decision (Azhari 2017a). Meanwhile, the Parliamentary Joint Committees began pushing, once again, for incinerators, including one in the highly populated capital (Masri and Dwarkasing 2019), and the process of land reclamation from these landfills continued (Azhari 2017b).

Were any proposed governmental plans impacted? Yes. After six months of the waste crisis, and after attempting to throw the waste from one marginalized area to another by enticing municipalities to accept a landfill, the government devised another plan: to dispose of the trash outside of Lebanon (at an inflated price). The government, with the support of the Prime Minister and the Council for Development and Reconstruction, told the public that a British company, by the name of Chinook, was to export the garbage to Russia. The government signed a $600 million contract with Chinook and paid $10 million as a first instalment. The campaign of We Want Accountability succeeded in revealing that, in fact, the documentation given to the government by Chinook was forged. The company claimed to have permission from the Russian 'Ministry of Natural Resources and Ecology' when such a ministry does not even exist in Russia.[11] Though the Prime Minister kept insisting the documents weren't forged, the attention given was enough to force the government to cancel the contract. Had activists not uncovered the fraud, and had attention not been given to the discovery, the taxpayers might have paid for the waste to be thrown into the Mediterranean Sea. In this case, a potentially damaging governmental plan was halted.

Individual transformations?

Did the protests transform individuals? Were new individuals encouraged to join the protests, and were they then radically changed by those protests? Do these individuals have the potential to influence or create transformation?

In her study on those for whom 2015 was their first protest, Carole Kerbage found no *marked* individual transformations.

> The motives behind first-time protestors' participation and withdrawal varied and took on class, regional, and sectarian dimensions ... They protested against their leaders and sectarian parties in August 2015, but were not necessarily liberated from their ties. They took to the streets, cheered against them and demanded their ousting; however, they quickly returned to them. (Kerbage 2017)

The findings of that study are supported by the parliamentary election of 2018, the first time that the Lebanese had voted for parliament in nine years: the number of votes to oppositional candidates were less than the number of the largest protest in 2015,[12] leading some to conclude that many protestors had reverted to their previous political allegiances.

However, individual change occurs in more than just drastic, radical ways; *gradual change, from one perspective to another more progressive one, may be more common.* In this regard, the protest movement definitely aided in moving individuals along on a *spectrum of change*, particularly for those individuals who were disenchanted with the system and who sought to explore alternatives. Without doubt, among the tens of thousands of people protesting together – disenfranchised working-poor protesting in the same space as middle-class bourgeoisie – conversations and interactions were likely to occur and potentially ignite new thought. One organizer told me of smaller changes she witnessed. Protestors had been making sexist chants in the protest beside her, so she approached them and talked to them about sexism and feminism. Later, she found those same individuals correcting others and reprimanding them for their sexist slogans.

Furthermore, the protests changed certain individuals' involvement in political parties: some joined political parties for the first time, choosing, perhaps naturally, the political parties that had been organizing for change, such as the Lebanese Communist Party and Citizens in a State.[13] Several individuals told me that they had never been affiliated with any political party, but because of the 2015 protests, they decided to join a secular political party. Given that those individuals are now organizers within oppositional political parties whose very mission is the transformation of the country, their individual change is likely to impact others as well.

Networks transformed?

Were lead organizers and networks of the protests transformed? Most of the networks were built on those that had been in place before 2015. Those networks were maintained or strengthened, while small, new networks were also built between organizers. Leading individual activists in 2015 had all been organizers in previous protests; they were not radically changed by the 2015 protests, though some may have been pushed further into, or away from, political organizing.

New organizations, though, were developed. For example, the protests encouraged already active individuals to form a coalition called Beirut Madinati, which ran in the 2016 municipality elections on a non-sectarian platform, campaigning on issues of garbage and water. They won 32% of the vote, a significant victory considering they were pitted against a coalition of the dominant sectarian political parties. However, they chose to sideline contentious political issues and structural inequalities in the city, in favour of an accommodating developmental programme, and to 'advance a consensual understanding of politics and social change that is more techno-moral and less contentious' (Khneisser 2019). The coalition did not survive intact beyond the municipal elections.

After the 2016 municipal elections, a new political party called Sabaa[14] announced itself as being led by experts, activists and concerned citizens seeking to 'organize the participation of citizens in public affairs' through 'the formation of a modern and advanced model for political action, following the latest technologies and latest political concepts' (Ayoub 2016). This party was built on a liberal discourse, and perceived ideological struggles as a thing of the past, 'celebrating all choices and projects available to come up with suitable solutions in all sectors'.[15] This new organization fractured after participating in the 2018 parliamentary elections.

Out of the desire for organized political work, made more relevant by the failures of policy change after the 2015 protests, another political party was organized, called Citizens in a State (Mouwatinoon wa mouwaitnat fi dawla).[16] The party's founders and first members, however, were not politicized by the waste crisis or the protests, since they were all either experienced politicians, political advisors or seasoned activists. In contrast to Beirut Madinati and Sabaa, this party adopted a clear political discourse, and 'explicitly departed from the civil society handbook [of other newly formed groups] by politicizing opposition to the sectarian system and its ensemble of practices' (Halawi and Salloukh 2020). This party has been participating in every election since 2015.

Societal transformations?

Did the protest discourse and the attention given to the waste crisis bring about societal environmental changes and the development of alternative visions of natural resource management? Some academic, technocratic attention was given to waste management, while the political elite also recognized the power of greenwashing. In May 2018, the municipality of Beirut declared the launching of an organic waste and recycling project, in coordination with the privatized sanitation company Ramco (*Daily Star* 2018). The timing of the launch, a week before the 2018 parliamentary elections, seemed disingenuous. The plan itself was also suspect. Members of the Waste Management Coalition, which includes local academic environmental health organizations, raised concern that part of the municipality's waste management plan included incineration, thus contradicting the idea of recycling as a solution to waste disposal since the thermal decomposition factories require paper and cardboard waste for fuel and for use in the organic waste incineration process (Hamdan 2018). The Beirut municipality was clearly trying to greenwash its attempts to push for an incinerator. This move was not surprising given that, at the CEDRE conference in Paris, the Lebanese government had asked for $1.4 billion to fund three incinerators in Lebanon. Samar Khalil, a professor at the American University of Beirut and member of 'The Health of Our Children Is a Red Line', noted, 'All indicators show that the [Beirut] government plan is not a sorting plan. If there were a real inclination toward sorting, the Beirut municipality would have launched awareness campaigns two years ago and expanded the waste treatment factories' (Hamdan 2018).

Meanwhile, shifts in public opinion on the issue remained nominal, with pockets of changes mostly limited to bourgeois circles. Certain designers, for example, during the Beirut Design Week in 2017, incorporated recycled and upcycled projects, claiming that the waste crisis 'gave weight' to their sustainability argument (Sabina 2017). Previous initiatives, such as the Green Glass Recycling Initiative, also gained popularity. There was also an increase in grassroots efforts to promote recycling (Stoughton 2016). However, none of these ventures gained the momentum to become mainstream; their presence and visibility were limited to particular neighbourhoods and particular economic classes.

Critical questions remain, and certain elements have yet to be understood to ascertain whether these changes have penetrated across and throughout the capital and the country. Exactly how many recycling centres are operating in Lebanon, and how many of them are run by grassroots organizations, by private for-profit companies and/or by municipalities? What impact, if any, have these centres had on the waste pickers who supplement their income by going through the garbage, looking for plastic, glass and other material to be sold to recycling centres? Furthermore, and most importantly, have these grassroots initiatives served to lessen the responsibility on the municipalities to fulfil their public duties of waste management, or have they drawn attention to the municipalities? Has there been a recognition that the waste management crisis is a political crisis, and thus can also be manufactured for further political gains or capitalized for political transformation?

As mentioned, the 2015 waste protests did not merely attempt to raise attention to waste mismanagement itself. Instead, the main cry from protestors throughout the country was about material grievances, including tenants' housing rights, unemployment, healthcare, education, electricity cuts, water shortage, the disappeared of the war (1975–91), rights of disabled people, and mass environmental protection and environmental health, rather than any technical issues about waste management (Kreichati 2017). Thus, if an alternative vision were to have been born or encouraged, it would have been the recognition of the critical importance of economic equity, intersectionality and holistic politics, and the rejection of technocratic solutions.

This narrative was most definitely hindered by the increased popularization and use of the term 'civil society' as a replacement for 'political thought' or 'political parties'. The term had entered the lexicon previously, particularly with the exponential growth of non-governmental organizations in Lebanon after the civil war at a rate of some 250 NGOs a year (Kingston 2014). The media attributed the title 'civil society' to the 2015 protests, further popularizing this (problematic) concept that opposition within a separate realm called 'civil society' 'is the powerhouse of oppositional politics and social transformation' (Harvey 2005). With the rise of the term 'civil society' came the decline of the word 'politics'. 'Talking politics' or 'doing politics' became, more and more, something ugly. What was the impact of this narrative on the potential for change and transformation?

Fuelled by mainstream media and popular culture's usage of this new term, all the various protest organizers were presented as one homogeneous 'civil society'. As one organizer lamented, 'we were presented as "being the same" despite our differences, and we were held responsible – by the media and the non-participating citizens – for the actions of other groups or individuals'.[17] Some protest organizers themselves encouraged this, by either calling themselves 'civil society' or claiming to use apolitical discourse.

This narrative had repercussions for the 2018 parliamentary elections. Under this one umbrella of 'civil society', electoral coalitions were encouraged in these elections. In addition to various opposition candidates, the largest national electoral coalition was formed in 2018: sixty-six candidates ran in nine districts in Lebanon. This coalition was only developed for the parliamentary elections and dissolved after the elections; its alliance was ephemeral. The use of the term 'civil society' instead of the more accurate 'political party' was revealed to create confusion.[18] Certain dominant, sectarian, right-wing political parties, such as the Lebanese Forces and the Phalange (Kataeb), Lebanon's chapter of the Muslim Brotherhood (Al Jama'a Al Islamiya), and even the problematic former director of the Lebanese Internal Security Forces Ashraf Rifi (who had supported local militias during the Tripoli armed clashes a few years earlier), all claimed to be 'civil society'. Meanwhile, among the self-identifying 'civil society activists' who were taking an interest in the political process, there was a declared distaste of political parties. Even during the elections themselves, when all were running for a seat in the parliament, the term 'political party' was used, by some, as an insult, while the term 'civil society' was used as a compliment. Media, both print and television, were complicit in that respect: they adopted the 'Civil Society vs Political Parties' view, thus marginalizing opposition parties, and, when there was more than one oppositional coalition, the media claimed that they were 'civil society' against 'civil society', concluding that the only option is to support the 'strong', and therefore to support the dominant, sectarian power structure.

Language can clarify or obfuscate. Avoiding the use of clear descriptions and accepting the popularized demonization of 'politics' encouraged technocratic and non-political behaviour, like separating waste management and environmental health issues from economic equality, environmental justice and political representation. This discrepancy in narrative was one of the main challenges for the organizers themselves.

Now What? Reverberations into the 17 October 2019 Uprising

The 2015 protests were minor in comparison to the uprising in October 2019. Instead of tens of thousands of people protesting in the capital, in October 2019 hundreds of thousands protested. Days prior to the protests, more than 100 fires destroyed forests throughout the country in a twenty-four-hour period, to which the government's response was glaring indifference and incompetence (Zaza 2019). With the embers still hot, and with ongoing public service failures still present, a government minister proposed a flat tax to be applied to WhatsApp calls, the cheapest and most common form of communication in the country. Within a few days, hundreds of thousands of people took to the streets in outrage (Masri 2019). These protests grew to perhaps the most comprehensive anti-government protests since the civil war, in terms of size, geographic spread and class diversity (Chehayeb and Sewell 2019). Throughout a period of months, streets were blockaded with burning tyres; certain MPs' offices were attacked in the south; public squares in cities were reclaimed; and tents set up within which political and general educational discussions were held.

Socio-economic grievances were again at the core of the protests (Bou Khater and Majed 2020), except this time there was a larger sense of revolt, and some even spoke of revolution. 'Calls for outright revolution were made', noted assistant professor in social psychology Rim Saab, 'and they gained unparalleled momentum, with hundreds of thousands of people across Lebanon and across sectarian and class lines taking to the streets over the days and weeks that followed, demanding an end to corruption, the ousting of the ruling elite and the fall of the current regime' (Saab 2020). A government fell (or chose to resign), and an economic and financial bankruptcy became vividly apparent. A hundred days later, another government was appointed. The 'new' government, however, was merely a facade for the sectarian leaders themselves who had appointed it, and the central economic problems were not addressed, so the protests continued. Some sectarian elites attempted to co-opt the protests (Nahas 2019b), and to depoliticize the discourse, while violence from the state continued to escalate. Months into the uprising, when the Covid-19 pandemic hit Lebanon and people were sequestered in their homes, security police destroyed the tents in the Beirut city square in an attempt to erase any memory of the protests. During the pandemic and the continuing financial and economic bankruptcy, Lebanon also suffered another blow: a literal explosion, one of the largest non-nuclear examples in human history. On 4 August 2020, an explosion ripped apart the Beirut port, killing hundreds, injuring thousands and destroying the homes of more than 300,000 people. The explosion might have moved some individuals who were already questioning the system and recognizing the incompetence, impotence and corruption inherent within it, but it did not change previously held opinions and narratives. The explosion, and the lack of accountability, did however weaken the euphoria of hope that had erupted in October 2019.

The uprising did contribute to several smaller victories, including most prominently the success of the Save the Bisri Valley campaign, which halted the destructive proposed World Bank-funded project, the Bisri Dam (Moughalian and Masri 2019). Many organizers, activists and social scientists consider that, although nothing was accomplished on the level of institutional politics and policy, there was a shift *here* in discourse and collective consciousness (LCPS 2020).

Though the October 2019 uprising was larger and more intense than the previous protests in 2015, many of the same internal and external organizational challenges remained. In addition to increasing state violence and economic pressure, there was also a lack of organized strategy and united objectives, and overall, a tendency for apolitical discussions. Calls for a united front, even though the organizers represented different political perspectives, were similar. More critically, the divided 2015 discourse discussed earlier reverberated here: political technical solutions versus political clarity, which had reached into the 2016 municipal elections and the 2018 parliamentary elections, continued in 2019. Certain organizers demanded a technocratic new government, as if the current and former ministers and parliamentarians did not have technical expertise, and as if public policies can be apolitical at all.[19] This discourse detracted from the needed political clarity in calling for a government composed of competent, experienced and courageous individuals, not beholden to sects or foreign influences, and focused 'on the structural problems of the system itself' (Bou Khater and Majed 2020). In 2019, as in 2015, the groups pushing for change in the sectarian system were themselves reluctant to organize among themselves, prompting fundamental questions: 'Can protests against [a] political system … be apolitical? Can they escape the need to cultivate proper organizational forms?' (Halawi and Salloukh 2020).

The need for political clarity and political strategy is greater now, in 2020, than it was in 2015. Lebanon is in the midst of an economic and financial bankruptcy caused by decades of financial engineering, a trade deficit of 86%,[20] and a sectarian leadership unable to develop public policies benefiting all its citizens.

The crisis comprises not only the loss of billions in individuals' bank deposits and savings and in public moneys allocated to social security, but also a threat to the fabric of society due to mass emigration. Based on demographic studies, since 1991, 50% of men and 40% of women between the ages of twenty and fifty have left the country. In other words, approximately 45% of Lebanon's productive population has emigrated since the end of the civil war. Before the 4 August explosion, 'some 66,000 Lebanese emigrated in 2019, twice as much as the previous year' – 80% of the population between twenty and forty years old. 'If the coronavirus epidemic has temporarily curbed the bleeding, the worsening of the economic crisis in the absence of specific measures will only reinforce the trend' (Babin 2020).

The events in Lebanon are, in the end, fundamentally part of a greater regional and global phenomenon: stepping away from real political organization and embracing liberal, demand-based campaigns. As carefully illustrated by Halawi and Salloukh (2020), 'what in the past was euphemistically labelled the ngo-ization of civil society confined collective action to demand-based campaigns. With this came a liberal rulebook on how to approach the political, thus assuming that collective action is outside the political.'[21]

The questions faced in Lebanon are present elsewhere. Our world is changing; how do we direct the change towards something positive (building an egalitarian economy and society), rather than being passive spectators of destruction and oppression? How do we break through the increasing sense of despair and hopelessness among citizens that lulls people into believing that there is no alternative to this broken, divisive and incompetent system? How do we create an imagining for a different reality?

Notes

1 The average cost for waste management in Lebanon was US$155/ton, while the worldwide average was US$75/ton.

2 For information on the environmental and public health impacts of the Bourj Hammoud landfill, refer to Moughalian 2016a.

3 For information on the environmental and public health impacts of Naameh landfill, refer to Moughalian 2016b.

4 Examples of fearmongering during that period including the following threats that activists received: 'If you go to the protests we'll confiscate your vehicle'; '... we'll beat you senseless'; '... we'll fire you from your job'; '... we'll have you arrested and let you rot in jail'; '... we'll cut off your financial support' (hospitals, universities, etc.).

5 There has only been one official census in Lebanon, in 1933.

6 By the United States and its allies.

7 Eighteen out of twenty banks have major shareholders linked to political elites and 'four out of the top 10 banks in the country have more than 70% of their shares attributed to crony capital' (Chaaban 2016).

8 Tulet Rehetkun/ طلعت ريحتكم

9 / بدنا نحاسب Badna Nhasseb

10 Personal conversation with Nael Kaedbey, lead organizer with We Want Accountability. August 2018.

11 The equivalent Russian office is called 'The Ministry of Natural Resources and Environment'.

12 While at one time the protests reached more than 100,000, the oppositional candidates in 2018 secured 70,000 votes.
13 'Mouwatinoon wa mouwaitnat fi dawla', mmfidawla.com
14 Literally the number 7 in Arabic.
15 As per their website, http://www.sabaa.org.
16 As stated earlier, I joined this party in 2016, was elected to serve as the President of the Council of Delegates in 2017, and ran as a candidate in the 2018 parliamentary elections. I was one of the party's main on-the-ground organizers in the October 2019 uprising.
17 Personal conversation with Nael Kaedbey, lead organizer with We Want Accountability. August 2018.
18 There is no legal difference between the two terms in Lebanon.
19 An allegedly technocratic government was formed and has shown itself incapable of leading the country forward.
20 $18 billion of goods are imported for every $3 billion exported.
21 For an eloquent discussion on how NGO-ization of society 'domesticated' opposition, see Plaetzer 2014. Also see Tagma et al. 2013.

References

Avelino, F. (2017) Power in Sustainability Transitions: Analysing Power and (Dis)empowerment in Transformative Change Towards Sustainability. *Environmental Policy and Governance*, 27(6): 505–20.

Ayoub, L. (2016) 'Sabaa', a New Political Party ... Without a Leader. *Assafir*, 20 October, http://assafir.com/Article/514789 (in Arabic). Accessed 13 April 2020.

Azhari, T. (2017a) Regulations Bypassed in Bourj Hammoud Landfill Project. *Daily Star*, 29 July, www.dailystar.com.lb/News/Lebanon-News/2017/Jul-29/414364-regulations-bypassed-in-burj-hammoud-landfill-project.ashx. Accessed 13 April 2020.

—— (2017b) The Lucrative History of Lebanese Land Reclamation. *Daily Star*, www.dailystar.com.lb/News/Lebanon-News/2017/Jul-19/413250-the-lucrative-history-of-lebanese-land-reclamation.ashx. Accessed 13 April 2020.

Babin, J. (2020) Le Liban menacé par une accélération de la fuite des cerveaux. *Le Commerce Du Levant*, 26 March, www.lecommercedulevant.com/article/29732-le-liban-menace-par-une-acceleration-de-la-fuite-des-cerveaux?fbclid=IwAR3Dn-ejPGnHoRe-1jiqyAiFGQqlyFjBUeIg-NzGcftM7JFdOQQrwOmpG_j0.

Baumann, H. (2013) The 'New Contractor Bourgeoisie' in Lebanese Politics: Hariri, Mikati and Fares. In Knudsen, A., Kerr, M. (eds) *Lebanon After the Cedar Revolution*. Oxford and New York, Oxford University Press.

—— (2016) Social Protest and the Political Economy of Sectarianism in Lebanon. *Global Discourse*, 6(4): 634–49. https://doi.org/10.1080/23269995.2016.1253275.

Bou Khater, L., Majed, R. (2020) *Lebanon's 2019 October Revolution: Who Mobilized and Why?* Working Papers. Asfari Institute for Civil Society and Citizenship. AUB. www.activearabvoices.org/uploads/8/0/8/4/80849840/leb-oct-rev_-_v.1.3-digital.pdf?fbclid=IwAR2Im5dzf-LXGsrS8AH2l-runsAAYrUhk4r0rBHUXka3Q-Gt7F0csHY6-6U.

Boutros, J. (2015) *Garbage Crisis in Lebanon – 1997: Same Policy, Repeated History. The Legal Agenda*, 16 October.

Chaaban, J. (2016) I've Got the Power: Mapping Connections between Lebanon's Banking Sector and the Ruling Class. *ERF Working Papers*, 1059 (October): 3, http://erf.org.eg/?p=14032.

Chahhal, N. (2015) The Movement Within a Class-based Analysis. Presentation at a public panel, 15 October. Organized by the Asfari Institute for Civil Society and Citizenship Held at the American University of Beirut.

Chehayeb, K., Sewell, A. (2019) Why Protestors in Lebanon Are Taking to the Streets. *Foreign Policy*, 2 November, https://foreignpolicy.com/2019/11/02/lebanon-protesters-movement-streets-explainer/. Accessed 13 April 2020.

Clark, J.A., Salloukh, B.F. (2013) Elite Strategies, Civil Society, and Sectarian Identities in Postwar Lebanon. *International Journal of Middle Eastern Studies*, 45(4): 731–49.

Daily Star (2018) Beirut Launches New Organic Waste and Recycling Initiative. 23 June, www.dailystar.com.lb/News/Lebanon-News/2018/Jun-23/454084-beirut-launches-new-organic-waste-and-recycling-initiative.ashx. Accessed 13 April 2020.

Finlay, A. (2010). *Governing Ethnic Conflict: Consociation, Identity and the Price of Peace*. Abingdon: Routledge.

Haddad, S. (2009) Lebanon: From Consociationalism to Conciliation. *Nationalism and Ethnic Politics*, 15(3–4): 398–416.

Halawi, I., Salloukh, B.F. (2020) Pessimism of the Intellect, Optimism of the Will after the 17 October Protests in Lebanon. *Middle East Law and Governance*, 12: 322–34.

Hamdan, H. (2018) Is Lebanon's New Recycling Project a Bunch of Garbage? *Al-Monitor*, 15 May, www.al-monitor.com/pulse/originals/2018/05/lebanon-beirut-municipality-hariri-paper-project-garbage.html.

Harvey, D. (2005) *A Brief History of Neoliberalism*. Oxford: Oxford University Press.

Horowitz, D. (1985) *Ethnic Groups in Conflict*. Berkeley: University of California Press.

Hume, T., Tawfeeq, M. (2016) Lebanon: 'River of Trash' Chokes Beirut Suburb as City's Garbage Crisis Continues. CNN, 25 February, http://edition.cnn.com/2016/02/24/middleeast/lebanon-garbage-crisis-river/. Accessed 10 April 2020.

Kerbage, C. (2017) *Politics of Coincidence: The Harak Confronts Its 'Peoples'*. Working Paper #41. Published by the Issam Fares Institute for Public Policy and International Affairs, American University of Beirut. www.aub.edu.lb/ifi/Documents/publications/working_papers/2016-2017/20170213_wp_hirak_english.pdf.

Khneisser, M. (2019) The Specter of 'Politics' and Ghosts of 'Alternatives' Past: Lebanese 'Civil Society' and the Antinomies of Contemporary Politics. *Critical Sociology*, April. https://doi.org/10.1177/0896920519830756.

Kingston, P.W.T. (2014) *Reproducing Sectarianism: Advocacy Networks and the Politics of Civil Society in Postwar Lebanon*. Albany, NY: SUNY Press.

Kreichati, C. (2017) Knowledge and the Trash: The Predominance of the Expert Model in the 2015 Beirut Protests. Thesis. American University of Beirut, MA, Lebanon.

LCPS (Lebanese Center for Policy Studies) (2020) Setting the Agenda: Has the October 17 Revolution Accomplished Anything at All? 17 October, http://lcps-lebanon.org/agendaArticle.php?id=197.

Leenders, R. (2012) *Spoils of Truce: Corruption in Postwar Lebanon*. Ithaca, NY: Cornell University Press.

Masri, R. (2019) Sectarian Leaders Are Facing a Historic Responsibility. Interview with the Arabic Hour. www.youtube.com/watch?v=VNTVv1Rx68o.

Masri, R., Dwarkasing, C. (2019) Beirut Incinerators Expansion Plans and Wastepickers Struggle, Lebanon. EJAtlas, https://ejatlas.org/conflict/as-the-plans-for-incinerators-in-beirut-lebanon-expand-wastepickers-recycle-to-survive.

Meraaby, T. (2017) Costa Brava Landfill, Lebanon. EJAtlas, https://ejatlas.org/conflict/costa-brava-landfill-lebanon.

Mouawad, J. (2017) *Unpacking Lebanon's Resilience: Undermining State Institutions and Consolidating the System?* IAI Working Papers 17 / 29 October. Foundation for European Progressive Studies.

Moughalian, C. (2016a) Bourj Hammoud Landfill, Lebanon. EJAtlas, https://ejatlas.org/conflict/bourj-hammoud-garbage-mountain.

—— (2016b) Naameh Landfill, Lebanon. EJAtlas, https://ejatlas.org/conflict/naameh-landfill-lebanon.

Moughalian, K., Masri, R. (2019) Bisri Dam, Lebanon. EJAtlas, https://ejatlas.org/conflict/bisri-dam.

Nagle, J. (2013) 'Unity in Diversity': Non-sectarian Social Movement Challenges to the Politics of Ethnic Antagonism in Violently Divided Cities. *International Journal for Urban and Regional Research*, 37(1): 78–92.

—— (2018) Beyond Ethnic Entrenchment and Amelioration: An Analysis of Non-sectarian Social Movements and Lebanon's Consociationalism. *Ethnic and Racial Studies*, 41(7): 1370–89. https://doi.org/10.1080/01419870.2017.1287928.

Nahas, C. (2019a) القرن موازنة :المحطة في نحّاس شربل, 31 May, www.youtube.com/watch?v=s5xrtLn90Gc.

—— (2019b) قدراً ليست مأساتنا :نحاس شربل, 14 November, www.youtube.com/watch?v=TiJ9buJXesY.

Nayel, M.A., Moghnieh, L. (2015) The Protests in Lebanon Three Months After: A Reading of Police Coercive Strategies, Emerging Social Movements and Achievements. *New Politics*, 7 November, http://newpol.org/content/ protests-lebanon-three-months-after. Accessed 26 February 2016.

Ofeish, S. (1999) Lebanon's Second Republic: Secular Talk, Sectarian Application. *Arab Studies Quarterly*, 21(1, Winter): 97–116.

Plaetzer, N. (2014) Civil Society as Domestication: Egyptian and Tunisian Uprisings Beyond Liberal Transitology. *Journal of International Affairs*, 68(1): 255–65.

Rodríguez, I., Inturias, M. (2018) Conflict Transformation in Indigenous Peoples' Territories: Doing Environmental Justice with a 'Decolonial Turn'. *Development Studies Research*, 5(1): 90–105. https://doi.org/10.1080/21665095.2018.1486220.

Saab, R. (2020) Setting the Agenda: Has the October 17 Revolution Accomplished Anything at All? Lebanese Center for Policy Studies, 17 October, http://lcps-lebanon.org/agendaArticle.php?id=197. Accessed 20 December 2020.

Sabina, R. (2017) Lebanese Designers Embrace Rubbish Following Beirut Trash Crisis. *Dezeen*, 26 June. www.dezeen.com/2017/06/26/lebanese-designers-embrace-recycling-after-trash-crisis-beirut-design-week/.

Safa, S.J. (2015) AUB Headed Research Warns of Alarming Carcinogen Levels Near Open Waste Dump Fires. AUB Office of Communications, 1 December, http://aub.edu.lb/news/2015/pages/carcinogen-waste-fires.aspx. Accessed 10 April 2020.

Saghieh, N. (2015) Beyond Sectarianism: Whom Does the Lebanese State Serve? *The Legal Agenda*, 32, 27 October.

Salloukh, B.F., Barakat, R., Al-Habbal, J.S., Khattab, L.W., Mikaelian, S. (2015) *The Politics of Sectarianism in Postwar Lebanon*. London: Pluto.

Stoughton, I. (2016) Lebanon's Rubbish Crisis Fuels Green Alternatives. *Al Jazeera*, www.aljazeera.com/news/2016/07/lebanon-rubbish-crisis-fuels-green-alternatives-160718083519422.html.

Sweepnet (The Regional Solid Waste Exchange of Information and Expertise Network in Mashreq and Maghrib Countries) (2014) *Country Report on The Solid Waste Management in Lebanon*. GIZ, April.

Tagma, H.M., et al. (2013) 'Taming' Arab Social Movements: Exporting Neoliberal Governmentality. *Security Dialogue*, 44(5/6): 375–92.

Temper, L., Walter, M., Rodríguez, I., Kothari, A., Turhan, E. (2018) A Perspective on Radical Transformations to Sustainability: Resistances, Movements and Alternatives. *Sustainability Science*, 13: 747–64. https://doi.org/10.1007/s11625-018-0543-8.

Westley, F., Olsson, P., Folke, C., Homer-Dixon, T., Vredenburg, H., Loorbach, D., Thompson, J., Nilsson, M., Lambin, E., Sendzimir, J., Banarjee, B., Galaz, V., van der Leeuw, S. (2011) Tipping Toward Sustainability: Emerging Pathways of Transformation. *AMBIO*, 40: 762–80.

Zaza, B. (2019) Lebanon Turns to Neighbors to Douse Forest Fires. GulfNews, 15 October, https://gulfnews.com/world/mena/lebanon-turns-to-neighbours-to-douse-forest-fires-1.67150002. Accessed 13 April 2020.

Zbeeb, M. (2012) $5 Million a Year: The Tip of the Sukleen Waste Pile. *Al-Akhbar*, 27 August.

Section 2

From Individual to Institutional Transformations

In this section of the book, the focus starts to shift towards an understanding of the scales of transformations. We present cases from Belgium, India and Argentina that address transformations that have taken place at different individual, collective and institutional scales. We look at how transformations scale up, scale out and scale deep, and at the internal conflicts that movements experience during their struggles. The Belgian case study looks inside the movement dynamics and at the personal level to understand the role that capabilities/internal relations and interactions in the movement have in bringing about transformations. The Indian case study will look at the role of gender dynamics in pastoral communities struggling to bring about transformative change. Finally, the Argentinian case study centres on the interaction between movement and state institutions in bringing about such transformation.

6

Free the Keelbeek from the Prison! A Deep Analysis of Individual and Collective Empowerment Within the Resistance Movement Against the Brussels Mega-prison Project

Jérôme Pelenc

Introduction

This chapter analyses a particularly strong social mobilization against a mega-prison project that is planned to be built on a 20 hectare natural site in Haren in the Brussels region (Belgium). The actors involved in the resistance are diverse (local inhabitants, local NGOs, activists who occupied the site, the syndicate of magistrates and a lawyers association, among others). This resistance is part of the largest European movement against unnecessary and imposed mega-projects. My aim in this chapter is to analyse the transformations that occur along the interactions between the individual and collective level within the movement. At the collective level, I first analyse how the resistance movement has developed collective capabilities to fight against the project. Then I analyse how the personal involvement in the collective movement of resistance impacts on individual well-being and agency. To do so I have assessed the impact of the involvement in the mobilization on individual capabilities (e.g. how the personal involvement impacts positively or negatively individual capability and agency). Finally, I have also conducted an analysis of the internal conflict of the resistance movement. I conclude by highlighting how this resistance movement has enabled transformation processes that occur at three levels: individual (impacting positively on individual capabilities and transforming subjectivities), collective (by the development of collective capabilities and the creation of new collective organizations) and societal (by repoliticizing issues of federal prison policy and regional land-planning).

The objective of this chapter is to analyse and understand the 'transformative potential' of place-based resistance movements through the study of the resistance movement opposing a mega-prison project in Brussels. The goal is to advance our understanding of the interactions between the individual and collective level through the analysis of this mobilization. This chapter *draws upon* and *continues* previous works that tried to better characterize groups' dynamics and collective action in the field of sustainability transformations through the lens of the capability approach (Pelenc et al. 2013, 2015, 2017).

According to Evans, collective action is required to create the collectivities that will, in turn, make it possible to foster individual and collective freedoms by establishing the link between the individual and the social level: 'Organized collectivities are fundamental to people's capabilities to choose the lives they have reason to value. They provide an arena for formulating shared values and preferences, and instruments for pursuing them, even in the face of powerful opposition' (Evans 2002: 56). According to David Vercauteren (2018), there is a need of developing a true 'micropolitics of group'. Such a micropolitics is still lacking. He explains, paraphrasing Simone de Beauvoir's 'One is not born a group, one becomes one'. He means that a group of interacting individuals is never given once and for all; the creation of a group and thus of collective practices and collective actions is a dynamic and ongoing process.

Social movements, collective action, etc. are generally geared towards challenging power structures. According to Rodríguez and Inturias (2018), unlike power of domination, which is known as 'power over', power of agency is usually known as the 'power to'. 'Power to' refers to the capacity to act, to exercise agency and to realize the potential of rights, citizenship or voice (Gaventa 2006). In the field of human development, agency is closely related to empowerment. According to Ibrahim and Alkire (2007), empowerment can be defined as the improvement of agency. This definition encompasses the institutional environment that offers people the opportunity to exercise their agency successfully (ibid.). According to Drydyk (2008), people are durably empowered when they exercise enhanced decision-making and can influence strategic life-choices and overcome barriers to agency and well-being freedom. When people lack agency, it is because the process aspect of freedom is blocked. The capability approach can be useful to investigate how an environmental resistance movement can improve individual and collective empowerment.

The capability approach (CA) has been praised by scholars from both political ecology (Forsyth 2008; Holifield 2015; among others) and environmental justice (Ballet et al. 2013; Schlosberg 2009; among others) for its analytical and normative power to tackle issues of justice. The CA fits well with EJ and PE, stating by essence that people should be free to choose the lifestyle they value the most (Sen 1999, 2009).[1] In sum they should be the protagonists of their own life, asserting their rights to have control over their life and territories and to shape their own future.

However, there exist very few studies that have tried to operationalize the CA framework to analyse interactions between individual and collective empowerment, and even fewer to characterize what Vercauteren calls 'group's micropolitics'. Being able to better understand this group micropolitics and more generally how the individual and collective level interrelate with each other is crucial for sustainability transformations. Therefore, this chapter aims to bring new conceptual and methodological tools to better investigate and understand the dynamics of resistance movements in the field of environmental justice and more widely in the field of social transformation.

To conduct my research I embraced John Holloway's recommendation about 'resistance studies' – that is to say, not doing 'studies about resistance' considering 'resistances' as an object of study, but rather 'to think our studies as resistance studies in the sense of being part of the struggle against capitalism' (2015: 13). More precisely, I adopted an 'activist-research' position that can be defined as 'scholarly engagement with activism and socio-ecological movements where researchers become scholar-practitioners practicing an applied, practical political ecology' (Demmer and Hummel 2017: 611). This position acknowledges the subjectivity of the research and embraces it. I started to work with the movement opposing the mega-prison in October 2015 through the mode of participant observation. While I became more and more socially embedded in the movement, I conducted twenty-two semi-structured interviews from January 2016 to March 2017 with the four categories of actors of the resistance described above (NGO members, local inhabitants, occupiers, activists who were not living permanently on the Keelbeek site). The interviews were conducted in a place jointly decided upon. For most of the interviewees, the interview was a special moment dedicated to reflexivity among the everyday conversations we had about the resistance. The interviews lasted between two and six hours; when necessary, we met two or three times. This study is part of a larger one. I have also conducted a power-cube analysis regarding how the resistance movement develops strategies to address spaces, forms and levels of hegemonic power (see Pelenc 2020) and helps the repoliticization of the environment; two participatory workshops on alternative land-planning; and two others on the process of collective writing of a book.[2]

The chapter is structured as follows: the first section describes the resistance movement against the mega-prison project in Brussels; the second analyses the collective capabilities that the movement has built up; the third analyses the impact that personal involvement in the resistance has on the individual well-being of the resisters; and the fourth provides some perspective to advance a group's micropolitics by offering a framework to identify the different types of internal conflicts that occur within the resistance movement.

The Resistance Movement Against the Mega-prison Project in Brussels

The Haren neighbourhood and the Keelbeek site

The mega-prison project (1,190 prisoners) was planned to be built in the Haren neighbourhood (Brussels region,[3] Belgium) on a natural site named the Keelbeek.[4] This green space provides a wide array of ecosystem services to local residents and more largely to the city of Brussels. The project and its consequences represent a typical environmental injustice (Schlosberg 2009) case where the distribution of cost and benefits is unfair, the participation of residents and local NGOs severely limited, and the diverse socio-ecological values are not recognized (see below). The Haren neighbourhood is located in the north of the Brussels region on the Flemish border. It is still considered a semi-rural area. This is a 'sacrifice zone' (Lerner 2010), considered to be the 'garbage dump' of Brussels because of the numerous polluting infrastructures that have been built in this peripheral area far from the city centre and far from the rich neighbourhoods to the south. The residents feel very 'enclosed' because of the numerous infrastructures that surround them (ring road, regional and international trains, international airport of Brussels, several dangerous industrial sites, bus depot, railway marshalling yard, etc.) and because of the lack of public transportation and services. The Keelbeek natural site, where the mega-prison project is being built, represents the biggest and most pleasant green space of the neighbourhood, an area clearly deprived of such spaces already. The Keelbeek featured a mosaic of ecosystems for a superficies of around 20 hectares, which included a green park, a cropland, natural areas with small wetland and a few protected animals and vegetal species. Above all, this green space offered an open landscape, quite a rare resource in a dense city. The Keelbeek was comprised of many land properties belonging to different private owners including individuals and business companies. The land has been bought by the federal state of Belgium through its public real estate agency (la Régie des Bâtiments).

The resistance movement

From 2008 to 2011, the local residents of Haren were starting to hear that a prison would be built in their area. At that time, a prison of 'regular' size (400 prisoners) was supposed to be built on an already artificialized site (an ancient warehouse near the Keelbeek). The residents had no problem with this first project. They discovered at the end of 2011/beginning of 2012 that it would be a mega-prison of 1,190 prisoners and that it would be built on the Keelbeek, destroying the entire natural/agricultural site. This triggered the mobilization. The residents started looking for some help with regional and national NGOs active in the field of environment- and land-planning as well as in the anti-prison sector. A turning point occurred on 17 April 2014 when 400 environmental activists claiming food sovereignty came to Haren and started illegally planting potatoes on the Keelbeek. The land was then occupied under the banner of 'Zone to Defend' (ZAD),[5] the famous label invented by the activists fighting against the airport project in Notre-Dames-des-Landes in France.

In early spring 2015, the 'Platform Against Prison Disaster' was created to regroup all the associations involved in the resistance together with the residents. The platform gathered anti-prison and pro-human rights NGOs as well as environmental ones. The platform represented, to some extent, a convergence of the struggle between the anti-prison movement and the environmental movement. The construction and environmental permits had been granted at the end of 2016 and the beginning of 2017. In March 2018 all trees of the Keelbeek had been cut down. After a few months, during which the project seemed paralysed, the preliminary construction works finally began in October 2018 after the eviction of the occupiers in August 2018. At the time of writing, the prison is being built, but a fraction of the movement is still alive and continues to develop activities against prisons in general.

I have identified four types of actors that have been involved in this resistance: 1) local residents; 2) local or national NGOs from both the environmental and justice/human rights sectors; 3) environmental and anti-prison activists who do not live permanently on the occupied site; and 4) activists who live permanently on the site. I will give more details about the movement in the section on power analysis.

Regarding the critique of the mega-prison project, the resistance movement has offered many arguments ranging from the cost and the size of this project to a radical critique of imprisonment policy and prisons in general, demonstrating that the more you build prisons the more you fill them. Ultimately, prisons are a crucial element of capitalism to maintain the class structure of society because poor, immigrant, illiterate people are over-represented there.

The resistance movement has also offered many arguments regarding the environmental impacts of the project (loss of many ecological functions; loss of local identity, landscape and sense of belonging). Last but not least, it is unacceptable for the people involved in the mobilization to lose a potential arable site of 20 hectares that could foster urban agriculture in order to improve food sovereignty. In a nutshell, given that there is no real argument from the government to demonstrate how this project will improve the prison/justice sector, the environmental destruction it entails appears entirely absurd and incoherent with the regional, national and global discourses around sustainability.

This case illustrates well the cumulating of social and environmental inequalities that prisons entail. Indeed, as has been demonstrated by the US Environmental Protection Agency, prisons are often located in polluted or degraded areas (Pellow 2017).[6]

To conclude, this 'unnecessary and imposed mega-project'[7] reveals the weaknesses of our supposed democratic institutions.

Four Collective Capabilities for Developing a Transformative Territorial Resistance

A brief presentation of the capability approach

Capabilities correspond to the various functionings that a person can chose to adopt, according to their values, in order to achieve expected lifestyles (Sen 1999). While Sen rejects the idea of a list of fundamental capabilities, i.e. of fundamental dimensions of well-being, others have provided such a list (see Pelenc 2017). Such fundamental dimensions can include for example subsistence (functionings related to food provision, shelter, etc.), protection (functionings related to feeling secure, etc.), affection (functionings related to being loved and having fruitful social relations, etc.), and so on. Drawing on Evans' pioneering work, some scholars of the CA have introduced the concept of collective capabilities to better characterize the freedom to act and choose found in a group (Dubois et al. 2008; Evans 2002; Ibrahim 2006). In a synthesis paper Pelenc et al. (2015)[8] define 'collective capabilities' as the real opportunities available to a group of interacting people to achieve a set of functionings that is defined collectively as valuable and that would have been impossible to achieve if acting alone. The collective capabilities of a group emerge from the interactions among the participants. They are an emergent property of the group and not just the aggregate of individuals' capabilities. For collective capabilities to emerge, the group of interacting persons will have to pool some resources and transform these resources – thanks to rights and conversion factors – into potential collective functionings, i.e. real opportunities for collective action (see Griewald and Rauschmayer 2014, and Pelenc et al. 2015 for examples).

Applying the capability approach to place-based resistance movements

To better describe the political and societal impacts entailed by local opposition movements, the geographer Léa Sébastien (2013) highlights the need to observe, through time, the evolution of the social landscape, the production of different types of legitimate knowledge, the role of place attachment and the political dimension of identities. She shows through the analysis of a resistance movement against a landfill in France how local opposition, far from being defined by the NIMBY label, can in fact enrich democracy through the constitution of four types of capital: social, scientific, patrimonial and political (see for further details Sébastien et al. 2019). From the work of Sébastien, I derive four types of collective capability in order to analyse the empowerment process that occurred at the collective level among the resistance movement against the mega-prison. These four collective capabilities are *being able to build alliances*, *being able to develop a counter-expertise and advance other legitimate knowledge*, *being able to develop place attachment* and *being able to develop a political agency*. Speaking of 'capability' instead of 'capital' helps to give a more dynamic view of the resistance and suggests a touch of 'freedom to act', whereas capital is more static and associated with the idea of accumulation. Finally, the concept of capability is broader than the one of capital as far as it encompasses the idea of 'resource or capital' to be converted into 'well-being achievements or greater freedom to act and choose'.

I now describe how these four collective capabilities have emerged and evolved during the mobilization against the prison. What follows is based on my participant observations.

Being able to build alliances

The mobilization was initiated by the Haren local inhabitants committee, who rapidly sought the help of a regional NGO specialized in supporting inhabitants' struggles in land use conflicts. The pre-existing social capital was very limited before the struggle because the Haren inhabitants committee was typically just facilitating local neighbourhood sport and cultural activities. Step by step, several improbable alliances have been built, gathering a huge variety of actors (from the syndicate of magistrate and Brussels president of the Bar, to anti-jail and ecological radical activists, and passing by environmental, agricultural, and social, regional and national NGOs such as the Belgian section of the International Observatory of prisons, Le Centre d'action laïque, Le Début des haricots, etc.). Although this diversity of actors, claims and methods to some extent represented a kind of 'convergence of the struggles' (*convergence des lutes*) it also generated some strong internal conflicts (see 'Individual Capabilities and Empowerment' below). The social density of the movement is hard to evaluate because it is characterized by continuous ups and downs, but we can identify three types of groups: the core group (between ten and twenty persons working on a regular basis); the midrange network of thirty to forty people supporting activities like movie screenings, debates and workshops on the occupied land; and a more extended network of 100 to 400 persons attending the festival and other similar events. Over the years of the conflict, four civil society/activist organizations have been created: 'The Free Keelbeek' collective, 'Platform Against Prison Disaster' 'Le forum potagistes' ('The Gardeners Forum') and the 'Haren Observatory'. Two more collectives that still exist have their roots in this movement, the CLAC (Collectif de luttes anticarcerale – a collective against prisons in general) and the platform Occupons le terrain ('Occupy the Land') which groups together several struggling spaces in French-speaking Belgium.

Being able to develop a counter-expertise and other legitimate knowledge

I have observed the development of a strong counter-expertise by local inhabitants helped by local NGOs and also a strong deconstruction/criticism of the prison system, an issue not usually confronted by an average member of the public. The movement deconstructed the environmental impact assessment,[9] the prison federal policy,[10] and legal and technical procedures; resisters learned about food sovereignty, urban agriculture, prison, land-planning, etc. In sum, opponents developed a 'critical system thinking' approach that helped them to link problems together. A vast variety of methods have been used – scientific, artistic, and so on (e.g. colloquium, festival, film-making, drawings, naturalistic exploration of the threatened land, etc.). I witnessed an interesting complementarity between those who bring empirical facts, those who have gained real-life experiences in the field, those who have legal skills or strategic and communication skills, and others who bring more emotions and energy for direct action. This cross-learning and systemic approach led the resisters to radically contest the compensation language used by the promoters.

Being able to develop place attachment

Place attachment was highly strengthened for local inhabitants, and outsiders 'fell in love' with the 'beauty of the site' despite it being surrounded by infrastructure and located nearby Brussels International Airport. The 'simple' fact of having a 'natural' (non-urbanized) and open landscape in the city with some remnants of a past rural identity played a great role in the place attachment. Most of the 'foreigners' did not know the Keelbeek before it was threatened with destruction. Thanks to the occupation, opponents discovered and co-produced many ecosystem services provided by the site, notably provisioning services (vegetable gardens, orchards, water, flower and fruit gathering, firewood, etc.). The 'living experience' of the occupation played a key role in the connection with nature. The most vivid example of this shared place attachment between inhabitants and foreigners is the celebration of the spring solstice in 2016 with representatives of the Kogi Indigenous people from Colombia who were passing by Brussels at an opportune moment. The sharing of place attachment between the inhabitants and the 'foreigners' was also built upon the claim of the farming inheritance of Haren, notably the cultivation of the chicory that originates in this area. Nevertheless, there were also tensions between the inhabitants of Haren and the 'occupiers' living on the Keelbeek, the former sometimes claiming that they were more legitimate than the Haren inhabitants regarding the management of the area and vice versa.

Finally, the struggle triggered, at least with the most involved actors, a reflection on the future of Haren orientated towards reclaiming lands to stop urbanization and starting urban agriculture projects in order to recover/recreate a rural identity. Conclusively, the place attachment that was present in the members of the inhabitants committee before the struggle was clearly strengthened through the battle and was shared largely with people that came from elsewhere.

Being able to develop political agency

(for more details see Pelenc 2020)

A large variety of actions were undertaken (from petitions, lawsuits, letters to politicians and encounters with them, to civil disobedience, direct action and illegal occupation, passing by drawings and short films, festivals, movie screenings and debates, leaflet distribution, etc.). Notably, thanks to the diversity of actors involved, I observed in the discourse of the inhabitants a scaling up of claims that began with the preservation of the local landscape and local mobility problems and ended with contesting federal prison policy. Another important political outcome was fostering, through the creation of the Gardeners Forum, the debate about the rapid urbanization process in the Brussels region. We also observed a thematic broadening of claims, from anti-jail attitudes to food sovereignty issues, passing by the 'right to the city' and a complementarity between legal and 'illegal' means of action. So, we can say that there was a significant convergence of struggles here. The Platform Against

Prison Disaster plays a role at the national level advocating for a moratorium on the prison policy, among other issues, and the Gardeners Forum plays a role at the regional level claiming the right of the citizens to actually participate in the making of the city, highlighting the importance of the conservation of natural/agricultural spaces for sustainability. Moreover, alternative projects to the mega-jail have been brought to the forefront by different members of the movement. Finally, the people involved felt a crude lack of democracy surrounding this project that led to strong distrust about 'democratic' institutions. They acquired the conviction that the supposed 'democratic' institutions will not necessarily guarantee public interest.

The description of these four collective capabilities illustrate the empowerment of the collective level (for more details about how these four collective capabilities are the result of a combination of different resources, rights, conversion factors and values, see Table 6.2 in the Appendix). In the following section I will explore how engagement in the resistance movement impacted on individual empowerment and well-being.

Individual Capabilities and Empowerment

In this section I analyse how personal involvement in the collective movement of resistance impacts on the individual well-being and agency of the resisters. To do so I have assessed the impact of involvement in the mobilization on individual capabilities (e.g. how the personal involvement impacts positively or negatively on individual capabilities and agency). I used a list of capabilities that I have developed elsewhere from a rapprochement between Max-Neef's fundamental needs (1991) and the capability approach (Pelenc 2017). This list comprehends ten fundamental dimensions of human well-being:

1. *Subsistence* (essential functionings to survive),
2. *Protection* (essential functionings to feel safe),
3. *Affection* (essential functionings to feel loved),
4. *Understanding* (essential functionings to understand other persons and nature),
5. *Participation* (essential functionings to participate in society),
6. *Idleness/leisure* (essential functionings for pleasant and playful entertainment),
7. *Creation* (essential functionings to create, to give life to things),
8. *Identity* (essential functionings to exist as a person, to belong to the human community and to the Earth),
9. *Spirituality* (essential functionings to development of spirituality[11]),
10. *Freedom* (essential functionings to have choices and responsibilities).

For each of these ten fundamental dimensions, I asked the interviewee to score the impact of their involvement in the collective action on their well-being (understood here from an agency perspective). The interviewee can score with + or ++ if the impact is positive or highly positive, and with – or – – if the impact is negative

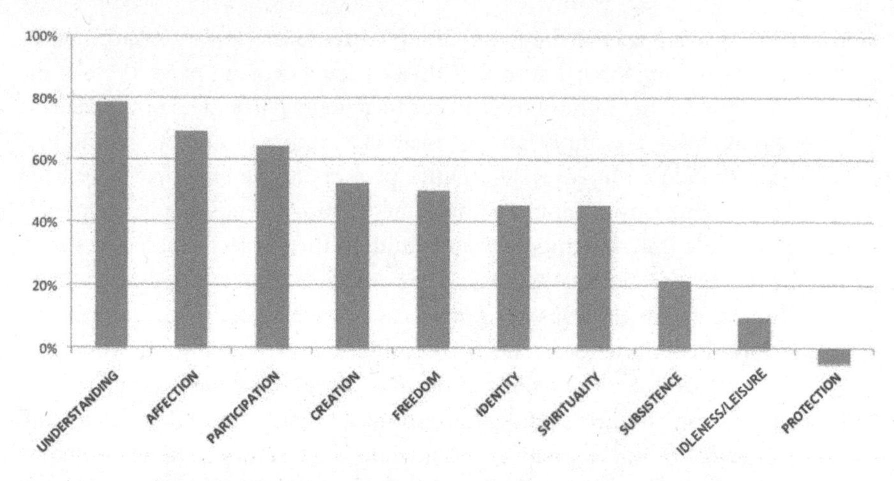

Figure 6.1 Impact of involvement in the collective action (the resistance movement) on the individual well-being of the resisters (aggregated results)

or highly negative. If their involvement in the collective action has no impact on a particular dimension, they wrote a 0. What really matters is not only the score but also the comments of the interviewee regarding the reason for it. Collecting both a score and a testimony allows a combination of both qualitative and quantitative results. The testimony helps justify the score and offers an interpretation of the dimension of well-being under discussion.

I first present the aggregated results and then present the results according to the different categories of resister (inhabitants, occupiers, NGO militants and activists).

A first look at Figure 6.1 shows that most of the dimensions have been impacted positively by involvement in the resistance. The most improved dimension (capability) is Understanding, followed by Affection and Participation. The least improved are Subsistence and Leisure, and Protection is impacted negatively. Let's now analyse these results in detail.

Understanding

For most of the actors involved in the struggle, their participation had a very positive impact on their Understanding capability. The following quotes illustrate this empowerment:

> It has improved my understanding of the imprisonment system and illustrated what a convergence between different struggles could be (NGO militant)
>
> It has improved my understanding of the imprisonment system, the issue of arable land and the history of the Haren neighborhood (Inhabitant of Haren)
>
> My involvement has improved my knowledge on the imprisonment system and the link between prison and social exclusion. It has changed our claims and geared them *toward* convergence, a mutual reinforcement of our claims with the other actors of the movement (Inhabitant of Haren)

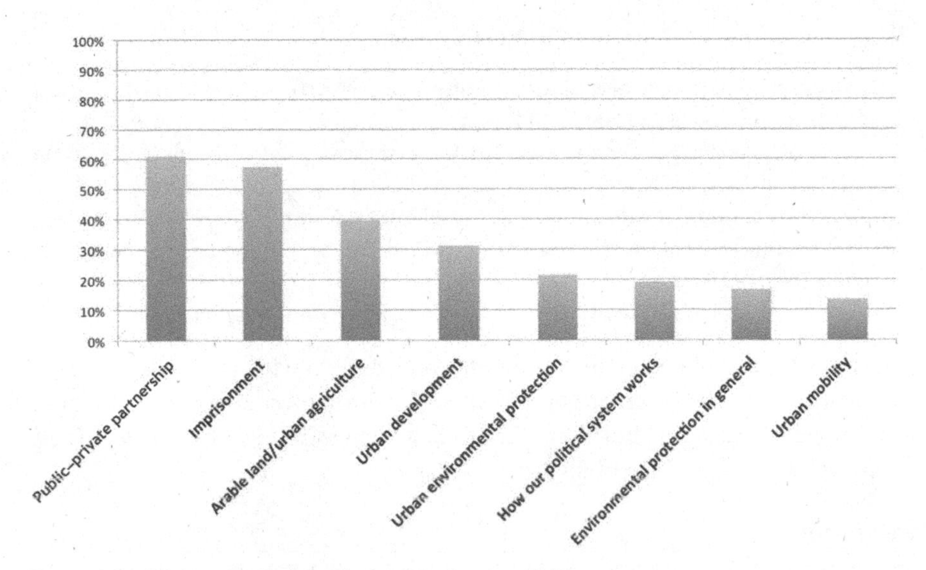

Figure 6.2 Disaggregating the Understanding capability

This echoes the 'Being able to develop a counter-expertise and other legitimate knowledge' collective capability I described above. We can form the hypothesis that there is a feedback loop process between the individual improvement of this Understanding capability and the need for the movement to be able to collectively develop critical knowledge to find arguments to protest against the project and to make broader claims. Given that this is the most improved capability, Figure 6.2 presents in more depth what kind of knowledge has been enhanced here.

This figure shows that it is the knowledge about public–private partnership (PPP) and imprisonment that have been the most improved by the resistance, followed by knowledge about food sovereignty/urban agriculture. This is very interesting because it illustrates cross-learning and the convergence of struggles.

Affection

The capability of Affection is the second most improved dimension of well-being for the individuals involved in the struggle. The social bonds formed during a battle like this are very strong. They are forged in the action, in the resistance, during the defeats and the victories, despite the conflicts and the tensions internal to the group (as we will see in the last section), and in face of powerful repression (as we have seen, involvement in the collective action has a negative impact on the capability of Protection). These bonds nurture the collective capability 'Being able to build alliances', but as mentioned by one of the interviewees they are also the site of possible tensions and conflicts, as in any human relations.

Some quotes from the resisters illustrate this result:

> Explosion of meetings, beautiful encounters, very enriching more than the regular friends or the family (Inhabitant of Haren)
>
> A lot of wonderful encounters but there are also a lot of conflicts inside the movement (NGO militant)
>
> I have become friends with people I would never have met elsewhere (Occupier)
>
> My involvement in the movement helped me to get out of my depression (Inhabitant of Haren)
>
> The struggle is an activity that has tied us together (Inhabitant of Haren)

These testimonies illustrate the 'social intensity' of the resistance. The ties that are created in such a context are strong and often transform the identity of participants. Depending on individual situations, the self may change during the struggle – as the evolution of the Identity capability shows.

Participation

The Participation capability is the third most improved dimension of well-being. One may well have expected that this capability would be the most improved, but it is interesting to see that Understanding and Affection are more impacted. From this capability onward we start observing a diversity of impacts on the individuals participating in the resistance that the aggregated results hide. Indeed, depending on the life course of the interviewee, the impact on the Participation capability varies according to the degree to which the person was previously involved in other struggles, or whether it is their first participation in a resistance movement or even in a social movement.

Some quotes illustrate this analysis:

> It is the very purpose of it [the affirmation of the Participation capability], it goes way beyond the defense of the Keelbeek, we have all been transformed, we will take the experience of the battle with us. This battle has changed us (Activist)
>
> Without this struggle, I would never have created Solid'Haren[12] (Inhabitant of Haren)
>
> This is the meaning of my life, resisting to a world of injustice I do not accept (Activist)
>
> We succeeded in blocking and delaying the project, that's quite something! (Occupier)
>
> Still not a brick and thousands of thousands of euros wasted! (Occupier)
>
> The movement is good but we are facing a wall (Inhabitant of Haren)
>
> Nope, no impact I was already engaged [already exercising my Participation capability] (Inhabitant of Haren)
>
> We gave a lot of energy and we had nothing back! (NGO militant)

I will not analyse all the capabilities in depth, but will just say a few words on the least improved ones. The Subsistence capability is strongly improved among the occupiers mainly because they could cover all their basic needs without money. Money was replaced on the ZAD by solidarity, mutual aid, and collecting food and material leftovers in different stores and markets. The vegetable garden launched on the Keelbeek also provided a bit of food during the first occupation. Above all, there was solidarity with inhabitants, who provided food, showers and internet access to those living on the Keelbeek with no running water and no electricity. Regarding the Idleness/leisure capability, this dimension often triggered a debate with the interviewee. For example, for the occupiers this dimension made no sense because once you live on the ZAD there is no difference between 'work time' and 'free time' – these categories of work and leisure do not apply as they do in regular society. Inhabitants also displayed different perspectives: some of them told me that the struggle was like a second job and took up all of their free time. Others thought that their involvement in the resistance could not be considered as leisure or free time, but instead a 'noble political cause'.

To conclude this subsection, I will discuss the only negatively impacted capability, i.e. Protection. The fact that this capability is negatively impacted illustrates well the risk-taking that the resisters felt aware of when confronted with state and capital power. There were very few positive testimonies on this subject; most of them are negative or neutral.

> Positive because I was able to feel the strength of the group (Activist)
>
> The network of solidarity enables to go deeper in the commitment (Activist)
>
> The threat and the protection are diffuse but we rather feel backed up by the inhabitants and the NGOs (Occupier)
>
> This is a rock and roll life, the solidarity depends on ever-changing conditions (Occupier)
>
> Going to the court is very frightening (Inhabitant)
>
> I feel in danger in my democracy (Inhabitant)
>
> There is a risk taking, you can be tracked by the police (NGO militant)
>
> Negative impact, I realized that decision-makers do not take good decisions (NGO militant)
>
> Now I'm afraid of being arrested and taken away by the police. Before, I was telling myself that I was not afraid to go to prison but now that I've been there, I mean in a cell during my custody, I changed my mind. (Occupier)

For most of them this feeling of insecurity and threat is unsettling because they strongly believe that they fight for the common good, for greater justice, and the judicial and physical repression make them feel instead like criminals.

This investigation of the interactions between individual and collective empowerment and capability would not be complete without an analysis of the tensions and conflicts that also arise during this kind of collective action.

Perspectives: Digging Even Deeper – Internal Conflicts Among Resisters, Towards a Group Micro-politics

In the two previous sections I have shown that involvement in the struggle is very empowering individually and collectively. However, it is also important to investigate the tensions, conflicts and difficulties that arise inside the resistance movement. For example, the 'emotional' cost of involvement can be very high. I noticed a significant level of burnout and separation of couples. 'Judicial and financial' costs also exist; some of the militants involved have been condemned to suspended prison terms and others burdened with financial fines – and the cost of a lawyer for legal battles is also very high. The feeling of empowerment and the energy of the movement go up and down – there is no linearity, with the mood depending on victories and defeats and on the fluctuation of the energy individuals are able to put in. For example, at the beginning when one starts the fight, there is an understanding of the parameters and a feeling of being strong enough to fight against the federal state, while also comprehending the vertiginous nature of the challenge. Moreover, the resister understands that the state is definitely not guided by the general interest. Then one starts fighting collectively, wins some battles, and so there is a feeling of great empowerment. And then come the defeats. On the legal side, but above all the physical side – it is very disempowering to see the police violently evict a camp, the destruction of all the gardens and facilities, the razing of the Keelbeek site with all the trees cut down. The resister then realizes that the state does not respect the laws and that they will start the construction works without permits or order the eviction without any legal basis. It is also an unsettling feeling to be the target of judicial threats; you feel like a criminal whereas you think you are fighting for justice.

Last but not least, it is worth mentioning all the internal conflicts that I also witnessed inside the movement. These conflicts are on the one hand necessary because the movement operates as an 'oppositional public space' (see Pelenc 2020) where different practices and political discourses can confront one another, but on the other hand they can generate a great fatigue among the resisters.

Table 6.1 attempts to capture the different types of conflicts that I observed inside the resistance movement (the structure of the table's inner/outer and individual/collective categories are inspired by Rauschmayer et al. 2011).

These observations are crucial to foster the self-reflexivity of the group. Indeed, these internal conflicts weaken the trust on which commitment to others and mutual aid rely. In turn commitment and mutual aid are the basis for the development of the four collective capabilities described above on which rely the political agency of the group.

Even if there were huge tensions between the promoters of legal and illegal actions, most of the actors interviewed recognized that without the occupation of the site the mobilization would never have been so strong, and that without the legal action through lawsuit the mobilization after the expulsion could have been finished. The lawsuits are also a backup option and provide credibility for those occupying the site. This particular struggle is a good illustration of the dialectics and perhaps the complementarity between legal and illegal actions. The eviction of the first ZAD in 2015 has been judged illegal and this allowed some of the activists to re-occupy the site one year later. In August 2018 the camp was evicted one more time and then the machines came and destroyed the Keelbeek land.

Table 6.1 The Different Categories of Conflicts Internal to the Resistance Movement

	Inner	Outer
Individual	Conflicts that arise within the self Self-questioning about the meaning of commitments in the battle, reflecting on one's position regarding prison or regarding our 'democratic' institutions, questioning personal responsibility: 'should I expose myself or not to physical or judicial problems?', questioning the meaning of work and everyday life compared to the exciting social moments of the mobilization ('should I go back to work on Monday or to the resistance movement?'), questioning legal vs legitimate actions, etc.	Conflicts that arise between the individual who is involved in the mobilization and their family, friends or work colleagues Conflicts with partners, friends or colleagues or even children regarding how involvement in the mobilization changes how the others perceive the involved individual, tensions regarding social role and social identity of individuals, etc.
Collective	Conflicts that arise within one of the particular groups of actors involved in the mobilization For example, tensions that arise inside the group of mobilized inhabitants regarding leadership and distribution of roles/tasks, etc., or within the group of occupiers regarding everyday life in difficult conditions (no drinkable/running water, no electricity, threat of expulsion, noise of the planes taking off just nearby the site, helicopter of the police flying over the ZAD at night, etc.).	Conflicts that arise between two (or more) different groups of actors involved in the mobilization For example, conflict or tensions between those who prefer to fight on the physical terrain through the occupation, often advocating for direct and illegal actions, and those who privilege the legal battle.

Conclusion: Three Levels of Transformation

The project has been delayed for six years (the prison was supposed to be operational in 2016 and is now supposed to be delivered in 2022; the construction work is currently finished), but this is not the main result to note.

Even if the battle has been lost, and the prison is built (2023), the resistance movement has entailed transformations at three levels:

1. At the individual level, I observed an improvement of individual capabilities (described in detail in 'Individual Capabilities and Empowerment' above). Only one interviewee out of twenty-two has seen his capabilities negatively impacted by his engagement in the struggle. Moreover, we have observed the emergence or the strengthening of radical subjectivities (there is clearly a process of

subjectivation during participation in the resistance – there is a 'before' and an 'after'; individual involvement in the resistance movement has transformed identities). Almost all the actors interviewed have seen their political positions transformed. This is particularly true for the inhabitant category of actors. Indeed, most of them said that before the struggle they were not politicized at all but thanks to their implication in this battle they developed strong critical discourse and critical practices and are now more able to position themselves regarding political issues – notably, of course, regarding the issue of prisons, land-planning and our supposed 'democratic' political system.

2. At the collective level (the resistance movement) I observed the creation of four collective capabilities (being able to build alliances; being able to develop counter-expertise and advance other legitimate knowledge; place attachment; and political agency) that enable the movement to fight against the project but also to propose alternatives directly by developing counter-hegemonic practices. More generally, the struggle has brought together different categories of people that would never have met elsewhere. This has formed friendships among different sociological profiles and different ways of life and practices of resistance. The individuals involved have experienced cross-learning and improved their understanding of the capitalist system. For many of the people involved, this struggle has played a key role in the decolonization of the imaginary. Indeed, this movement showed that a very motivated and diversified group of people could challenge a federal state project that involved billions of euros and transnational corporations. More concretely for those living directly on the Keelbeek site, by illegally settling there they demonstrated that another way of life was possible based on self-organization, solidarity and mutual aid, and respecting the ecological integrity of the Keelbeek site.

At the societal level, I observed the repoliticization of the federal imprisonment policy and the regional level of the land-planning policy. More widely this conflict has shed light on the weakness of the Belgian 'democratic' system. This process of repoliticization is possible thanks to the creation of what Oskar Negt calls an 'oppositional public space' (see Pelenc 2020). Finally, I have observed the creation of several collectives that illustrate the scaling up of the claims and of the network, from the Free Keelbeek Collective and the Haren Observatory which focus specifically on this battle, to the Platform Against Prison Disaster and the Gardeners Forum which gather together struggling territories in the Brussels region, to the 'Occupons le terrain' platform which does the same for struggling territories in French-speaking Belgium, and finally to 'CLAC', a collective that struggles against all prisons in Belgium.

Appendix

Table 6.2 A Synthesis of the Four Collective Capabilities

Being able to build new alliances	
Capability's parameters	**Components**
Resources	Pre-existing individual social capital
Conversion factors	The trust that is built among the group through the actions undertook together + the recognition from the public acquired through different actions + the recognition given by integrating new actors with high credibility
Rights	Right of association
Values	We are stronger together than alone
Results	The creation of five more or less formal civil society organizations

Being able to develop a counter-expertise and advance other legitimate knowledge	
Capability's parameters	**Components**
Resources	Individual knowledge, skills and expertise
Conversion factors	Complementarity between the individuals involved in the resistance (in terms of skills and expertise but also in terms of energy, availability, etc.)
Rights	Access to information; this right has been seriously restricted
Values	Distrust of the information given by the authorities, self-confidence in their own judgement
Results	Development of counter-expertise, critical approach to hegemonic discourse, counter-hegemonic discourse

Being able to renew and share an attachment to the threatened place	
Capability's parameters	**Components**
Resources	Haren semi-rural identity + the chicory heritage + the Keelbeek as natural/agricultural space + Haren past history (heritage)
Conversion factors	Occupation of the land
Rights	Cross the border of legality by illegal occupation of the land
Values	Difference between what is legal and what is legitimate; it is legitimate to occupy the Keelbeek to save it, even if it is illegal
Results	Renewed place attachment for the locals, shared attachment with the outsiders

Being able to develop a political agency	
Capability's parameters	**Components**
Resources	The three previous capabilities
Conversion factors	The fluctuating political context
Rights	Legal/illegal, legitimacy/illegitimacy dialectic, criminalization
Values	This project is toxic for society; it is not improving the common good/general interest
Results	The project has been delayed but above all a debate has been created around this project and more largely around imprisonment policy in Belgium and land-planning in the Brussels region (see Pelenc 2020 regarding how this movement fostered repoliticization)

Notes

1 This can be a backfiring argument because some could claim the right to maintain Western life at the cost of others, but many scholars have framed the CA into the suitability discourse (see among many other publications the special issue of the *Journal of Human Development and Capabilities* edited by Lessmann and Rauschmayer 2013). Here we understand the claim for being able to choose one's meaningful lifestyle as the right or the freedom to demand respect and defend different ways of life and relation to the environment.

2 https://niprisonnibetonlelivre.be/.

3 Brussels is both a municipality and a region. The region of Brussels is comprised of nineteen municipalities, among which is the municipality of Brussels.

4 Today, the prison is built; the construction work began in autumn 2018.

5 See Mauvaise Troupe 2016.

6 For further details see the 'Prison Ecology Project': https://nationinside.org/campaign/prison-ecology.

7 The term comes from the network of resistance movements opposing mega-infrastructure projects. This movement has been developing in Europe since 2010 (see Robert 2014 and Collectif Des plumes dans le goudron 2018).

8 For a complete presentation of the concept of collective capability see Pelenc et al. 2015.

9 Through discovering species, hydrography, questioning the model to calculate car traffic, understanding of the legal and political system, understanding of the architectural plans of the mega-jail and critique of them, understanding the 'consultation' procedure and critique of it.

10 Studies have demonstrated that the more jails we build, the more prisoners we have. The construction of jails has never been a solution to jail overpopulation. Reducing the number of prisoners is a political choice.

11 This should be read rather in a secular sense than one referring to organized religions.

12 Solid'Haren is a neighbourhood association created by some inhabitants involved in the resistance movement. Solid'Haren's goal is to foster social cohesion and transformation towards sustainability in the area through the implementation of a diversity of community projects such as participatory vegetable gardens, collective composting, etc.

References

Ballet, J., Koffi, J-M., Pelenc, J. (2013) Environment, Justice and the Capability Approach. *Ecological Economics*, 85: 28–34.

Collectif Des plumes dans le goudron (2018) Résister aux grands projets inutiles et imposés – De Notre Dame des Landes à Bure. Paris: Editions textuel, Collection Petite encyclopédie critique.

Demmer, U., Hummel, A. (2017) Degrowth, Anthropology, and Activist Research: The Ontological Politics of Science. *Journal of Political Ecology*, 24(1): 610–22.

Drydyk, J. (2008) Durable Empowerment. *Journal of Global Ethics*, 4(3): 231–45.

Dubois, J-L., Brouillet, A-S., Bakhshi, P., Duray-Soundron, C. (2008) Repenser l'action collective: une approche par les capabilités. Paris: L'Harmattan.

Evans, P. (2002) Collective Capabilities, Culture and Amartya Sen's Development as Freedom. *Studies in Comparative International Development*, 37(2): 54–60.

Forsyth, T. (2008) Political Ecology and the Epistemology of Social Justice. *Geoforum*, 39(2): 756–64.

Gaventa, J. (2006) Finding the Spaces for Change: A Power Analysis. *IDS Bulletin*, 37(6), 23–33.

Griewald, Y., Rauschmayer, F. (2014) Exploring an Environmental Conflict from a Capability Perspective. *Ecological Economics*, 100, 30–9.

Holifield, R. (2015) Environmental Justice and Political Ecology. In Perreault, T., Bridge, G., McCarthy, J. (eds) *The Routledge Handbook of Political Ecology*. London and New York: Routledge.

Holloway, J. (2015) Resistance Studies: A Note, a Hope. *Journal of Resistance Studies*, 1(1): 12–17.

Ibrahim, S. (2006) From Individual to Collective Capabilities: The Capability Approach as a Conceptual Framework for Self-help. *Journal of Human Development*, 7(3): 397–416.

Ibrahim, S., Alkire, S. (2007) *Agency and Empowerment: A Proposal for Internationally Comparable Indicators*. OPHI Working paper. Oxford Poverty and Human Development Initiative, Oxford University.

Lerner, S. (2010) *Sacrifice Zones: The Front Lines of Toxic Chemical Exposure in the United States*. Cambridge, MA: MIT Press.

Lessmann, O., Rauschmayer, F. (2013) Re-conceptualizing Sustainable Development on the Basis of the Capability Approach: A Model and Its Difficulties. *Journal of Human Development and Capabilities*, 14(1): 95–114.

Mauvaise Troupe (2016) *Defending the ZAD*. L'Eclat. https://mauvaisetroupe.org/spip.php?article143.

Max-Neef, M.A. (1991) *Human Scale Development: Conception, Application and Further Reflections*. New York: Apex Press.

Pelenc, J. (2017) Combining Capabilities and Fundamental Human Needs: A Case Study with Vulnerable Teenagers in France. *Social Indicators Research*, 133(3): 879–906.

—— (2020) Resistance as Repoliticization: The Resistance Movement Against the Mega-Prison Project in Brussels. In Murru, S, Polese, A. (eds) *Resistances: Between Theories and the Field*. London: Rowman & Littlefield.

Pelenc, J., Lompo, M.K., Ballet, J., Dubois, J.L. (2013) Sustainable Human Development and the Capability Approach: Integrating Environment, Responsibility and Collective Agency. *Journal of Human Development and Capabilities*, 14(1): 77–94.

Pelenc, J., Bazile, D., Ceruti, C. (2015) Collective Capability and Collective Agency for Sustainability: A Case Study. *Ecological Economics*, 118: 226–39.

Pellow, D. (2017) Environmental Inequalities and the U.S. Prison System: An Urgent Research Agenda. *International Journal of Earth and Environmental Sciences*, 2(140). https://doi.org/10.15344/2456-351X/2017/140.

Rauschmayer, F., Omann, I., Frühmann, J. (eds) (2011) *Sustainable Development: Capabilities, Needs, and Well-Being*. London and New York: Routledge.

Robert, D. (2014) Social Movements Opposing Mega-projects: A Rhizome of Resistance to Neoliberalism. Master's thesis, KTH Stockholm. http://stophs2.org/wpcontent/uploads/2014/12/D-Robert-Social-movements-against-UIMP.compressed.pdf.

Rodríguez, I., Inturias, M.L. (2018) Conflict Transformation in Indigenous peoples' Territories: Doing Environmental Justice with a 'Decolonial Turn'. *Development Studies Research*, 5(1): 90–105.

Schlosberg, D. (2009) *Defining Environmental Justice: Theories, Movements, and Nature*. Oxford: Oxford University Press.

Sébastien, L. (2013) Le nimby est mort. Vive la résistance éclairée: le cas de l'opposition à un projet de décharge, Essonne, France. *Sociologies pratiques*, 2: 145–65.

Sebastien, L., Pelenc, J., Milanesi, J. (2019) Resistance as an Enlightening Process: A New Framework for Analysis of the Socio-political Impacts of Place-based Environmental Struggles. *Local Environment*, 24(5): 487–504.

Sen, A.K. (1999) *Development as Freedom*. Oxford: Oxford University Press.

—— (2009) *The Idea of Justice*. Cambridge, MA: Harvard University Press.

Vercauteren, D. (2018) *Micropolitiques des groupes. Pour une écologie des pratiques collectives*. Paris: Éditions Amsterdam.

7

Raika Women Speak

Meenal Tatpati and Shruti Ajit[1]

Raika men are as straight as a cow, but Raika women are as cunning as a fox.

– A Raika proverb

Introduction

Proverbs such as these tend to encompass the resilience and business acumen of women, but often fail to take into account the articulations of the women themselves, about their own lives and ways of being. We came across this proverb while trying to initiate an action research project with the pastoralist Raika community in the Indian state of Rajasthan, to document their worldviews on development. The community has been in the limelight for trying to resist changes in their way of life due to the grazing restrictions placed on their traditional grazing land that falls under the Kumbhalgarh Wildlife Sanctuary.

The articulation of the community's vision of well-being and development has been overwhelmingly represented both in discourse and practice by men. The women of the community are navigating the sedentarization of a nomadic existence, because of which various changes have taken place within and around the community. Shrinking common lands due to exclusionary conservation policies, privatization of grazing commons and breakdown of the interrelationships among the various communities in the landscape who were earlier interdependent on each other are having a profound effect on the Raika women. At the same time, they are also negotiating their agencies within a highly patriarchal society.

It was our attempt with this study therefore to provide a gendered discourse to the already documented work on the community and understand, in the face of state/society-imposed 'development' discourse, how women whose lives are interlinked with the livestock they keep view them and navigate life through them. Along with this, the study also glances at some traditional practices and worldviews that have permeated through generations within the community, and how they impact the women of the community.

Developing the Methodology

The initial methodology included structured individual and focus-group interviews with women, as well as sessions to explore mapping of commons and future-building exercises. However, it became clear during the initial phase of fieldwork that arranging group meetings with the women would be difficult because the women are extremely busy managing their households and herds, added to which was the cultural inhibition on women holding meetings. Therefore, the methodology was modified to formulate a comprehensive question guide[2] to record life stories of individual women and their hopes and dreams for their families and the community. Four villages in Pali district were identified based on familiarity with the women to be interviewed as well as the proximity of the villages to the Kumbhalgarh Wildlife Sanctuary. Eighteen women were identified, because they were the most willing to speak to us. They were between the age of 15 to 90 years. We stayed with Raika families, helped with household chores and participated in celebrations such as births, marriages and religious festivals, which provided valuable insights into their daily lives as well as forging an intimacy which helped the women overcome their hesitation in speaking to and discussing their lives with strangers. Observations on the landscape and pastoralism were discussed while accompanying women and men on their grazing duties inside the Kumbhalgarh Wildlife Sanctuary.

All the women interviewed, except two, are from families that are now sedentarized and living in their native villages, and herd flocks of largely sheep and goat (some exceptions include cattle like buffalo and camel) around the village commons for seven to eight hours every day. The families of two women migrate with their flock for six to nine months of the year.[3] These families migrate from their villages just before winter towards the neighbouring state of Madhya Pradesh and return to the villages during the monsoon. Most of the families of the women interviewed have flocks of sheep and goats, while four families rear only goats and buffaloes and one family has camels. We could not observe interactions of the Raika women with other communities, limiting somewhat the understanding of the commons within the landscape. The paucity of secondary literature on women in pastoralism in India also hampered research, but enriched the case for further engagement in this area.

Location of the study

The study was conducted with women from the Godwar Raika community in Hiravav settlement (Sadri municipality), Ghanerao *panchayat*[4] (Desuri *tehsil*[5]), Latara and Dungarli (Bali *tehsil*) villages around Kumbhalgarh Wildlife Sanctuary (KWLS) in the Pali district of Rajasthan (Figure 7.0). Spread across 610.528 square kilometres of the Udaipur, Pali and Rajsamand districts of Rajasthan, the KWLS is an important sanctuary that includes parts of the Aravalli mountain range and serves as a barrier to the extension of the Thar Desert to the east along with other forest corridors (Foundation for Ecological Security 2010).

Pastoralism in India: Where Are the Raika Placed?

About 35 million people in India are engaged in pastoralism.[6] In most parts of the country, agriculture and livestock rearing have closely evolved together and ensure

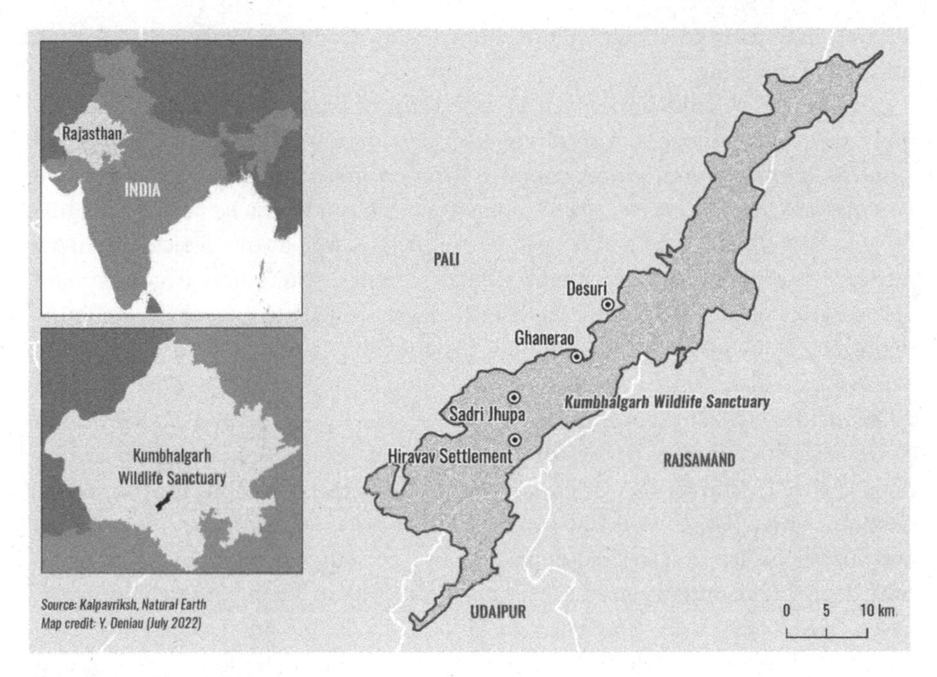

Figure 7.0 Map of the study area

the optimal use of the same landscapes (Sharma et al. 2003). Pastoralists contribute to livestock genetic diversity, upkeep of natural ecosystems and the overall rural economy.

In the dry and semi-arid parts of Rajasthan, an intricately woven resource-use system existed in village societies whereby diverse caste groups practised specialized livelihoods. These systems were based on complex interrelations[7] and the diversified use of landscape as a common property resource: forest land, village *gauchars* (land categorized as grazing land), stony and gravel lands, and agricultural fallows, all of which were accessible to pastoralists for migratory routes and grazing through negotiations with other castes. The Raika emerged as a specialized camel and small ruminant breeding pastoralist caste group. Oral history suggests that they may have settled in western Rajasthan from Iran or Baluchistan, and eventually moved eastwards, engaged by different kingdoms in Rajputana to care for state camel *tolas* (herds) (Kohler-Rollefson 2014). The origin myth of the Raika tells that Lord Shiva created them to become caretakers of camels (Raika Samaj Panchayat 2009). The Raika are concentrated in and around Jodhpur and in Godwar, the fairly fertile and forested region between Marwar and Mewar. Records suggest that the Godwar Raika around KWLS started rearing goats and sheep some 200–250 years ago and undertook long-term migrations with their herd towards the late nineteenth century due to *chappania akaal* (1899–1900), a severe drought that affected parts of north India including Rajasthan. This migration continued until the late 1950s but only in times of severe drought (Robbins 1998). Traditionally they traded meat, milk, wool and dung.

Today, many Raika have given up any form of pastoralism and are increasingly sedentary.[8] Those that continue with pastoralism sell meat and dung and combine it with crop production, either through sharecropping or on their own land parcels. The factors leading to sedentarization can be traced back to colonial policies that aimed to dismantle both pastoralism as well as the intricately woven resource use system in Rajasthani village societies. The colonial government viewed grazing as wasteful and legislated complex codes to ensure the scientific management of forests and commons for timber production and agricultural expansion, which created revenue for the colonial state (Agarwal and Saberwal 2006; Balooni 2002; Gooch 2009). Princely rulers such as the Maharajah of Kumbhalgarh shared this perspective and in the late 1800s colonial foresters acting on his behalf in the territory of the Godwar Raika began restricting grazing access (Robbins et al. 2007).[9] Post-independence, this focus on agricultural expansion and timber production for 'development' and 'growth' has continued. Concerns over depleting commons and forests have given way to an increasingly 'fortress-based' model of conservation, whereby large tracts of land are declared protected areas and all activities including grazing, lopping, collection of forest produce and traditional fire management practices are heavily regulated or banned completely through legal pronouncements.

In independent India, the forests around KWLS continued being used for commercial extraction of timber but were declared a Wildlife Sanctuary in 1971. Since 1975, parts of the KWLS began to be closed off for establishing nurseries, and grazing charges began to be levied (Kohler-Rollefson 2014). Three years after the KWLS was finally notified in 1986, the government of Rajasthan issued orders banning all livestock from forests. After protests by Raika as well as other livestock breeders in villages around the sanctuary, some forests were reopened but protected forests remained closed, thereby reducing grazing areas significantly. The final blow came in 2004 when the Rajasthan Forest Department refused to issue grazing permits for the KWLS after the Supreme Court passed an order prohibiting the removal of any forest produce.[10] Since 2006, all forms of grazing and lopping have been completely banned from the KWLS. Ironically, this was also the year when the Scheduled Tribes and Other Traditional Forest Dwellers (Recognition of Forest Rights) Act 2006 (also called the Forest Rights Act or FRA) was introduced and eventually passed by parliament in 2007. Meant to address the 'historic injustice' meted out to communities dependent on forests for their livelihoods, it recognized grazing as a right and made provisions for recording and vesting rights of use and entitlements over grazing for both settled and transhuman communities, and traditional seasonal resource access for nomadic and pastoral communities.[11]

Village agricultural fallows are another important source of grazing commons available for the Raika. Since agriculture was rain-fed, fields would be available for almost five months of the year (January to May) in the dry season to graze. However, this important source of fodder has also dried up recently due to the expansion of land taken up for cropping, double-cropping on former community fallows due to increased irrigation networks, and invasion of scrub species (*Prosopis juliflora*) into community grazing lands (Robbins 2004). These factors, combined with the blanket ban on grazing in the forests (although its implementation is ad hoc and arbitrary with illegal fines being developed on the spot by foresters[12]) and other forest management decisions such as the alleged rescue and release of leopards (*Panthera pardus*) into the area, which villagers claim has increased leopard numbers, has caused a considerable restriction in access to grazing resources.[13]

In the last 250 years, the Raika have managed to adapt to these periods of transformation. They have developed breeds of sheep suitable for sedentary environments and migration (Raika Samaj Panchayat 2009), adjusted their flock size and breeds to overcome periods of droughts and crisis, diversified their livelihood options, and even agitated politically and legally challenged adverse policies. They have forged alliances with various civil society organizations and used the law, including the FRA.[14] However, the uncertainties of livestock rearing, the increasing loss of pastureland and the risks that confront them while accessing it are all leading towards a trend of many younger Raika wanting to switch over to a more settled life, presenting a major challenge to Raika communities today.

At the same time, in many pastoralist communities, women, who were traditionally silenced, have begun to speak out more and articulate their vision for the future while emphasizing the need to hold on to some aspects of traditional pastoralist lifestyles. Women pastoralists around the world are advocating for the recognition of their contribution within pastoralist societies, access to grazing lands and commons secured through policies and legislation, their need for education and appropriate healthcare, and appropriate modern technologies (Women Pastoralists 2012). However, the voices of women within the Raika community have not been heard so far, either through academic research or within the arenas the community has used to politically take up its cause. We believe that the articulations put forth by women could bring out a new dimension to the issues faced by the community, and their vision of the future needs to be part of the collective envisioning process within the community that wants to continue maintaining a pastoralist lifestyle. Further, these articulations when shared with other women within pastoralist communities in India can start a dialogue on well-being.

Raika Women as Pastoralists

Taking care of animals is like taking care of children. One cannot do it without one's wife.

– A Raika elder, explaining why he could no longer keep his flock after the demise of his wife

Pastoralism is a family-supported operation with fixed and well-defined gender roles. Men are chiefly responsible for grazing, selecting breeds, the sale of animals, and making political and community decisions, and women are responsible for looking after the young and sick animals, the family and household, and processing animal products. Within the Raika community too, it is believed that women play a nurturing role, while men are engaged in production and political decision-making (Ellen 2001; Kohler-Rollefson 2017). Women are responsible for household activities, including cooking, cleaning, fetching water for washing clothes and consumption, taking care of guests, and other such work that often goes unnoticed. Within families that have land parcels, women also farm.

Most of the women we interviewed aged 15–35 were born into families where nomadic pastoralism was given up when they were very young, and some of them are married into families where day pastoralism is practised. These sedentarized families have sold off a large part of their *evad* (ancestral herd) consisting of sheep and goats, and now have herd sizes of up to 100 or less heads. There is wider diversification of livelihoods within a household, with sons migrating to cities and towns to seek alternate 'jobs' or to pursue higher education. In economically weaker households, family members are involved in supplementing their income through MGNREGA[15] and other jobs when available.

> I used to get impatient with the sheep at times, since they would keep interfering with my cleaning, and would hit them with the broom. My father-in-law told me that touching a broom to the sheep meant that they would surely fall ill.
>
> – A young woman who was married into a household with sheep, but did not have sheep in her maternal home

Young children, especially daughters, are taught basic household chores and rituals of everyday animal care like stall-feeding livestock, cleaning the pens, fetching water and feeding it to the livestock and the like, deemed women's work. As they become older, they are expected to take up more tasks related to animal rearing and birthing, and processing of animal products. This may entail adaptation to the changing animals being reared. These skill sets are learned through observation as well as through instruction from older women and men in their families.

Interviewees were of the opinion that their mothers had done far more household work when they were their age. The present generation admits that they would not be able to travel long distances like their mothers or grandmothers used to do while migrating with sheep and goats. The women in sedentary households observed that while they now have some time in the afternoon for rest, those who migrated would often have a lot of work tending to the young and sick animals left behind in the pen, looking after the children, unloading and setting up the temporary settlement and packing up again. However, inversely younger women in sedentarized households preferred grazing the livestock and cattle to staying back at home, as they felt that they had more time to themselves when they took the animals to graze. They felt that it was less work than household chores. Owing to the diversification of livelihoods from pastoralism, women are expected to contribute towards pastoral, agro-pastoral and other livelihood needs, while also taking on some of the responsibilities performed by men earlier, including grazing and lopping fodder responsibilities.

As implied at the beginning of the chapter, there is a general acknowledgement of Raika women being the 'controllers' of economic transactions within the family, including the sale of animals through interactions with traders and middlemen, the sale of *mingna* (sheep and goat droppings) which is a valuable source of manure, especially in sedentary households, the sale of milk and milk products, and the purchase of goods for the family (Kohler-Rollefson 2017). However, from the interviews we conducted with the women from sedentary households, we observed that only older women above the age of 45 carried out actual transactions of animals in the absence of their husbands. Younger women were unaware of the prices of the animals and did not participate in their sale or purchase. Where livestock flock sizes were up to fifty heads of sheep and goat, the sale of *mingna* and milk and ghee (within the village) was controlled by the women. However, if a household shifted to rear cattle instead of sheep or goats, the sale of milk was controlled by men as the surplus milk was sold in dairies in the nearest market, whereas the income from the sale of *mingna* was lost. Influential groups within the community have also started advocating for sedentarization and rearing of stall-fed cattle. This will affect the economic autonomy of women to some extent.

Raika women interact with a host of other communities within the village. They sell small quantities of milk and milk products when surplus, to members of other communities within the village. They also undertake additional day-labour (*majuri*) whenever it is available and required for the family. *Mingna* is traded with traders and farming communities from outside the village. Raika women healers are sought after to heal many ailments, especially the reproductive ailments of women from within the village and surrounding villages, as well as minor ailments of cattle and livestock.

Access to and Knowledge of Commons

> When I go out to graze the *evad*, I observe the landscape to see where there is sufficient fodder. I keep those routes in mind and go there the next day.
>
> – A woman from Ghanerao

In Rajasthan's ecosystem, commons are pastures, forests and wastelands used by the community as a whole, without exclusive private rights. These include community grazing land including uncultivable and cultivable fallow lands, village forests and woodlands, private croplands that are available for grazing after harvesting of crops, waste-dumping grounds, community ponds, water wells, migration routes, etc. (Jodha 1986). Previously, nomadic Raika used to migrate outside the villages in search of commons due to large flock sizes. With increasing sedentarization, they are now wholly dependent on *gauchars*, commons, agricultural fields and forest land surrounding their villages. Largely landless to begin with, Raika families have also begun to purchase small plots of agricultural land, and are moving towards agro-pastoralist livelihoods.

The nomadic Raika travelled in groups ranging from eight to twenty herding units, mostly families with a herd. These herding units are close-knit social units. The women who are part of these units would stay back at the camp, taking care of the young animals in the herd and the camp itself. However, it seems like sedentarization has also made pastoralism a family-centric business. One observes a married couple or a father and daughter grazing their herds around the villages. Therefore, the women are now also very closely involved in grazing duties, while almost exclusively taking up all the household chores as well.

All women interviewed had a wide-ranging knowledge of the common resources around their village. While ten of them still access the commons for grazing, collection of firewood and medicinal herbs, four women aged 20–35 do not graze animals and four above sixty years have stopped going into the forests due to their age. Here we examine the changes impacting the different commons and the women's relationship to these spaces.

For the Raika women, commons like *gauchars*, village agricultural fallows and *orans* (patches of forests preserved in the name of local deities or saints) are important spaces as they take their animals there for grazing and to collect firewood or fodder. Women we spoke to felt that all these spaces needed to be opened up and restored, with reasonable restrictions placed on their use. They described the forest as a 'peaceful' place that allowed them some time to themselves when the animals were grazing. However, the creation of the Kumbhalgarh Wildlife Sanctuary, the resulting restrictions placed on access and the increasing population of carnivores are all contributing to these spaces becoming inaccessible and unsafe.

Forests[16] *and* orans

All the women reiterated that forests are important for the well-being and survival of their flock and families, for grazing, firewood, medicinal herbs and food. Women above forty-five also spoke of forests as being repositories of food and fodder during times of drought. Women who herd spoke of the synergistic relationship between wild animals and the Raika, explaining that the wild animals would attack sheep and goats, but this was essential for the survival and good health of their flock, as well as for the survival of the forests and the wild animals themselves. They recalled how the community would take special care of the forests during the drier months to control the spread of wildfire. Certain restrictions are placed on women regarding access to parts of the *oran*. For example, women are forbidden from entering the area where the sacred fire or *dhuni* is lit when they are menstruating. However, for Raika women, *orans* were very often the only source of refuge for families while migrating, and especially for pregnant women and ewes who give birth there. So these restrictions were not followed strictly by the community. The women from one village revealed that since they did not have a *gauchar*, forests were their chief source of fodder. However, women fear going into the forests for grazing and lopping due to fear of wild animals and thieves. Some women were of the opinion that wild animals, especially leopards, were becoming more aggressive towards humans and were not afraid of people. They attributed this change to greater protection measures from the forest department towards the wildlife sanctuary. They believe that the wild animals do not have access to enough wildlife to eat and hence the carnivores are becoming fearless. Women healers reported loss of medicinal herbs and plants from nearby areas and said that they had to go deeper into the forests to collect them. Women reported that while the forest officers did not harass them with fines as they did to the men, they were asked to refrain from entering enclosed nurseries and regularly turned out from forests.

Gauchars

Women considered *gauchars*, as important sources of fodder and also safer places to graze livestock compared to the forests, particularly so in the last decade. All the women reported loss of *gauchar* land due to encroachment. Some village *panchayats* have started actively protecting the *gauchar*, and protection activities include demarcating areas that are kept enclosed for regeneration. Since women cannot always go into the forests, they sometimes need to graze their herds stealthily in the enclosed areas of the *gauchars* for fear of fines by village committees. Neither Latara nor Ghanerao *panchayats* have a single Raika representative in the ward and they are thus unable to bring up grazing issues.

Agricultural fallows

Women above the age of forty-five remember agricultural fallows as an important source of fodder for livestock and food for the Raika during dry winter and summer months before the monsoon. Farmers used to provide for the services of the Raika in kind and give food grains like *bajra* (pearl millet), *jowar* (sorghum) and wheat, and sometimes vegetables, in exchange for the flock fertilizing their fields while grazing. However, privatization of land and changing crop patterns mean that these fallows are now available for barely two months of the year and the Raika have to travel further from the villages to search for fallows. Women now have to negotiate with farmers, help clear land and lop overgrown trees on the fields to prepare the land for the next agricultural cycle in exchange for grazing their flock on the land. In one instance, a younger girl was stung by bees while clearing the tree branches in one of the fields that her family had gone to graze in. The younger women also expressed unhappiness at being driven away by farmers.

The herd as a commons

> I have been surrounded by animals since being born. Whenever I have to travel to places without animals I don't feel at home. It feels like a part of me is missing.
>
> – Raika woman from Hiravav village

Like pastoral societies around the world, the Raika value their animals as *dhan* or wealth. Women, especially those who graze sheep and goats and are used to looking after the young and sick animals in the family, frequently describe a sense of attachment to their herds. They keep sacred herds, herds for the daughter and sacred breeding rams. Herds, while essentially viewed as a single family unit's *dhan*, are also shared with other members of the community in times of crisis, or if one loses one's herd to pestilence or thieves, or if one has given up the pastoral lifestyle but wants to re-establish a herd.

> *A cow for the daughter to drink milk in her* sasural *[husband's household].*
>
> – A Raika phrase

While the herds are passed down from father to son, the *dhamena* (daughter's herd) is usually given to the daughter of the house when she is sent off to her *sasural* (marital home). The fathers and brothers either pledge a cow (a more recent phenomenon) or separate a few animals from each of their herds as a gift to the daughter. Other families from the village or from the immediate family also contribute to this herd. Even as the herd breeds and grows, it remains the daughter's herd in her marital home.

Each Raika family can also set aside a particularly strong ram from a good breeding pedigree as a sacred ram, which is dedicated to Lord Shiva and cannot be sold. Within the Raika community, rams with a good pedigree are also shared for breeding. There is a strong disapproval of selling female animals, which was likened by many women to selling one's daughter for profit.

Agency of Raika Women

A common theme overarching this study is the change from a nomadic to a sedentarized way of being. This has not only impacted the way women interact with the commons but also the agency of the women in various aspects of their life, including participation in formal and informal decision-making bodies (as also in issues of commons), access to formal education, the institution of marriage, and personal and reproductive health.

Marriage as an institution

Within the Raika community, marriages and relationships are a community concern. Marriage is also a way in which families ensure the continuing of pastoralism in one way or the other. The Raika practise the paying of a bride price, and other forms of marriages have evolved as a means of avoiding or mitigating the paying of bride price. One such is the *Ata-Sata* (literally 'one in exchange for the other') where a male and female from one family are pledged in marriage to female and male members from another family. This eliminates the bride price that both families had to pay for the respective brides. Another system, common earlier but losing popularity now, is the *ghar jamai*, where a groom from a poor family works for the bride's family for seven to twelve years, learning the trade from the father-in-law and offering his services to prove his worth in lieu of the bride price. Families in which such systems cannot be arranged have to pay an enormous bride price.

Marriage is decided among elders in a family. These systems are rigid, and the community socially boycotts families whose children find partners from other communities. Divorces need deliberation, approval and finally the payment of huge fines to the community. Despite this, the women we spoke to observed that divorces were common and attributed this to the non-consensual nature of marriages. Women were unanimous about the need for young people to have the freedom to choose their own partners. They also felt that they should be able to get married after getting to know the person. One young woman interviewed was married relatively later in life (at the age of twenty-seven) and had insisted on meeting and talking to her prospective groom before marriage. However, this was an extremely rare instance in the community. Women felt that it was very important to be able to choose someone who would treat them with respect. However, they did not consider alliances being formed outside their community as favourable or desirable. This can be associated with the strong caste system prevalent in the villages, where such a move would invite ostracization.

Personal and reproductive health

Personal and reproductive healthcare is seldom addressed while researching issues specific to women. Women from nomadic pastoralist communities rely on the knowledge passed on from one generation to another and have limited access to state-sponsored healthcare. Raika women use various terms to indicate the arrival of menstrual periods, including *kapda-aana* (often used by older women) whereas the younger Raika women call it 'MC' (menstrual cycle), as it is referred to in school. Younger women (15–35) were told about reproductive cycles by peers and friends, and close female relatives, and invariably hid the onset of menses from their families because attaining puberty means being sent off to one's marital home. Some young mothers have started explaining the process to their girl children.

In the past, women relied on traditional knowledge for reproductive and maternal well-being. Midwives, traditionally called *dais*, would assist women throughout pregnancy, home births and postnatal care. Nomadic women shared their experiences of giving birth while migrating from place to place – they were assisted by other women in the group or sometimes even managed on their own. Traditional midwifery knowledge was passed down from one generation to the other. Currently, Rajasthan has a high maternal mortality rate driven by the weak health of mothers and high rates of female infanticide and foeticide. The state has launched various programmes to encourage hospital births to help reduce these rates. Younger Raika women are opting to use these services. They acknowledge that over time the transmission of traditional knowledge around reproductive health has reduced and there are fewer *dais*. One older Raika woman observed that cases with complications during traditional childbirth were on the rise, increasing the preference for hospital births. On the other hand, cases of death due to negligence by hospital staff have also been reported. There is thus a divided opinion regarding prenatal care options. Expectant mothers continue to be guided by elders and even those living in cities often come back to be surrounded by the family, rest and be well fed during pregnancy. Pregnant women also often consult the *dai* for issues during pregnancy, provided a *dai* still resides in their village or close by.

Access to education

The Raika place increasing emphasis on educating their children, induced by the gradual breakdown of the pastoralist way of life (Kavoori 2007). Most sedentary Raika families are now keen to educate both their sons and daughters. All interviewees aged 15–35 except one have gone to school. The Raika are mobilizing around the issue of reserved jobs and education.[17] Raika families seek educated grooms for their daughters since it indicates a 'stable, settled' life and better economic prospects. These factors have led to aspirations among younger Raika men to leave nomadic/semi-settled pastoralist lifestyles and get jobs or start their own businesses in cities and towns. Since the state does not provide accessible education to nomadic families, Raika families practising permanent or semi-permanent migration enrol their children into formal schooling and leave them in the village with sedentary nomadic families or aged family members. All women felt that girls needed to complete high school to be able to interact with the larger world and to navigate the changes that are rapidly taking place around them. Some women mentioned that girls needed to be allowed to pursue higher education in the face of a rapidly changing environment. Most people in the community feel that illiteracy is the primary reason the state has been able to manipulate and curtail their rights over forests. Most of the younger women expressed that being educated would help them address this inequality of power.

Young Raika girls we spoke to perceive education as an important tool to navigate through the various oppressive institutions that they interact with on a day-to-day basis, and as a means to secure a better livelihood outside day-labour jobs. However, women who have successfully finished high school are now being discouraged from interaction with the outside world, including going to the nearest town to pursue higher education, for they are still confined within the informal institutional barriers of the community where it is predominantly the men who decide the fate of the women. In economically weaker Raika families, in the absence of young men (leaving the house for better job prospects), the lives of younger Raika girls are tied down to pastoralism. For example, in one household, a twelve-year-old girl would accompany her father to graze the family flock on the days she did not have to attend school, and tend to the young sheep and goats and procure lopping for them most afternoons when she came back from school. She was recently pulled out of school to help the family with grazing responsibilities.

The Raika caste Panchayat[18]

The Raika *samaj* (community) Panchayat is a caste-based institution that deals with conflict resolution within the community. The Panchayat addresses failed marriages, divorces, and land and other disputes between two parties within the community which they cannot resolve themselves. The Panchayat consists of male members from six, twelve or twenty-four villages, out of whom four are elected to preside over the meeting and pronounce arbitration. Women cannot participate and if they are involved in the dispute, men represent them. Any decision that the Panchayat takes is taken by men and often the gendered nuances and arguments within each are left out. People are discouraged from going to court to seek redress.

Most older Raika women feel that the judgement that the *samaj* gives usually benefits the larger community as a whole. While some of them may not agree with the judgement, they feel that they can do very little about it. The threat of being removed from the community has led to women depending on the men to decide their fate with very few or no avenues of representation. Some of the young women as well as men we spoke to are of the opinion that there are increased incidences of corruption and low accountability, and many believe the judgements pronounced are often unjust. Younger women are of the opinion that they should be allowed to participate in customary institutions, including the Raika Panchayat, to get direct representation.

Participation in aspects of Raika struggle over forests

While Raika women have stepped out of their homes in large numbers during protest demonstrations over prohibition of grazing in KWLS, except for the lone voice of Dailibai Raika,[19] no other woman has represented the community in leading these struggles. They are also not actively present in village-level institutions. While Hiravav, Latara and Ghanerao had applied for Community Forest Resource (CFR) rights under the FRA,[20] none of the women, except Dailibai Raika, were aware of this. Women, while aware of village-level forest management committees or *gauchar* protection committees in the villages, revealed that it was not possible to express their concern regarding access to these resources and that it was the men in the family who generally represented these concerns to the committees or state institutions.

Conclusion

In the past the Raika have adapted to continue their pastoralist lifestyle, in the face of privatization of land, fragmentation of commons and a politics of control over natural resources, whether it is by diversifying into other jobs within a household or through legal measures, through community mobilization over issues of education, livelihood, land and grazing rights, or through electoral processes where representatives from the community are attempting to procure seats in the state assembly to take up the issues of the community at the state level, or through dialogues with pastoralists in other states (like Gujarat). Currently, the community stands at a crossroads. Within this context, the present study has attempted to understand the lives of women within the community. It is a preliminary attempt to understand the transformations taking place within and outside it.

While women over the age of forty felt a great sense of pride in identifying themselves as the group that specialized in their *khandani dhando* (traditional business) and the acumen of breeding animals, they also expressed a sense of helplessness about the younger generation carrying forward the legacy of the business, due to tremendous pressure from external factors towards giving up their pastoralist lifestyle.

Women below the age of forty identified the pastoral lifestyle as something that is being done out of compulsion (*majburi*) to feed their families because they have neither the skill nor the education required to pursue other occupations. However, they also reiterated that it was better to pursue their *khandani dhando* than do manual labour, because it afforded them some dignity and the ability to take their own decisions about their livestock and also put their children through school. Some of the women also revealed that the younger generation that could not pursue higher education or secure good jobs in cities worked for minimum wages in cramped spaces away from their families, and learning the skills of rearing animals could give them a dignified life closer to their communities, if they chose to return.

The Raika have been undergoing transformations for centuries. These were induced by both natural and human-made causes, which would compel the community to adapt to the situation at hand, to ensure that their way of life continued. In the case of drought in the early 1900s, the Raika left their semi-sedentarized settlements and crossed state boundaries to access fodder. There are families who continue to migrate eight months a year to another state. The impact of drought meant that there was a shift in the roles and responsibilities within a household, the breeds of sheep, goats and camels they kept, forging new alliances and seeking new commons. When state policies became exclusionary and conservation policies at a national level refused to acknowledge the role of pastoralists in the landscape, the Raikas living around KWLS found ways to negotiate, so that they could continue grazing in the wildlife sanctuaries, whether through challenging these legislations in courts or by rallying against the declaration of national parks.

These transformations have impacted members of the community (based on gender, age, economic status, etc.) very differently. Understanding a community's notion of well-being from a gendered lens throws open a worldview that is not vastly different from what the men in the community have to articulate to outsiders. From this case study, it has emerged that the contemporary challenges that the Raika face have impacted the women most of all, as they have altered their role within the household as well as within the community. The effects on Raika women of fragmentation of land and changes in the pastoral way of being are not immediately visible and therefore these narratives are left out from larger discourses. The above articulations attempt to shed light on how the lives of these women are still linked to pastoralism and the commons. The Raika woman today is even more intimately involved with taking care of her flock, as men migrate to cities for jobs. Young Raika girls are taught how to take care of the animals and have limited access to education as both work and the community obstructs their right to individual choice. It is still the men who are addressing their concerns on a policy level and women are often left out from these conversations, though they are equally impacted by the loss of commons and exclusionary conservation policies.

While there is a large shift within the community towards leaving pastoralism as a way of being, there is also resilience among the Raika women, who associate their dignity with the animals. They are finding ways to express dissent with the existing systems, for example by stealthily grazing their flock in village *gauchars* where there is prohibition on grazing put into place through committees in which the Raika are not represented.

In conclusion, it is imperative to have the voices of pastoral women represented in the articulation of what 'development' means for the Raika community as a whole.

Notes

1 Both authors are members of Kalpavriksh.
2 Please contact the authors for the questionnaire.
3 To understand the migration patterns of the Raika, please see Agarwal 1992.
4 The *panchayat* is the formal local self-governance system in India. It is divided into wards represented by ward members directly elected by the village assembly or the *gram sabha*. The *panchayat* is chaired by the *sarpanch* who is the president.
5 A state is usually divided into districts. Districts are divided into *tehsils* and each *tehsil* consists of several blocks. These are all official administrative units in India.
6 See https://centreforpastoralism.org/about-us/.
7 Also called the caste system, this has created a hierarchical structure, leading to historical and continuing oppression of certain castes which needs to be acknowledged.
8 There are no official census records of this diversification of livelihoods among the Raika. This is reported by local organizations like Lokhit Pashu Palak Sansthan (LPPS) and the Raika themselves.
9 This was confirmed by Raika herders when they spoke of *haqdari* (record of rights) receipts being issued to them for grazing before 1986.
10 In 1996, the Supreme Court (SC) passed an order prohibiting the removal of any type of forest produce including dried and green wood, grass, leaves, etc. from Protected Areas across India. In March 2000, the Sadri forest office of the KWLS received the order and they issued their own order in Hindi which mentioned prohibition of 'grazing', although this was not mentioned in the original SC order. In 2004, the Central Empowered Committee of the SC, taking cognizance of the 1996 order, issued its own

orders to all state governments to implement the SC order while also enumerating grazing as one of the activities prohibited in Protected Areas.

11 See the preamble and Sec 3(1)(d) of The Scheduled Tribes and Other Traditional Forest Dwellers (Recognition of Forest Rights) Act 2006.

12 Personal communication with many Raika herders and reiterated in Robbins et al. 2009). Also corroborated by the Range Officer of Sadri, who reported that fines were levied on the spot as per the 'damage' done and 'assessed by the foresters'. He could not explain how the damage was quantified by the foresters.

13 Most Raika herders including women healers and graziers who enter the forests spoke of a decline in good fodder species and medicinal herbs from areas which were easily accessible before, especially due to an increase in canopy-like species such as *Lantana camara* and *Prosopis juliflora* which did not allow 'good' fodder to grow, and the introduction of leopards, bears and crocodiles into the area by foresters in 'cages' thereby increasing the number of wild-animal attacks on livestock while grazing. This was however not corroborated by the Range Officer of Sadri.

14 LPPS has filed Community Forest Resource Rights (Sec 2(a) and 3(1)(i)) claims under the FRA for some villages (Latara and Sadra of Latara Gram *panchayat* in Bali *tehsil*, Gudajatan village of Mandigarh Gram *panchayat* and Joba village in Desuri *tehsil*) where Raika herders stay. These claims were submitted to the Sub Divisional Level Committees during 2012–13, but they were not given any proof of receipt. The villages have not received any communication regarding the status of the claims.

15 The Mahatma Gandhi National Rural Employment Guarantee Act is a labour law that that aims to secure livelihoods in rural areas, by providing a minimum of 100 days of wage employment to adult members of every household within a village who volunteer to do unskilled manual labour.

16 Here, 'forest' is land legally defined as forest under the Indian Forest Act 1927. Most of the forest land that these villages use is within the KWLS.

17 In 2008, a law was enacted to provide four communities, including the Raikas, with the Special Backward Class (SBC) status (an affirmative action agenda of the state to ameliorate the effects of the caste system) which reserved 5% of seats. This was implemented in 2009, but was put on hold after the High Court issued an order of the violation of exceeding the 50% ceiling as per the Indian Constitution. In September 2015, a bill was passed in the Rajasthan State Assembly as the Special Backward Class (SBC) Reservation Act to reinstate the 5% reservation to the communities, which was struck down again by the High Court in December 2017.

For more details see www.financialexpress.com/india-news/5-castes-including-gujjars-re-included-in-obc-list-in-rajasthan/677292/.

18 In this case, the term *panchayat* is used to denote a non-state, community-led arbitration and decision-making institution.

19 Dailibai Raika is the only vocal female voice from within the Raika community who is demanding the right to graze as well as the safe passage for pastoralists in forests. She has been associated with the Lokhit Pashu Palak Sansthan, a local organization supporting the Raika.

20 The significance of CFR can be found in Sec 5 of the Forest Rights Act which empowers communities to 'protect forests, wildlife and biodiversity, and to ensure protection of catchments, water sources and other ecologically sensitive areas'. When read with

Section 3(1)(i) of the Act and Rule 4(1)(e) and (f) of the Amendment of 2012 (which elaborates on the constitution of a committee which can perform these functions as well as prepare conservation and management plans for its CFR), Sec 5 creates a space for forest-dwelling communities to practise forest management and governance by using their own knowledge systems and institutions and integrating them with modern scientific knowledge.

References

Agarwal, A. (1992) *The Grass Is Greener on the Other Side! A Study of Raikas, Migrant Pastoralists of Rajasthan*. International Institute for Environment and Development.

Agarwal, A., Saberwal, V.K. (2006) India Seminar, www.india-seminar.com/2006/564/564_a_agarwal_&_v_saberwal.htm. Accessed 29 May 2017.

Balooni, K. (2002) Participatory Forest Management in India: An Analysis of Policy Trends amid 'Management Change'. *Policy Trend Report*, 87–113.

Ellen, G. (2001) *Sheep Husbandary and Ethnoveterinary knowledge of Raika Sheep Pastoralists in Rajasthan, India*. Deventer, The Netherlands: Wageningen University.

Foundation for Ecological Security (2010) *Assessment of Biodiversity in Kumbhalgarh Wildlife Sanctuary: A Conservation Perspective*. Anand, Gujarat: Foundation for Ecological Security.

Gooch, P. (2009) Victims of Conservation or Rights as Forest Dwellers: Van Gujjar Pastoralists Between Contesting Codes of Law. *Conservation and Society*, 239–48.

Jodha, N.S. (1986) The Decline of Common Property Resources in Rajasthan, India. *Population and Development Review*, 11(2): 247–64.

Kavoori, P.S. (2007) Reservation for Gujars: A Pastoral Perspective. *Economic and Political Weekly*, 3833–5.

Kohler-Rollefson, I. (2014) *Camel Karma: Twenty Years Among India's Camel Nomads*. New Delhi: Tranquebar Press.

—— (2017) Purdah, Purse and Patriarchy: The Position of Women in the Raika Shepherd Community in Rajasthan (India). *Journal of Arid Environments*. https://doi.org/10.1016/j.jaridenv.2017.09.010.

Raika Samaj Panchayat (2009) *Raika: Bio-Cultural Protocol*. Sadri, India/Rajasthan: The Raika Samaj Panchayat with help from Natural Justice, Lokhit Pashu-Palak Sansthan and League for Pastoral Peoples and Endogenous Livestock Development.

Robbins, P. (1998) Nomadization in Rajasthan, India: Migration, Institutions and Economy. *Human Ecology*, 26(1): 87–112.

Robbins, P. (2004) Pastoralists Inside-Out: The Contradictory Conceptual Geography of Rajasthan's Raika. *Nomadic Peoples*, 8(2): 136–49.

Robbins, P.F., Chhangani, A.K., Rice, J., Trigosa, E., Mohnot, S.M. (2007) Enforcement Authority and Vegetation Change at Kumbhalgarh Wildlife Sanctuary, Rajasthan, India. *Environmental Management*, 40(3): 365–78.

Robbins, P., McSweeney, K., Chhangani, A.K., Rice, J.L. (2009) Conservation As It Is: Illicit Resource Use in a Wildllife Reserve in India. *Human Ecology*, 559–75.

Sharma, P. V., Koller-Rollefson, I. and Morton, J. (2003) Pastoralism in India: a scoping study. New Delhi: DFID.

Women Pastoralists (2012) *MERA Declaration*. https://landportal.org/node/8047.

8

Transformative Environmental Conflicts: The Case of Struggles Against Large-scale Mining in Argentina

Mariana Walter and Lucrecia Wagner

Introduction

This chapter explores an intriguing case of environmental mobilization, Argentina's anti-mining movement. This movement has contributed to the cancellation or suspension of about half of the contentious projects they have opposed. Argentina's national governments have openly supported mining activities as a strategic policy during the past two decades. Nevertheless, actors critical to mining activities have led to the approval of regulations/laws restricting large-scale mining activities in nine out of twenty-three national provinces (including two restrictive laws that were later reverted). This process of mobilization and institutional change is quite unique when compared to other Latin American and worldwide environmental mobilization processes (Scheidel et al. 2020). In this chapter we argue that while the most visible outcomes of mining mobilization are apparently institutional, changes in institutions were possible due to the multi-scalar mobilization of actors that successfully deployed strategies targeted on the different forms of power (institutions, actors/networks and narratives/discourses) that were supporting mining development in the country. This process has led to far-reaching transformations in Argentina's social and environmental narratives and policies.

According to the Latin American Observatory of Mining Conflicts, in 2021, Mexico (58), Chile (49), Peru (46) and Argentina (28) had, in this order, the largest number of mining disputes in the region (OCMAL 2022). In comparison with the other countries in this ranking, mining is quite recent in Argentina. Between 1997 and 2017, the extraction of metal ores (in tonnes) multiplied by ten in Argentina (UNRP 2018). Nevertheless, the high increase of ore material extraction is part of a larger Latin American (and global) trend. Mining booms started in the context of (regional and global) economic liberalization reforms and skyrocketing metal ore prices that fostered a shift of the international geography of mining investment towards the developing world during the 1980s and 1990s (Bridge 2004). This trend remains to date (Karl and Wilburn 2017). Such extractive pressures will likely increase given the strategic role of Latin America for the global provision of copper, lithium and nickel, among other key metals and minerals for the energy transition (USGS 2020).

While mining activities have a long history in the Latin American region, the size of material extraction pressures and exports, particularly regarding mineral ores, has no historical precedent (León et al. 2020). Mining exploration and exploitation pressures are fostering a growing number of mining conflicts in the region and (deadly) violence against environmental defenders (Global Witness 2020). Furthermore, hegemonic energy transition policies are becoming a relevant source of extractive pressures (IEA 2021) and socio-environmental conflicts (Deniau et al. 2021).

As we explain in the outline of our methods, this research is nurtured by more than eighteen years of engaged research conducted by the authors with mobilized groups in Argentina. In the first stage of the research, we conducted the first nationwide systematic analysis of all large-scale mining conflicts in Argentina (Wagner and Walter 2020; Walter and Wagner 2021). In this chapter we examine in more detail how this mobilization process has targeted different forms of power, fostering socio-environmental transformations. We developed, in collaboration with the Environmental Justice Atlas (https://ejatlas.org/) – a global database on environmental justice struggles – a systematic identification and analysis of all large-scale mining conflicts in the country from 1997 (when large-scale mining began) to 2018. We examined trends related to the history of mining activities and conflicts in the country, the mobilized actors, their strategies and the outcomes of their struggles. We explored how the actors and strategies mobilized in mining struggles in Argentina have targeted and impacted key forms of (visible and invisible) hegemonic power. Based on the Conflict Transformation Framework (Rodríguez and Inturias 2018), we claim that actors and networks were able to move across scales to impact institutions (e.g. mining regulations), actors/networks (e.g. creating unexpected alliances with local governments and heterogenous actors) and socio-environmental discourses (e.g. challenge sustainable mining narratives). Furthermore, we claim that the transformative outcomes of social mobilization went beyond mining concerns, to address broader socio-environmental matters.

After this brief introduction, the second section presents our conceptual framework for approaching mining conflicts and power. The third section explains our methodological approach. The fourth section outlines the results that we discuss in the fifth, and in the final section we offer our conclusions.

Mining Struggles, Outcomes and Transformations

Mining is among the most conflictive activities in the world (Scheidel et al. 2020). Literature examining mining struggles around the globe identifies different concerns that mobilize communities against mining activities. Environmental impacts and risks are usually central in these disputes (Conde 2017). According to the US EPA, mining is one of the main sources of heavy metals and pollutants to the environment (EPA 2020). However, communities mobilizing against large-scale mining are not only reacting to perceived or visible environmental impacts but also to a lack of information, lack of representation and participation in decision-making processes concerning their development path, a lack of monetary compensation, and distrust of the state (e.g. lack of independence, control capacity) and mining companies. Mining struggles are encompassed by different discourses that range from compensation and other market-embedded measures, to broader claims regarding post-material and socio-ecological alternatives (Conde 2017). The defence of livelihood in mining conflicts should not be understood only as the protection of a source of subsistence and income, but also as the protection of its embedded meanings, values and identities (Bebbington et al. 2008). In Latin American contentious politics, there is an inseparable relationship between the material and the cultural in livelihoods (Escobar 2001).

In their review on mining struggles research, Conde and Le Billion (2017) identify that dependency on mining companies, political marginalization, and trust in institutions tend to reduce the likelihood of resistance. However, large environmental impacts, lack of participation, extra-local alliances, and distrust towards the state and extractive companies tend to increase resistance. Many of the factors leading to resistance signalled by Conde and Le Billion (2017) are present in Latin America. Many communities distrust governments and companies. The notion of the 'right to decide' has been raised as a key concern of communities where projects are seen as imposed (Muradian et al. 2003, Vela-Almeida et al. 2021). It has become common for anti-mining groups to prevent or boycott public hearings, as these are seen as an empty requisite for project approval (Jahncke Benavente and Meza 2010; Walter and Urkidi 2017). In this vein, Vela-Almeida et al. (2021) conclude that large mobilizations, legal actions, calls for binding consultation, and forms of blockades are successfully used by affected communities to shape decision-making. While these actions from below achieve certain outcomes, they are only temporarily successful, as long-term decisions surrounding extractive governance and underlying structural inequalities remain unaffected. Nevertheless, these actions can create a path for questioning political participation outside the existing structural constraints and for questioning established social orders and building emancipatory tools.

While not all countries suffer the same environmental impacts from mining activities as Latin America, the internet and regional networks of activism are making them visible beyond borders. Supra-local networks systematizing and spreading knowledge and strategies of resistance among anti-mining local groups are key in the spread of anti-mining consultations in Latin America (Walter and Urkidi 2017).

Outcomes of mining struggles

The study of the outcomes of social mobilization and more precisely the factors or conditions under which social mobilization is able to enact (successful) collective action is a key enquiry of social movement studies (e.g. Bosi and Uba 2009; della Porta and Rucht 2002; McAdam et al. 2001; Tarrow 2011; Van Der Heijden 2006). Social movement studies highlight 'political opportunity structures' and the 'repertoires of contention' as key elements to consider the particular outcomes of social mobilizations.

Political opportunity structures are features of a political system that facilitate or constrain collective action. Studying these features allows us to understand the feasibility and outcomes of certain strategies in certain contexts. An international study offers some insights on the political opportunity structures related to mining projects and contestation. Özkaynak et al. (2021) studied 346 cases of mining struggles from the EJAtlas (including construction materials) and found that preventive mobilization and conflicts confronting smaller companies are more likely to stop projects than reactive mobilization (when projects are already active) against larger companies.

Furthermore, Argentina presents a particularity when it comes to mining and environmental governance institutions that became key political opportunity structures. The political decentralization fostered in Argentina (and Latin America) in the 1990s led to the development of semi-direct democratic instruments for provinces and municipalities that can conduct local plebiscites or referenda on environmental issues. These instruments have allowed social movements to promote such events (e.g. Esquel and Loncopué in Argentina).

The main features of the institutions that regulate mining activities are similar in all Latin American countries. However, while approval of mining projects – based on the revision of the environmental impact assessment (EIA) report – is usually centralized in the national government, in Argentina it is decentralized to provinces. There are laws regulating mining activities and environmental protection at the national and provincial levels. But provinces own and manage mineral resources and are in charge of the review, approval and control of mining projects. Civil society actors can present non-binding allegations to EIA reports and express their views in a public hearing where the technical document is presented and discussed for approval. In this vein, the provincial level is a key scale of mobilization for local groups against mining activities and the study of these processes.

Repertoires of contention refer to the different forms of protest or action mobilized by social movements which are often influenced by national and local contexts and trajectories (Tilly 2002). As we explore in this chapter, movements mobilized against mining activities in Argentina deploy diverse strategies in alliance with different actors, a feature that seems relevant to understand the outcomes of the social movement. In their analysis of the EJAtlas global dataset, Scheidel et al. (2020) found that if mobilized groups combine strategies of preventive mobilization, diversification of protest and litigation, they can significantly increase their success rate (ability to stop projects) to up to 27%.

Transformations: power and scales

The Introduction of this book outlined key definitions regarding different types of power involved in social transformations and the role of scales in these processes. As reviewed by Rodríguez and Inturias (2018), power of domination can manifest in the form of visible, hidden (Foucault 1971) and invisible/internalized forms (Gaventa 1980; Lukes 1974). In this chapter we highlight how the actors, the networks and the strategies mobilized in mining struggles in Argentina have impacted these three key forms of power: 1) institutions and legal and economic frameworks (visible power); 2) people and networks (hidden power), and 3) discourses, narratives and ways of seeing the world (invisible power). We recall below some of the definitions referred to in the Introduction.

The strategies and outcomes of social mobilization can address the visible or less visible faces of power domination. 'Visible' forms of power are manifested through decision-making bodies where issues of public interest, such as legal frameworks, regulations and public policies, are decided. *Institutional or structural* power refers to the public spaces where social actors deploy their strategies to assert their rights and interests (Rodríguez and Inturias 2018). The 'hidden' face of power refers to how some actors create barriers to participation, exclude matters from the public agenda or drive political decisions 'behind the scenes' to maintain their privileged position in society. Power of domination is deployed by *people and networks* (Long and Van Der Ploeg 1989). These actors are organized to ensure that their interests and worldviews prevail over those of others. This form of power works in an 'invisible' way through discourses, narratives, worldviews, knowledge and behaviours that are assimilated by society (Foucault 1971). These structural forms of power are 'materialized' in state institutions, the market and civil society and give rise to structural biases in the relationships and the consequent asymmetrical power relations. This form of power is known as *cultural power*. Invisible power (narratives/discourses) and hidden power (in actors/networks) often act together, one controlling the world of ideas and the other controlling the world of decisions (Rodríguez and Inturias 2018).

In this vein, distinguishing between the power concentrated in institutions, people and culture allows a better understanding of relationships of power and domination in socio-environmental conflicts and in the perpetuation of environmental injustices (Temper et al. 2018a). As was argued in the Introduction, the challenge for overcoming violence and injustice (Young 1990) and therefore for achieving conflict transformation is to generate strategies to impact on the three areas in which power is concentrated: 1) institutions and legal and economic frameworks, 2) people and their networks, and 3) discourses, narratives and ways of seeing the world. The final outcome of the struggles in terms of achieving the desired transformation would depend on knowing how and when to impact on each one of the types of hegemonic power. In this chapter we explore, with the EJAtlas data on Argentina's large-scale struggles, how these different forms of power were targeted and impacted by the diverse actors, networks and their strategies.

Moreover, by studying historical, local and collective mobilization processes we adopt a multi-scalar approach that allows us to capture how social groups and networks move across scales, challenging power inequalities. Such an approach also allows us to problematize the question of what is transformative or not, as what could be considered an unsuccessful struggle locally (i.e. contentious activities take place) can trigger powerful processes of organization and institutional changes that go beyond space, time or specific issues of contention (Walter and Urkidi 2017; Temper et al. 2018a).

Methodological and Conceptual Approaches

Our methodological approach combines an analysis of general trends – based on the EJAtlas dataset – with a review and analysis of mining conflicts and their history in the country. This is possible given the manageable size of the sample and the authors' research background. Both authors' PhD thesis and more than eighteen years of research has focused on mining conflicts in Argentina and Latin America. The authors' research emerged from their field collaboration with mobilized groups in Chubut and Mendoza and later collaborations with other Argentinean and Latin American movements in mining contestations. Many of the cases developed for this chapter are the result of these collaborations. Furthermore, the authors have been part of diverse collective publications co-developed with social movements and engaged scholars on mining challenges and risks (e.g. Deniau et al. 2021; Machado et al. 2011).

The identification and analysis of large-scale mining struggles in Argentina was conducted in collaboration with the Environmental Justice Atlas. The first author of this chapter is member of the Direction and Coordination Team of the EJAtlas and the second is an active collaborator. The EJAtlas is a collaborative mapping effort launched in 2011 that was co-designed by activists and academics (Temper and Del Bene 2016; Temper et al. 2015). The EJAtlas is today the largest global database on environmental conflicts with cases co-developed between engaged researchers, mobilized groups, students, journalists, etc. (Temper et al. 2018b). The EJAtlas defines socio-environmental conflicts as 'mobilizations by local communities, social movements, which might also include support of national or international networks against particular economic activities […] whereby environmental impacts are a key element of their grievances' (www.ejatlas.org). The EJAtlas documents social conflicts related to claims against perceived negative social or environmental impacts that follow these criteria. There has to be first, an economic activity or legislation with actual or potential negative environmental and social outcomes; second, a claim and mobilization by environmental justice organization(s) that such harm occurred or is likely to occur as a result of that activity; and third, a reporting of that particular conflict in one or more media stories (in some cases where such media coverage is difficult, exceptions can be made). For each case information is given regarding the location, features of the contested activity, the commodity at stake, the mobilized actors, their grievances, their strategies and the outcomes of the conflict.

In order to conduct this research, the authors undertook a review and update of all large-scale mining conflicts registered in the EJAtlas in Argentina regarding gold, copper, lead, iron, lithium and uranium up to December 2018. Moreover, a review of academic literature, media coverage (national, provincial and local newspapers) and non-governmental organization (NGO) sources was conducted in order to identify missing cases. New cases were identified, documented and added to the EJAtlas dataset. As a result, a total of thirty-eight cases were identified. A final review of all cases was conducted in order to ensure consistent and comparable categories among cases (e.g. how Indigenous, ethnic/racial groups or women groups were coded). The EJAtlas dataform structure identifies different types of actors, strategies, impacts and outcomes, which were used for the analysis. We classified strategies and outcomes considering the main forms of power that these addressed. This classification was adjusted and discussed with the support of Prof. I. Rodríguez and based on her work in the field of power and conflict transformation studies (see Rodríguez et al. 2015).

Mining Struggles in Argentina

The systematic search for large-scale mining conflicts led to the identification of thirty-eight struggles from 2002 (first ore mineral extraction conflict) to December 2018. These are not necessarily all *existing* conflicts, but those that adjust to the EJAtlas criteria referred to in the methodological section that excludes, for instance, labour and private conflicts. Figure 8.1 presents the geographical location of the identified conflicts, distinguishing those conflicts that led to the cancellation or suspension (temporarily) of mining projects. The map also signals those provinces where mining activities were restricted by specific legislation as the result of social mobilization.

Most conflicts are related to gold- and copper-mining projects (twelve cases each), followed by uranium and lithium projects (four conflicts each), lead (three cases), iron, silver and potassium (with one each). This distribution resonates with the weight of these metals in the economy. In 2017, gold was the main metal exported by Argentina (3.7% of total exports value), followed by silver, zinc, copper, iron and lead (OEC 2019). Argentina is currently the third worldwide lithium exporter and there are increasing pressures to open new mines in the north of the country, as lithium carbonate is a strategic mineral for electric batteries used in cars and phones. In fact, in March 2022, the EJAtlas listed seven conflicts related to lithium-mining registered in Argentina, three more than those considered in this initial study.

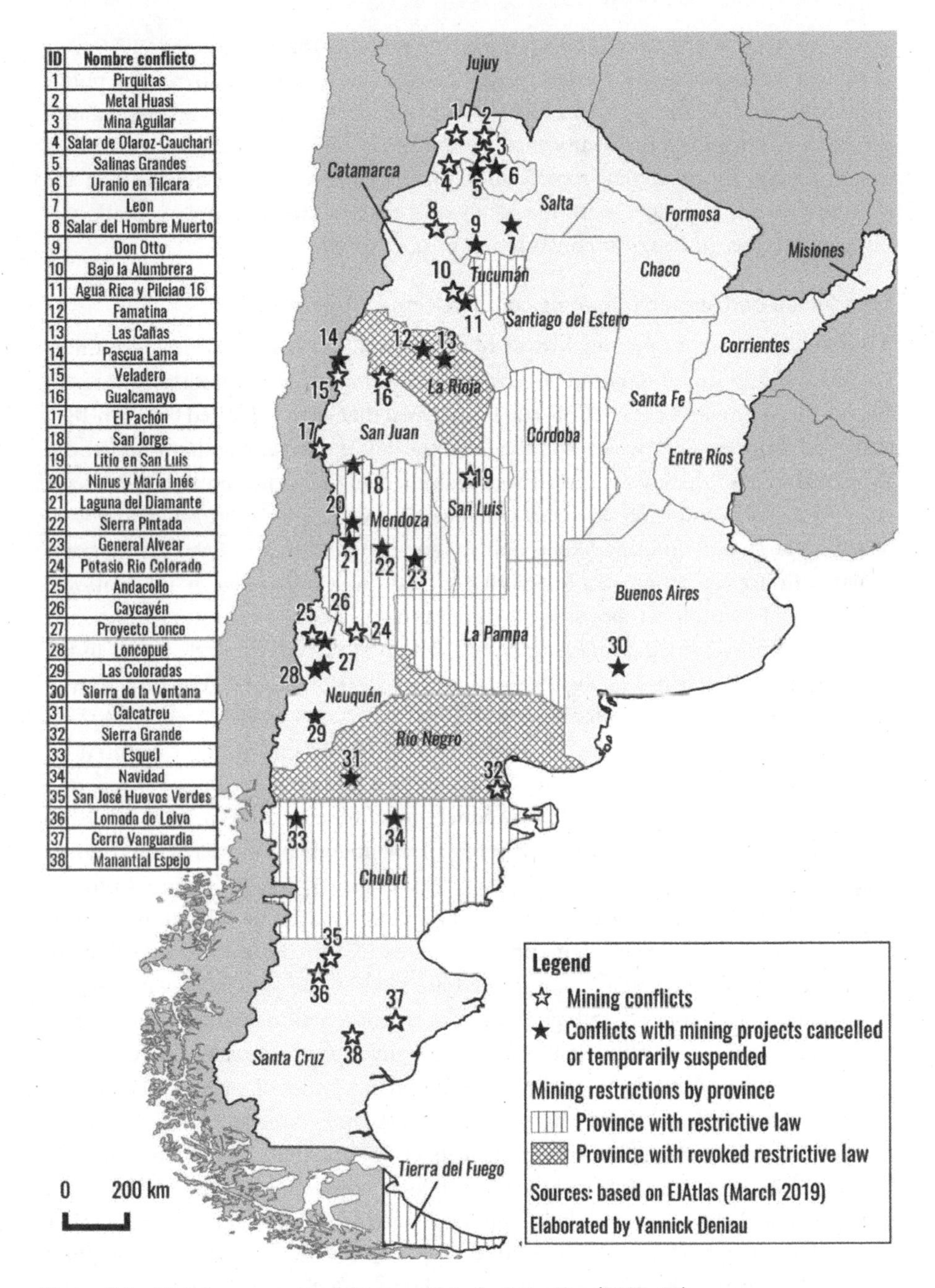

ID	Nombre conflicto
1	Pirquitas
2	Metal Huasi
3	Mina Aguilar
4	Salar de Olaroz-Cauchari
5	Salinas Grandes
6	Uranio en Tilcara
7	Leon
8	Salar del Hombre Muerto
9	Don Otto
10	Bajo la Alumbrera
11	Agua Rica y Pilciao 16
12	Famatina
13	Las Cañas
14	Pascua Lama
15	Veladero
16	Gualcamayo
17	El Pachón
18	San Jorge
19	Litio en San Luis
20	Ninus y María Inés
21	Laguna del Diamante
22	Sierra Pintada
23	General Alvear
24	Potasio Rio Colorado
25	Andacollo
26	Caycayén
27	Proyecto Lonco
28	Loncopué
29	Las Coloradas
30	Sierra de la Ventana
31	Calcatreu
32	Sierra Grande
33	Esquel
34	Navidad
35	San José Huevos Verdes
36	Lomada de Leiva
37	Cerro Vanguardia
38	Manantial Espejo

Figure 8.1 Main large-scale mining conflicts in Argentina (2003–18)

Note: Conflicts are identified by the name with which they are publicly known.

More than 50% of the mining conflicts in Argentina registered before 2019 were of high or medium intensity (twenty-eight cases). This means that there were public and visible mobilizations that in four cases resulted in violence or arrests (against protestors). Furthermore, twenty-three of the thirty-eight conflicts were identified as preventive, meaning that mobilization started before the development of the mining project, and eight were the result of a reaction to the implementation of mining activities (during the construction or operation).

Emergence and spread of mining contestation in Argentina

Argentina does not have a long history of either small- or large-scale metal-mining. Historically, the largest mines in Argentina were Famatina and Capillitas, located in the northern provinces of La Rioja and Catamarca that were exploited between 1850 and 1914. Low workers' wages and the exploitation of natural resources influenced the crises of these historic mines. These crises led to the abandonment of ore-rich sites for decades and even did away with regional economic approaches that were based on large-scale mining (Rojas and Wagner 2017). In the 1990s, the approval of new mining legislations (in line with a global process fostered by international development institutions) boosted the development of open-pit mines with the use of new technologies (e.g. cyanide leaching). The first three large-scale mines in the country started activities in the 1990s. Between 1997 and 2017, the extraction of metal ores (in tons) multiplied by ten (UNRP 2018). It was however in 2002, in Esquel, that large-scale mining contestation emerged and spread in the country.

The spread of anti-mining mobilizations in Argentina has been related to a cascade effect from two key conflicts: Esquel (gold-mine conflict in Patagonia) and La Alumbrera (copper and gold mine in one of the poorest provinces of the country, related to environmental accidents). These key cases triggered a learning process about the impacts of mining activities and the power of social mobilization (Svampa et al. 2009). The Esquel mining conflict and its 'No to the mine' movement in Chubut province (Patagonia) is considered the birth of the anti-mining movement of Argentina. This conflict emerged in the midst of a national political, economic and social crisis. Esquel municipality had a 25% unemployment rate and 20% of the population was under the poverty line. In this context, national, provincial and local governments were highly supportive of new investments (Walter 2008). The Esquel conflict placed the discussion of mining – its environmental, social and economic impacts, as well as the question of the right of local populations to choose their own development path – on the national political agenda (Walter and Martinez-Alier 2010). A municipal plebiscite was held in March 2003, the second vote of its kind in Latin America after the Tambogrande local consultation in Peru, in 2002. The Esquel plebiscite resulted in an 81% rejection of the gold- and silver-mining project located 7 km from the city. Esquel was also the first public mining conflict that reached the national media and led to the formation, in 2003, of a National Network of Communities Affected by Mining, with the participation of communities from six provinces in Argentina, all opposed to mining projects in their areas (Walter and Martinez-Alier 2010). Another result of the Esquel conflict was the sanction of a provincial law (Chubut) that restricted mining activities. The Esquel movement also engaged in the task, that remains today, of systematizing and disseminating national and international information related to large-scale mining impacts, especially regarding the use of cyanide. They created a website (noalamina.org) and have published materials and books on the matter. These resources were very important in other conflicts that followed – for instance for the anti-mining mobilization that emerged in Ingeniero Jacobacci, a very small town in Río Negro province where a gold mine project planning to use cyanide was proposed. This project was rejected by the local population, and thanks to this social mobilization, a provincial law restricting metalliferous mining was also sanctioned in Río Negro province in 2005.

That same year, in 2005, a conflict against mining appeared on the streets of another province, Mendoza. The neighbours of San Carlos, an agricultural town located in the central oasis of this province, rejected a metalliferous mining project planned near the Diamante lagoon, a natural protected area also considered a water reserve. After that, neighbours from different towns of Mendoza province rejected other metalliferous mining projects, including the reopening of a uranium mine, Sierra Pintada. These processes led to the organization of a provincial network of self-convened neighbour assemblies called Mendocinean Assemblies for Pure Water (AMPAP), one of the strongest organizations against mega-mining in Argentina. In 2007, a coordinated demonstration in different locations in Mendoza was key to the sanctioning of provincial law 7722, which protects Mendoza water from metalliferous mining activities (Wagner 2014). The same year, similar laws were sanctioned in La Rioja, Tucumán and La Pampa provinces. La Rioja province boasts one of the most important local movements against mining: the neighbour assemblies of Famatina and Chilecito. With the slogan '*El famatina no se toca*' ('Don't touch Famatina'), the neighbours organized a road blockade on the way to the mine, blocking the arrival of supplies to the mining camp (see Sola Álvarez 2016). One year later, Córdoba and San Luis provinces also sanctioned laws restricting mining, driven by local mobilizations in rejection of mining activities. Finally, a law restricting mining was sanctioned in Tierra del Fuego province, in 2011. However, the laws of Río Negro and La Rioja were later reverted or cancelled under the pressure of private and public mining promoters. Currently, there are seven laws restricting mining in Argentina (see Figure 8.1).

A strong mobilization also emerged in Catamarca province, where the first metalliferous mega-mine of the country, La Alumbrera, started activities in 1997. The lack of benefits perceived by local communities and the negative environmental impacts of this project drove the neighbours of Andalgalá and other towns to organize, preventing the opening of other new projects like 'Agua Rica' and 'MARA'.

As mining and other environmental struggles were spreading in the country, several neighbour assemblies from Argentina got together to form the 'Citizens

Assemblies Union' ('Union de Asambleas Ciudadanas' in Spanish), recently renamed as the Community Assemblies Union (UAC) (aiming to make clear the inclusion of Indigenous and peasant groups). This network was created in 2006 to promote the gathering of a diversity of struggles, not only against mining, but also against plunder and pollution ('*contra el saqueo y la contaminación*') (UAC 2018).

UAC members have successfully worked with other environmental organizations and foundations in the country to develop provincial and national environmental regulations and laws. Besides provincial mining restrictions, these networks led to the approval of some laws that aim to establish national minimum environmental standards ('*Leyes de presupuestos mínimos*'). Two laws were sanctioned thanks to environmental mobilizations and the involvement of scientists: the Glaciers Law, which catalogues and protects glaciers (from mining activities in particular) and the Native Forest Law, which aims to stop deforestation and the displacement of Indigenous communities. Currently, the environmental movement in Argentina is struggling to set a law to protect wetlands, the frontier of urban development near Buenos Aires and other sites (Langbehn et al. 2020).

In this sense, we would like to stress the role of local anti-mining movements in the formation of a larger national environmental network of movements that has fostered the transformation of environmental institutions, creating alliances and legitimating counter-hegemonic narratives that went beyond specific mining matters to address larger environmental and development issues.

In the sections that follow, we complement this narrative account of mining mobilizations in Argentina and the transformations they have fostered with evidence provided by the EJAtlas on the different outcomes, actors and strategies involved in these struggles.

Social Mobilization: Impacting Institutions, People/Networks and Cultural Power

Mobilization outcomes: social mobilization and institutional change

While each conflict identified and registered in the database is unique, with its own history, geographical setting and contexts, looking at them together offers an opportunity to identify patterns across localities in resistance movements. This approach complements the historical and interconnected approach offered in the previous section and provides further evidence on the impact of social mobilization.

Figure 8.2 presents the frequency of the different outcomes registered in the EJAtlas for the thirty-eight cases studied. There is usually more than one outcome per conflict but some are more frequent than others. We have grouped outcomes considering the impact on institutional and actor/network forms of power. For institutional forms of power impacted we distinguish outcomes on institutions (regulations, laws), justice (court outcomes), project modifications (its decision-making, technical assessment processes, etc.) and negative institutional outcomes (i.e. corruption). If we consider the actor/network forms of power, 'strengthening of participation'

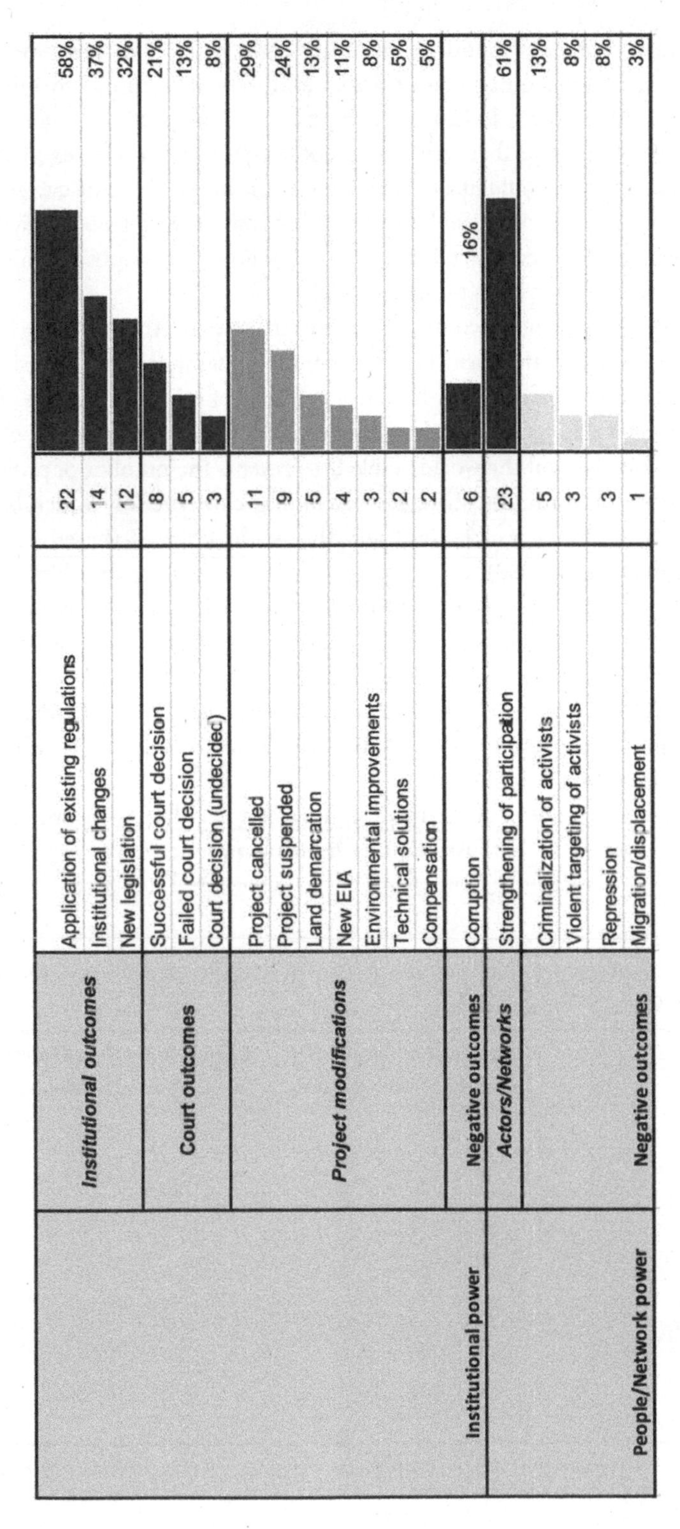

Form of power	Outcome category	Outcome	N	%
Institutional power	*Institutional outcomes*	Application of existing regulations	22	58%
		Institutional changes	14	37%
		New legislation	12	32%
	Court outcomes	Successful court decision	8	21%
		Failed court decision	5	13%
		Court decision (undecided)	3	8%
	Project modifications	Project cancelled	11	29%
		Project suspended	9	24%
		Land demarcation	5	13%
		New EIA	4	11%
		Environmental improvements	3	8%
		Technical solutions	2	5%
		Compensation	2	5%
	Negative outcomes	Corruption	6	16%
People/Network power	*Actors/Networks*	Strengthening of participation	23	61%
	Negative outcomes	Criminalization of activists	5	13%
		Violent targeting of activists	3	8%
		Repression	3	8%
		Migration/displacement	1	3%

Figure 8.2 Main outcomes of large-scale mining conflicts in Argentina and their impact on forms of power

Source: Own elaboration based on EJAtlas (2019).

Note: We classify 'failed court decision' as a court outcome but it could also be considered a 'negative outcome'.

is the most frequent outcome found in the thirty-eight cases. However, we also highlight some negative outcomes against actors and networks critical to mining activities, such as failed court decisions, corruption, criminalization, repression, assassinations or displacements that relate to the coercive forms of power exerted by states and corporations. We would like to signal that while in the outcome categories of the EJAtlas it is difficult to visualize the cultural/discursive transformations led by social movements, we explore their relevance in the following sections examining the strategies and actors involved in mining struggles.

One of the most significant outcomes of mining struggles in Argentina has been the stopping of projects (in the form of cancellations or temporary suspensions) in twenty out of the thirty-eight cases. This means 52.6% of mining struggles. This effect at the project level is very high when compared to mining struggles in other Latin American countries and the world. Table 8.1 presents the number of projects stopped as an outcome of mining struggles (excluding construction minerals) in other Latin American countries (9–30%) and the world (26.5%) (based on the EJAtlas database in October 2020).

Different factors could be considered to understand why a company or state stops a mining project – for instance, international metal price fluctuations; speculation; national, political and social policies; stability, etc. However, when we consider regional and global trends (Table 8.1; Özkaynak et al. 2021; Scheidel et al. 2020), the percentage of project cancellations/suspensions occurring in mining conflicts in Argentina clearly stands out.

Another unique outcome of Argentina's mining conflicts has been on formal institutions: the approval of new regulations limiting mining activities. This is

Table 8.1 Mining Conflicts in Latin American Countries and the World Where Projects Were Stopped (Cancelled or Suspended)

Country	Number of mining cases registered in the EJAtlas	Cases where the extractive project was stopped
Argentina	38	20 (53%)
Bolivia	13	2 (15%)
Chile	25	7 (28%)
Ecuador	14	2 (14%)
Mexico	22	2 (9%)
Peru	44	13 (30%)
EJAtlas worldwide mining conflicts	647	172 (26.5%)

Source: Based on EJAtlas database in October 2020 (Walter and Wagner 2021). Note that the level of coverage of mining conflicts varies across countries.

certainly an outcome of mobilization that removes incentives for private investment in some areas of the country. Since the beginning of mining contestation, nine provinces have approved laws restricting mining activities (two were reverted).

The impact of environmental justice movements (EJM) on institutional power is multifold. EJMs have not only fostered new regulations but have also demanded the application of existing environmental regulations (very progressive on paper, but not in practice). In some cases, regulations were applied for the first time as a result of mobilization strategies. Gutiérrez (2015) highlights how environmental organizations legitimized and revitalized environmental rights that seemed dormant before the environmental mobilization started at the beginning of the 2000s. Merlinsky (2013) also underlines the institutional and judicial productivity of environmental conflicts in Argentina. However, as we argue in this chapter (i.e. 'Emergence and spread of mining contestation in Argentina', above) mining mobilization has contributed to relevant institutional outcomes that go beyond specific mining controversies and institutional forms of power.

In Walter and Wagner 2021 we discuss some key political opportunity structures present in Argentina and the role of social movements and networks in navigating them. We argue that the approval of laws restricting mining activities can be seen as the result of the work of a social mobilization that was able to navigate the particularities of provincial politics and contexts. In Argentina, the environmental authority is national, but the jurisdiction of mining activities and environmental permits are provincial. This means that, while in most countries the technical and political daily management of mining is centralized in national ministries in the capital of the country (e.g. Chile, Peru), in Argentina it takes place at the provincial level. Hence, mining decision-makers are located closer to the sites of conflict and face more political accountability. This has provided a key political opportunity for movements to push for law compliance, project modifications and institutional transformations. However, these same decentralized institutional structures have also made local movements located in conservative and repressive provinces (the poorest provinces located in the north of the country) (Cerutti 2017; Möhle 2018) more vulnerable, although violence against activists is spreading to all provinces.

Previous research has also pointed out that socio-environmental contestation starting at the early stages of project development is more likely to stop these activities (Conde 2017; Özkaynak et al. 2021; Scheidel et al. 2020). This is the case in twenty-three out of the thirty-eight cases registered in Argentina that are preventive (60.5%). Usually small capital junior companies are involved at early stages of project development, as is the case in most examples identified in Argentina. Previous research has pointed out that junior companies usually don't have the experience or resources to deal with potential resistance negotiation with local communities, leading to more conflict and distrust (Conde and Le Billion 2017; Özkaynak et al. 2021).

Diversity of strategies to confront different forms of hegemonic power

Scheidel et al. (2020) concluded, based on their analysis of the global EJAtlas dataset, that if actors combine strategies of preventive mobilization, diversification of protest and litigation they can significantly increase the success rate to up to 27%. In our case study, the actors mobilized against mining activities in Argentina have employed a large diversity of strategies, including litigation, with successful outcomes.

Figure 8.3 presents the different forms of mobilization identified (or not) in the EJAtlas. This can be considered a picture of the repertoire of contention of movements mobilized against large-scale mining activities in Argentina. We have ordered the different strategies registered in the database considering the main forms of power that these strategies address, although in practice these are intertwined. We distinguish between those strategies targeted (mainly) at: *1) institutional, legal and economic frameworks* (formal allegations to EIAs, local referenda/consultations, etc.), *2) people and networks* (collective action, reaching out for support, creation and expansion of networks), and *3) narratives/discourses* that challenge cultural power embedded in hegemonic discourses, narratives and ways of seeing the world. Figure 8.3 illustrates how EJMs deploy interrelated strategies to transform visible and invisible forms of power. The most frequent strategies of mobilization aim to 1) build alternative narratives and discourses, 2) create networks of support and social mobilization, and 3) target institutions.

Strategies addressing cultural forms of power are highly relevant in mining struggles. The most frequent strategy relates to the creation of media-based activism channels. This strategy aims to support the mobilization of actors and the dissemination of bottom-up counter-narratives. The use of alternative media, blogs and social networks is a key strategy for local actors contesting mining activities in Argentina. It is present in 92% of our cases, but in only 47% of the global EJAtlas dataset (Scheidel et al. 2020). Most of these alternative media channels were created in the context of the conflicts. One key case is the creation of the 'No a la Mina' website ('No to the Mine'), by the Asamblea de vecinos autoconvocados (Autoconvened Neighbours Assembly) of Esquel, as a central outcome of the first large-scale metal-mining conflict in Argentina. This is currently a key source of information nationally and internationally. Moreover, the approval in 2009 of the National Law 26,522 of audiovisual communication services, in the context of a national debate about freedom of expression and fewer information monopolies, allowed for the spread of local and alternative communication media. This law facilitated the creation of alternative local radio and communication channels that gave voice to social movement concerns, in contrast to existing media channels. Moreover, in the context of many conflicts, local assemblies were the ones creating and managing these alternative media channels (like Radio La Paquita, in Uspallata, Mendoza), or were given spaces to voice their narratives and information (like Radio Kalewche, in Esquel). The existence of a national network of community

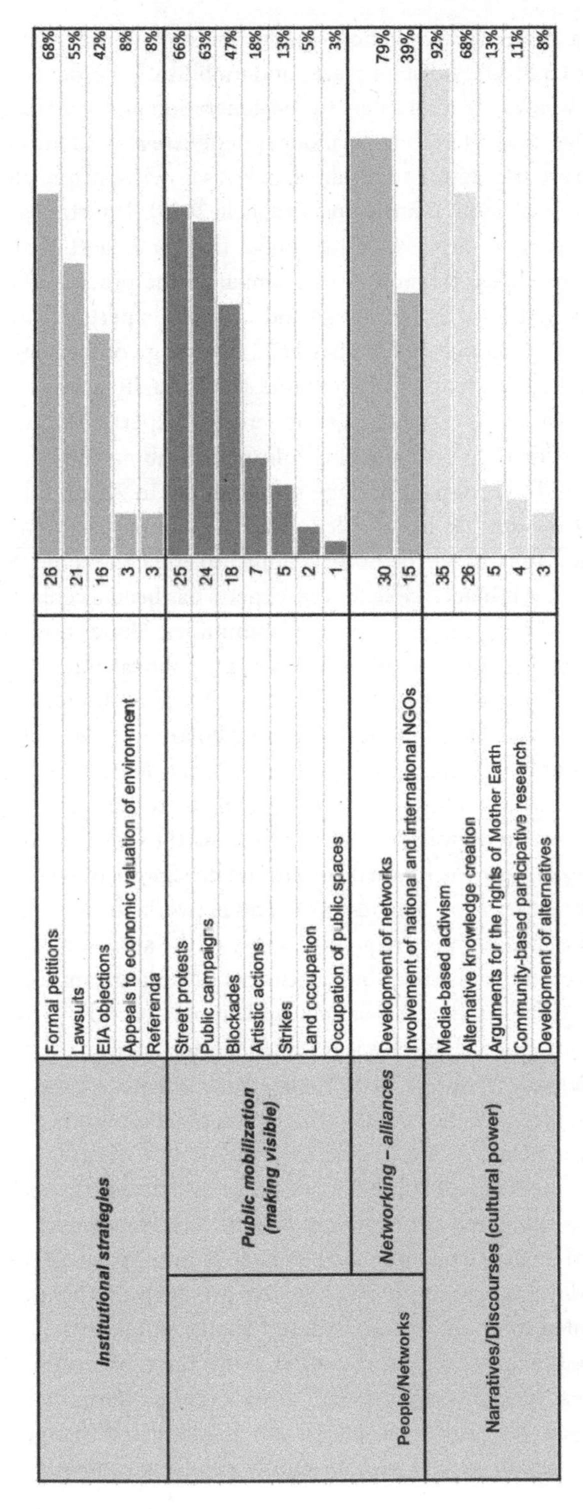

Figure 8.3 Forms of mobilization and their impact on forms of power

Source: Own elaboration based on EJAtlas (2019).

radio stations also allowed the circulation of information and different discourses around the country, connecting distant places and mobilization processes.

Alternative knowledge creation refers to the generation and systematization of alternative knowledge, also referred to as 'independent expert wisdom' or 'counter-expert wisdom' (we are translating the Spanish *saberes* as wisdom), related to local knowledge (Machado et al. 2011; Svampa and Antonelli 2009). This strategy is present in 68% of our cases and in only 36% of the global EJAtlas dataset (Scheidel et al. 2020). Alternative knowledge creation is also similar to the processes of 'activism mobilizing science' where local organizations and scientific experts co-produce new knowledge (Conde 2014; Conde and Walter 2022). Mining conflicts are spaces of dispute about scientific knowledge. Local organizations usually have members that are scholars or scientists or have the support of 'certified' experts. Mobilized groups have devoted much effort into systematizing information about the environmental and social impacts and experiences of large-scale mining in Argentina and other countries. This information has been widely circulated among actors challenging hegemonic expert accounts. Since the Esquel conflict in 2003, the claim that there is a technical rationale that is inaccessible to non-experts has been a central argument held by companies and governments against communities. Under this framework, local movements are labelled as irrational or led by political intentions (Walter 2008). As a response, 'knowledge' has become a central field of dispute. Documents such as technical reports, critical studies of mining environmental impact assessments and publications by critical scholars become central elements in the struggle. Actors demand to access EIAs to conduct reviews of the technical documents. In this vein, counter-knowledge strategies are closely related to institutional forms of power, as they aim to dispute the rationale of mining development narratives.

Strategies targeting the power held in people and networks are also central in the studied conflicts. The second most frequent strategy of the anti-mining movement in Argentina has been the construction of networks of support and collaboration across the country. These networks connect and, more recently, cross different environmental mobilization issues. This is reflected, for instance, in the creation of a network of local assemblies involved in different environmental struggles that meet regularly in different places of the country (the Union of Community Assemblies – UAC). This strategy appears in 79% of our cases and in 49% of cases in the global EJAtlas dataset (Scheidel et al. 2020). Networks have broadened the reach of local collective action, upscaling and coordinating actions. For instance, different local assemblies organized in 2011 a coordinated blockade of mining trucks transporting inputs to the La Alumbrera mine located in the province of Catamarca. Local assemblies coordinated intermittent blockades of routes and streets, accompanied with the handing of brochures and information about the mobilization to drivers and pedestrians. These blockades were called '*cortes de rutas informativos*' or 'informative route cuts'. Note how this coordinated action was also a channel to spread counter-hegemonic narratives. It is only in moments of high pressure and stakes that movements have conducted permanent blockades. These action and dissemination strategies were among other public mobilization strategies such as marches, protests and public campaigns (e.g. to support regulatory reforms).

Strategies targeting institutional powers are also frequent. Formal petitions, lawsuits and EIA objections were present in 68%, 55% and 42% of our registered cases and in 58%, 44% and 26% of the cases registered in the global EJAtlas dataset (Scheidel et al. 2020). The organization of referenda/consultations is present in three cases and is led by social movements and local governments that deploy a formal participation institution in a counter-hegemonic manner (Walter and Urkidi 2017).

Bottom-up and multi-scalar mobilization

The characteristics of mobilized actors also offer valuable information about all the actors, spaces and dynamics that have led to transformative processes. Figure 8.4 dissects the frequency of participation of different social actors in mobilizations against mining projects in Argentina (following the EJAtlas form options). We note that categories of actors are not mutually exclusive, but are actor characteristics. For example, Indigenous farmers can be counted under two groups (Indigenous peoples and farmers), showing intersectionality between different actor characteristics. We have grouped the actors (characteristics) registered in the database in four categories. The first category includes local grassroot actors and organizations. These are the most frequent actors. The second category groups professionalized actors, given their high level of activist organization (international NGOs) or professional formation (e.g. lawyers, chemists). The third category lists institutionalized groups, such as political parties, unions or the church. Finally, we have gathered together vulnerable groups, considering those highly exposed to intersectional discrimination (e.g. women, Indigenous groups) and those vulnerable groups that are directly affected in their (urban or rural) livelihoods by mining projects.

Figure 8.4 indicates a clear prominence of actors coming from the local scene, both organized movements (EJMs) and neighbours who mobilize or support the struggle without being nucleated in an organization. In contrast to this, only ten of the thirty-eight cases involve the presence of international organizations. This indicates that the participation of these actors is lower than local actors. In the largest global analysis of environmental conflicts to date and based on the overall EJAtlas database, Scheidel et al. (2020) show that local organizations (involved in 69% of cases) and neighbours (67%) are the two most frequent actor groups mobilizing to defend their environment. However, in Argentina these actors are identified in up to 95% and 84% of cases, indicating a significantly higher relevance of local actors.

As signalled in Figure 8.2, one of the most significant outcomes of mining struggles has been the strengthening of participation (i.e. new mobilized groups and actors). While in the overall EJAtlas dataset this outcome was flagged in 29% of cases (Scheidel et al. 2020), in Argentinean mining conflicts it appears in 60.5% of cases. Previous research (Schiaffini 2003; Sola Álvarez 2012; Wagner 2014; Walter and

Characteristics		Actors mobilized against mining activities in Argentina	Frequency of involvement	
			Cases	Percentage
Local grassroots organizations/groups		Neighbours	36	95%
		Local organizations	32	84%
		Social movements	23	61%
		Recreational users	7	18%
Professionalized actors		Scientists and other professionals	18	47%
		International NGOs	10	26%
Institutionalized groups		Government and political parties	22	58%
		Religious groups	8	21%
		Trade unions	7	18%
Vulnerable groups	Highly exposed to intersectional discrimination	Indigenous communities	18	47%
		Women collectives	5	13%
		Ethnically discriminated groups	1	3%
	Rural context	Farmers	19	50%
		Landless peasants	3	8%
		Pastoralists	7	18%
		Artisanal miners	2	5%
		Fisher people	-	0%
	Urban context	Informal workers	3	8%
		Industrial workers	2	5%

Figure 8.4 Social actors mobilized in mining conflicts in Argentina

Source: Own elaboration based on EJAtlas (2019) and Scheidel et al. (2020) proposed grouping of actors.

Martinez-Alier 2010), including fieldwork conducted by the authors of this chapter, highlights how local assemblies mobilized against mining were spaces of gathering and exchange between actors with previous political militancy (political parties, human rights organizations, unions, teachers associations, cultural and artistic collectives, among others) and actors with no previous experience in social organizations. The use of the terms 'autoconvened neighbours' and 'assemblies' chosen by the members of local movements to name themselves reflects their intention to establish horizontal structures of participation and decision-making, thus contrasting top-down technocratic procedures. Assemblies have horizontal structures, where decisions are made by consensus. There are no permanent or paid representatives, diverse social and environmental issues are addressed, they are composed by heterogenous participants and they have autonomy (Alvarez 2017). This form of organization became widespread during the 2001 socio-economic crisis in Argentina and later consolidated as the most common form of organization not only for grassroots movements, but also for the multi-scalar networks articulating local groups in Argentina (i.e. UAC, National Assembly of Communities). Assemblies can be seen as a challenge to hegemonic power in their aims and form, where horizontal and participative forms of deliberation and decision-making are put into practice (Alvarez 2017).

The third most frequent actor category in Argentinean mining controversies is local governments and political parties. These actors are the first to receive pressure from social mobilization and are strategic to impacting formal institutions (e.g. promoting laws or breaking hegemonic political discourses). In some cases, the mayors (local executive power) initially tend to assume a favourable position to extractive projects, but when the social opposition increases, they change their position and align themselves with the rejection expressed by the local population – even when this leads them to confront the provincial or national government (Walter and Urkidi 2017). Recent research on institutional transformations driven by social movements mobilized against mega-projects in Latin America has highlighted the importance of networks generated between civil society actors and different sectors of government, along with active mobilization (Silva et al. 2018).

In her study of environmental demands related to mining extractivism in Argentina, Delamata (2013: 87) states that these demands have exerted pressure on, and sometimes have been able to align, the general antinomy that transverses states between their dependence on investments and businesses, and the construction of political power of governments based on their popularity.

Figure 8.4 indicates, more broadly, the diversity of actors involved in mining struggles, pointing as well to the diverse forms of knowledge and strategies they provide to the mobilization process. For instance, scientists or professionals provide technical and scientific support to communities, especially on the environmental impacts of mining projects, where scientific or technical knowledge becomes central to the legitimation – or rejection – of mining projects. This group is present in almost half of the cases and has been key for the development of legal and technical strategies. Furthermore, as mentioned in above, scientists and EJMs have collaborated at the national level to foster broader environmental regulatory reforms.

Diverse vulnerable groups are also active in mining controversies and bring their knowledge and networks into these struggles. For instance, there are those actors whose livelihood can be directly affected by mining projects, such as farmers, Indigenous communities, pastoralists, trade unions and recreational users. In Argentina, the category 'farmers' includes different kinds, ranging from the farmer who produces on a small scale to large agricultural entrepreneurs, depending on each conflict, the regional configuration and the kind of predominant agricultural activity. The tension between agricultural activity and metalliferous mining activity is one of the main topics of debate related to mining impacts, evidenced by the fact that in half of the conflicts, the agricultural sector plays an important role.

Indigenous communities are also relevant actors in environmental conflicts (present in 47% of cases). Since the conflict over the gold mine in Esquel (province of Chubut) and Jacobacci (province of Río Negro), the presence of Mapuche communities has been increasingly relevant. They played a central role in mining conflicts in the province of Neuquén, especially in Loncopué, where the second community consultation on mining (after Esquel) of Argentina occurred (Wagner 2019a). Another example is the conflict related to the Navidad project, on the Patagonian plateau, where Mapuche-Tehuelche communities participated in the conflict. Indigenous communities are also playing a relevant role in conflicts over lithium-mining in the north of the country.

Religious groups are also relevant in mining conflicts, mainly the Catholic Church. Some parish priests have integrated with EJMs, and others have collaborated in denouncing the situation of inequality and injustice experienced by local communities who have rejected large-scale mining. The Pope Encyclical Letter called 'Laudato SI: on care for our common home' was an important document in supporting these actions by parish priests in communities affected by mining projects.

In summary, a central feature of mobilized actors has been the prevalence of local (i.e. neighbours, local organizations, social movements, farmers, Indigenous communities) and heterogeneous (i.e. access to diverse knowledge) actors, usually organized in the form of local assemblies and national networks of assemblies that work in alliance with different institutionalized and professionalized actors. In the section that follows we discuss how the mentioned actors and strategies are impacting different forms of visible and invisible power, and conclude by describing their role in diverse social and environmental transformative processes that went beyond the local level and mining policies.

Bottom-up power transformations

As described in this chapter, anti-mining assemblies have played a central role in the organization of a national network of socio-environmental movements (UAC) that has successfully impacted different forms of institutional power. The most salient impact has been the development of new regulations. Social mobilization has led not only to mining restrictions at the local and provincial level, but also to the approval at the national level of a 'glacier protection law' that maps and protects glaciers from activities such as mining. The Glacier Law was the result of national mobilization that had the active support of NGOs and scientists. Social mobilization has also campaigned for the approval of national laws to protect forests and wetlands in the country. Furthermore, social movements have demanded the application of existing (but not used) rights and laws, succeeding in putting into practice 'sleeping' legislation and introducing environmental issues to judicial courts. For instance, municipal consultations regarding mining activities were conducted for the first time in the country (i.e. Esquel in 2003), inspiring other communities to use this form of participation (Loncopué and others). This form of participation has also spread to other environmental struggles, as in the case of local consultation regarding a dam in Misiones (Wagner 2019b).

Assemblies have also introduced debates regarding mining and its impacts within different institutions such as universities and provincial legislatures (i.e. new regulations). For instance, there has been an intense debate within Argentina's national universities about their acceptance (or not) of funds from the La Alumbrera mining project. While some universities have accepted these funds, others have rejected them, a process that has led to intense debates among social movements, university administrations, professors, researchers and students (Andreychuk 2009).

We posit that in order to understand the shift in environmental and mining discourses and policies occurring in Argentina, it is central to observe the cultural transformations that occurred within local and inter-local assembly spaces. Local assemblies became a space of individual and collective empowerment, where local actors – many of whom had no previous mobilization experience – could express their views and feel listened to and recognized. Assemblies made it possible for marginalized actors to become protagonists (eg. women, Indigenous communities). Moreover, as local actors and their discourses spread and appeared on the TV, in newspapers and on the radio, local actors gained further internal and external recognition. We would like to note that we are not necessarily romanticizing these spaces, or denying the existence of internal tensions, but highlighting the role that these spaces had in the individual and collective, local and supra-local legitimation and dissemination of counter-hegemonic narratives and worldviews.

Furthermore, assemblies challenge hegemonic participatory and decision-making institutions, as these are spaces based on different forms of (non-hierarchical) organization, decision-making and networking, where a different way of debating, deciding and doing is performed. Assemblies allow for different forms of viewing and living the world, facilitating dynamics of solidarity and collaboration (within and among assemblies), as well as a space to learn (from local participants and allies), co-produce and disseminate counter-hegemonic narratives and discourses (Skill et al. 2017).

Figure 8.5 Argentinian national deputy offers a T-shirt featuring the slogan 'Water is worth more than gold' to Pope Francis

Source: Proyecto Sur.

The involvement of institutionalized groups, especially governments and political parties, but also the church and unions, had a significant impact on hegemonic powers, and their actors, networks and discourses. Social movements pressured governments and political parties to make transparent and public decisions about mining activities, succeeding in breaking the apparent unified support towards mining development policies and discourses.

'Experts'/scientists, churches and unions also became in many cases influential allies in transforming institutions and legitimatizing counter-hegemonic narratives. The Catholic Church was very important in putting the mining conflict on the public agenda, especially in small villages. However, this actor and its role remains poorly studied in the literature. Figure 8.5 shows Pino Solanas, a national deputy (close to the anti-mining movement), giving to the Pope in 2013 a T-shirt featuring the slogan 'Water is worth more than gold' (from Argentina's assemblies). This is an image that has travelled throughout Argentina and the world. This photograph illustrates the diverse and influential alliances built by anti-mining networks in Argentina and beyond.

The cultural transformations led by anti-mining and environmental movements are also becoming part of the public space, in slogans, murals, graffiti and on bodies (e.g. T-shirts). Slogans such as '*No a la mina*' ('No to the mine') and '*El agua vale más que el oro*' ('Water is worth more than gold') are nowadays ubiquitous in social protests in Argentina and Latin America (Figure 8.6). However, the narratives of protest not only denounce the negative impacts of polluting activities (the 'no to') but also recover and build positive and inspiring narratives born from local values, collective resistance and the recognition of other worldviews and knowledge (the 'yes to'; see Figures 8.7 and 8.8). The development of artistic

Figure 8.6 Graffiti with slogans in Esquel ('Water is worth more than gold', 'No to the mine')

Source: noalamina.org

Figure 8.7 Mural with the words 'Loncopué municipality free of mega-mining by decision of its people' painted on the town's church wall

Source: Lucrecia Wagner, 2016.

Figure 8.8 Mural in Lago Puelo (near Esquel), the region where anti-mining mobilization started in Argentina and where the first local consultation was conducted

Source: Artistas autoconvocados in Lago Puelo, documented in: https://peakd.com/hive-148441/@greengalletti/engesp-art-and-protest-unite-in-the-struggle-for-the-defence-of-the-environment-el-arte-y-la-protesta-unen-en-la-lucha-por-la-d.

expression and of alternative narratives to extractivism have been largely debated in national encounters of the Union of Assemblies of Communities (UAC 2018). The use of the arts contributes to organizational strategies of social movement and also raises environmental and political awareness and enhances public participation in decision-making (Sanz and Rodríguez-Labajos 2021). There is an emerging scholarly interest in exploring the role of aesthetic expression and artists in environmental protests in Argentina (Antonelli et al. 2021, 2022; Artilugio 2022; Merlinski and Serafini 2020).

Final Remarks and Conclusions

Argentina's anti-mining movement has succeeded in cancelling or suspending about half of the contentious projects they have opposed, and in approving regulations/laws restricting large-scale mining activities in nine out of twenty-three national provinces (including two restrictive laws that were later reverted). This process of mobilization and institutional change is quite unique when compared to other Latin American and worldwide environmental mobilization processes. In this chapter we have argued that while the most visible outcomes of mining mobilization are apparently institutional, institutional transformations were made possible by the mobilization of diverse actors. These actors successfully deployed strategies that impacted on different forms of power (institutions, actors/networks and narratives/discourses) that were supporting mining development in the country. Moreover, we claim that movements critical to large-scale mining have contributed to significant transformations that go beyond mining debates and policies to influence broader socio-environmental institutions (activation and creation of new environmental rights and laws, new participatory institutions), networks (national networks of collaboration and support on broader socio-environmental matters, diverse alliances with institutionalized and professionalized actors) and narratives (development and dissemination of social, cultural and environmental counter-hegemonic narratives).

Nevertheless, the question of how stable these transformations are could be asked, especially if we consider that mineral deposits and extractive interests remain. As we have detailed in this chapter, metal-mining is recent in Argentina and has a marginal relevance in its social and economic history. Large-scale metal-mining activities started in 1997 and since 2002 a powerful network of resistance has emerged and successfully stopped the installation of several projects. In most of the cases where projects were suspended/cancelled, this outcome persists as a result of continued social mobilization. One example is the massive mobilizations that took place in Chubut and Mendoza provinces in 2019, when there was an attempt to modify their two provincial laws restricting metal-mining activities (Merlinsky and Wagner 2019, Ullacia 2022). Even in the provinces where provincial laws restricting mining were reverted (La Rioja and Río Negro), mining projects were unable to begin their activities given the resistance of local actors. In fact, in provinces with unfavourable political opportunities for social movements, such as Catamarca, movements maintain their 'NO' to mining, notwithstanding the harsh repression they are suffering (Fontenla 2022). The slogan 'We will keep walking' ('*Seguiremos caminando*') can be read on the flag of Andalgala, a town of Catamarca that has counted more than 600 rallies to demand the halting of MARA's mining project (Cerutti 2021). In summary, large-scale metal-mining history is short in Argentina (less than three decades) and has been marked from its early stages by a history of resistance that has been sustained by a network of movement and alliances. We should perhaps revisit in the future the permanence of the transformations explored in this chapter; however, we can say today that mining resistance is leaving a permanent mark on Argentina's social and environmental history.

References

Alvarez, L. (2017) Asambleando el mundo. La experiencia de la Unión de Asambleas de Comunidades en las luchas socioambientales en Argentina. *Debates en Sociología*, 45: 113–40.

Andreychuk, L. (2009) La Alumbrera, 40 universidades y un 'dilema ético' por fondos. *El Litoral*, 30 December 2009. www.ellitoral.com/index.php/diarios/2009/12/30/educacion/EDUC-01.html.

Antonelli, M., Fobbio, L., Wagner, L. (eds) (2021) Dossier 'Estética, política y naturaleza: lenguajes y experiencias ecopoéticas'. *Heterotopías*, 4(8). https://revistas.unc.edu.ar/index.php/heterotopias/issue/view/2405.

—— (eds) (2022) Dossier 'Tomo II 'Bord(e)ados de vida. Marcas de condiciones de existencia en tiempos de violencias extractivas'. *Heterotopías*, 5(9). https://revistas.unc.edu.ar/index.php/heterotopias/issue/view/2536.

Artilugio (2022) Convocatoria #8 Eje temático 'Prácticas artísticas, crisis planetaria y vidas posibles'. https://revistas.unc.edu.ar/index.php/ART/call-for-papers.

Bebbington, A., Humpreys Bebbington, D., Bury, J., Lingan, J., Muñoz, J.P., Scurrah, M. (2008) Mining and Social Movements: Struggles over Livelihood and Rural Territorial Development in the Andes. *World Development*, 36(12): 2888–905.

Bosi, L., Uba, K. (2009) Introduction: The Outcomes of Social Movements. *Mobilization: An International Quarterly*, 14(4): 409–15.

Bridge, G. (2004) Contested Terrain: Mining and the Environment. *Annual Review of Environment and Resources*, 29(1): 205–59.

Cerutti, D. (2017) *Comunidades en resistencia frente a violencias (en)tramadas en América Latina. Megaminería y control social en un espacio subnacional: San Juan, Catamarca y La Rioja*. Tesis de Doctorado, CEA-Universidad Nacional de Córdoba, Córdoba.

—— (2021) Andalgalá: conjurar el agua y la memoria. *La Tinta*, 27 May 2021. https://latinta.com.ar/2021/05/andalgala-agua-memoria/.

Conde, M. (2014) Activism Mobilising Science. *Ecological Economics*, 105: 67–77. https://doi.org/10.1016/j.ecolecon.2014.05.012.

—— (2017) Resistance to Mining: A Review. *Ecological Economics*, 132: 80–90.

Conde, M., Le Billon, P. (2017) Why Do Some Communities Resist Mining Projects While Others Do Not?, *The Extractive Industries and Society*, 4(3): 681–97.

Conde, W., Walter, M. (2022) Knowledge Co-Production in Scientific and Activist Alliances: Unsettling Coloniality. *Engaging Science, Technology, and Society*, 8(1): 150–70.

Delamata, G. (2013) Actualizando el derecho al ambiente. Movilización social, activismo legal y derecho constitucional al ambiente de «sustentabilidad fuerte» en el sector extractivista megaminero. *Entramados y perspectivas*, 3(3): 55–90. https://publicaciones.sociales.uba.ar/index.php/entramadosyperspectivas/article/view/150.

della Porta, D., Rucht, D. (2002) The Dynamics of Environmental Campaigns. *Mobilization*, 7: 1–14.

Deniau, Y., Herrera, V., Walter, M. (2021) *Mapping Community Resistance to the Impacts and Discourses of Mining for the Energy Transition in the Americas* 2nd edn. EJAtlas/MiningWatch Canada. https://miningwatch.ca/publications/2022/3/4/mapping-community-resistance-impacts-mining-energy-transition-americas.

EJAtlas (Environmental Justice Atlas) (2019). https://ejatlas.org/. Accessed June 2019.

EPA (2020) *TRI National Analysis 2019*. USA: Environmental Protection Agency.

Escobar, A. (2001) Beyond the Search for a Paradigm? Post-development and Beyond. *Development*, 43(4): 11–14.

Fontenla, M. (2022) Crónica desde Andalgalá en días de represión. *ANRed*, 9 May 2022. www.anred.org/2022/05/09/cronica-desde-andalgala-en-dias-de-represion/.

Foucault, M. (1971) The Order of Discourse. In Young, R. (ed.) *Untying the Text: A Poststructuralist Reader*. London: Routledge and Kegan Paul.

Gaventa, J. (1980) *Power and Powerlessness: Quiescence and Rebellion in an Appalachian Valley*. Chicago: University of Illinois Press.

Global Witness (2020) Environmental Activists. www.globalwitness.org/en/campaigns/environmental-activists/. Accessed 6 July 2020.

Gutiérrez, R. (2015) Teoría y praxis de los derechos ambientales en Argentina. *Temas y Debates*, 30: 13–36. https://doi.org/10.35305/tyd.v0i30.320.

IEA (International Energy Agency) (2021) 'The Role of Critical World Energy Outlook Special Report Minerals in Clean Energy Transitions'. www. iea.org/reports/the-role-of-critical-minerals-in-clean-energy-transitions.

Jahncke Benavente, J., Meza, R. (2010) *Derecho a la participación y a la consulta previa en Latinoamerica*. Lima: Fedepaz Muqui Miserer CIDSE.

Karl, N.A., Wilburn, D.R. (2017) Annual Review 2016: Exploration Review. *Mining Engineering*, 69(5): 28.

Langbehn, L., Schmidt, M., Pereira, P. (2020) Las leyes ambientales en el ojo de la tormenta. Un análisis comparativo en torno a la legislación sobre glaciares, bosques y humedales en Argentina. In Merlinsky, G. (ed.) *Cartografías del Conflicto Ambiental en Argentina 3*. Buenos Aires: CLACSO-CICCUS.

León, M., Lewinsohn, J.L., Sánchez, J. (2020) *Balanza comercial física e intercambio, uso y eficiencia de materiales en América Latina y el Caribe*. Serie Recursos Naturales y Desarrollo, 200 (LC/TS.2020/150). Santiago: Comisión Económica para América Latina y el Caribe (CEPAL).

Long, N., Van Der Ploeg, J.D. (1989) Demythologizing Planned Intervention: An Actor Perspective. *Sociologia Ruralis*, 29(3/4): 226–49.

Lukes, S. (1974) *Power: A Radical View*. London and New York: Macmillan.

Machado, H., Svampa, M., Viale, E., Giraud, M., Wagner, L., Antonelli, M., Giarracca, N., Teubal, M. (2011) *15 Mitos y Realidades de la mineria transnacional en la Argentina*. Buenos Aires: Editorial El Colectivo-Ediciones Herramienta.

McAdam, D., Tarrow, S., Tilly, C. (2001) *Dynamics of Contention*. Cambridge: Cambridge University Press.

Merlinsky, G. (2013) Introducción. La cuestión ambiental en la agenda pública. In Merlinsky, G. (ed.) *Cartografías del Conflicto Ambiental en Argentina*. Buenos Aires: CLACSO-CICCUS.

Merlinsky, G., Wagner, L. (2019) La memoria del agua. Megaminería y conflictos ambientales en Mendoza. *Espoiler*, 30 December 2019. http://espoiler.sociales.uba.ar/2019/12/30/la-memoria-del-agua-megamineria-y-conflictos-ambientales-en-mendoza/.

Merlinsky, G., Serafini, P. (eds) (2020) *Arte y Ecología Política*. Ciudad Autónoma de Buenos Aires: IIGG-CLACSO.

Möhle, E. (2018) ¿Cómo se decide sobre el territorio? Gobernanza de conflictos mineros. Los casos de Andalgalá, en Catamarca, y Famatina, en La Rioja (2005–2016). Tesis de Maestría. Universidad Nacional de San Martín – Georgetown University, Buenos Aires.

Muradian, R., Martinez-Alier, J., Correa, H. (2003) International Capital versus Local Population: The Environmental Conflict of the Tambogrande Mining Project, Peru. *Society and Natural Resources*, 16: 775–92.

OCMAL (Observatorio de Conflictos Mineros de América Latina) (2022) www.ocmal.org/. Accessed March 2022.

OEC (Observatory of Economic Complexity) (2019) Exports: Argentina, https://oec.world/en/profile/country/arg/#Exports.

Özkaynak, B., Rodríguez-Labajos, B., Erus, B. (2021) Understanding Activist Perceptions of Environmental Justice Success in Mining Resistance Movements. *The Extractive Industries and Society*. https://doi.org/10.1016/j.exis.2020.12.008.

Rodríguez, I., Inturias, M.L., Robledo, J., Sarti, C., Borel, R., Cabria Melace, A. (2015) Engaging with Environmental Justice Through Conflict Transformation: Experiences in Latin America with Indigenous Peoples. *Revista de Paz y Conflictos*, 8(2): 97–128.

Rodríguez, R., Inturias, M.L. (2018) Conflict Transformation in Indigenous Peoples' Territories: Doing Environmental Justice with a 'Decolonial Turn'. *Development Studies Research*, 5(1): 90–105.

Rojas, F., Wagner, L. (2017) 'Desarrollos' fallidos en la minería histórica. Famatina y Chilecito, apuntes para pensar el presente socio-ambiental. *Trabajo y Sociedad*, 28: 281–307. www.unse.edu.ar/trabajoysociedad/28%20ROJAS%20Y%20WAGNER%20CONICET%20Mineria.pdf.

Sanz, T., Rodríguez-Labajos, B. (2021) Does Artistic Activism Change Anything? Strategic and Transformative Effects of Arts in Anti-coal Struggles in Oakland, CA. *Geoforum*, 122: 41–54.

Scheidel, A., Del Bene, D., Liu, J., Navas, G., Mingorría, S., Demaria, F., Avila, S., Roy, B., Ertör, I., Temper, L., Martinez-Alier, J. (2020) Environmental Conflicts and Defenders: A Global Overview. *Global Environmental Change*, 63. https://doi.org/10.1016/j.gloenvcha.2020.102104.

Schiaffini, H. (2003) 'El agua vale más que el oro': la constitución de fuerzas sociales en torno al conflicto minero en Esquel, 2002–2003. Tesis de Licenciatura, Departamento de Ciencias Antropológicas-UBA, Buenos Aires.

Silva, E., Akchurin, M., Bebbington, A. (2018) Policy Effects of Resistance Against Mega-Projects in Latin America: An Introduction. *European Review of Latin American and Caribbean Studies*, 106: 23–46. https://doi.org/10.32992/erlacs.10397.

Sola Álvarez, M. (2012) Conflictos socioambientales en torno a la megaminería metalífera a cielo abierto. El caso de Famatina, La Rioja, Argentina. Tesis de Maestría, FADU-UBA, Buenos Aires.

—— (2016) Estados subnacionales, conflictos socioambientales y megaminería. Reflexiones a partir del análisis de la experiencia del Valle de Famatina, Argentina. *Sociedad y ambiente*, 9. https://doi.org/10.31840/sya.v0i9.1632.

Skill, K., Ullberg, Susann B. (2017) Asambleas socioambientales en la Argentina: activismo como agenciamiento. *Etnografías Contemporáneas*, 3(4): 200–24.

Svampa, M., Antonelli, M. (2009) *Minería transnacional, narrativas del desarrollo y resistencias sociales*. Buenos Aires: Editorial Biblos.

Svampa, M., Sola Álvarez, M., Bottaro, L. (2009) Los movimientos contra la minería metalífera a cielo abierto: escenarios y conflictos. Entre el efecto Esquel y el efecto La Alumbrera. In Svampa, M., Antonelli, M. (eds) *Minería Transnacional, narrativas del desarrollo y resistencias sociales*. Buenos Aires: Biblos.

Tarrow, S.G. (2011) *Power in Movement: Social Movements and Contentious Politics*. Cambridge: Cambridge University Press.

Temper, L., Del Bene, D. (2016) Transforming Knowledge Creation for Environmental and Epistemic Justice. *Current Opinion in Environmental Sustainability*, 20: 41–9.

Temper, L., Bene, D., Martinez-Alier, J. (2015) Mapping the Frontiers and Front Lines of Global Environmental Justice: the EJAtlas. *Journal of Political Ecology*, 22: 256–78.

Temper, L., Walter, M., Rodríguez, I., Kothari, A., Turhan, E. (2018a) A Perspective on Radical Transformations to Sustainability: Resistances, Movements and Alternatives. *Sustainability Science*, 13(3): 747–64.

Temper, L., Demaria, F., Scheidel, A., Del Bene, D., Martinez-Alier, J. (2018b) The Global Environmental Justice Atlas (EJAtlas): Ecological Distribution Conflicts as Forces for Sustainability. *Science*, 13: 573–84.

Tilly, C. (2002) *Stories, Identities, and Political Change*. Lanham, MD: Rowman and Littlefield.

UAC (Unión de Asambleas Ciudadanas) (2018) *Construyendo caminos colectivos en defensa de nuestros territorios*. https://asambleasciudadanas.org.ar/wp-content/uploads/2018/04/CuadernilloUACAbril2018.pdf.

UNRP (2018) International Resource Panel (by CSIRO. Dr James West and Mirko Lieber). www.resourcepanel.org/global-material-flows-database.

Ullacia, Martín. 2022. *No fue No. Una crónica del Chubutazo*. Remintente Patagonia, Trelew.

USGS (2020) *United States Geological Survey*. www.usgs.org.

Van Der Heijden, H.-A., (2006) Environmental Movements and International Political Opportunity Structures. *Organization and Environment*, 19: 28–45. https://doi.org/10.1177/1086026605285452.

Vela-Almeida, D., Gonzalez, A., Gavilán, I., Fenner Sánchez, G.M., Torres, N., Ysunza, V. (2021) The Right to Decide: A Triad of Participation in Politicizing Extractive Governance in Latin America. *The Extractive Industries and Society*, 9(1). https://doi.org/10.1016/j.exis.2021.01.010.

Wagner, L. (2014) *Conflictos socioambientales: la megaminería en Mendoza, 1884–2011*. Buenos Aires: Editorial de la Universidad Nacional de Quilmes.

—— (2019a) Propuestas de inversiones chinas en territorio mapuche: resistencias a la minería metalífera en Loncopué. *Estudios Atacameños*, 63: 315–39. https://doi.org/10.22199/issn.0718-1043-2019-0028.

—— (2019b) Consultas comunitarias en Argentina: respuestas participativas frente a mega-proyectos. *Tempo e argumento*, 11(28): 181–211. http://dx.doi.org/10.5965/2175180311282019181.

Wagner, L., Walter, M. (2020) Cartografía de la conflictividad minera en argentina (2003–2018). Un análisis desde el Atlas de Justicia Ambiental. In Merlinsky, G. (ed.) *Cartografías del conflicto ambiental en Argentina 3*. Buenos Aires: CICCUS-CLACSO.

Walter, M. (2008) Nuevos conflictos ambientales mineros en Argentina. El caso Esquel (2002–2003). *Revista Iberoamericana de Economía Ecológica*, 8: 15–28. https://redibec.org/ojs/index.php/revibec/article/view/280/152.

Walter, M., Martinez-Alier, J. (2010) How to Be Heard When Nobody Wants to Listen: Community Action Against Mining in Argentina. *Revue Canadienne d'Études du Développement/Canadian Journal of Development Studies*, 30(1–2): 281–301. https://doi.org/10.1080/02255189.2010.9669292.

Walter, M., Urkidi, L. (2017) Community Mining Consultations in Latin America (2002–2012): The Contested Emergence of a Hybrid Institution for Participation. *Geoforum*, 84: 265–79. https://doi.org/10.1016/j.geoforum.2015.09.007.

Walter, M., Wagner, L. (2021) Mining Struggles in Argentina. The Keys of a Successful Story of Mobilisation. *The Extractive Industries and Society*. https://doi.org/10.1016/j.exis.2021.100940.

Young, I. (1990) *Justice and the Politics of Difference*. Princeton, NJ: Princeton University Press.

Section 3

Enacting Counter-hegemonic Alternative Politics, Economics and Worldviews

This section of the book focuses on counter-hegemonic alternatives that movements have carried out to bring about just transformation. These range from alternative forms of democracy and economies to cultural transformations in knowledge systems, discourses and worldviews. Case studies from Bolivia, India (from two regions) and Canada (at a national scale) will explore how communities are resisting extractive activities like mining and forestry by defining their own governance space, forms of territorial management and productive activities. Furthermore, this chapter upends traditional understandings of innovation to argue that reclamation and recreation of traditions and ancestral ways of being together with new social institutions provide a more convincing path to transformative change than high-tech solutions. At the same time, it shows that transformation can be internally incoherent, or contradictory, even if there are areas of transformation. It will help explore how the line between types of transformations is blurred, and difficult to separate.

9

The Monkoxi from Lomerío, Bolivia: On the Road to Freedom Through *Nuxiaká Uxia Nosibóriki*[1]

Mirna Inturias, Iokiñe Rodríguez, Miguel Aragón, Elmar Masay and Anacleto Peña

Introduction

In 1990, an Indigenous march demanding territorial rights walked 650 kilometres from the Bolivian lowlands to the highland capital of La Paz. So began a long-term nationwide Indigenous mobilization that ended up transforming the model of the state in Bolivia from 'monocultural' to 'plurinational'. The approval of a new political constitution in 2009 marked the birth of this new plurinational nation-state, with a new legal framework that grants Indigenous peoples their rights to the property of their territories, to maintain their own forms of culture and language and, most significantly, to their self-determination and to exercise their own forms of justice and government.

The Monkoxi from Lomerío were among those marching in 1990 for the dignity and territories of Bolivia's Indigenous peoples. They are one of the Bolivian lowlands' most emblematic Indigenous nations in terms of their political strength and organization, and their determination to open the way for a more plural democratic model in Bolivia. In addition to fighting for constitutional reforms, since the late 1900s they have persistently worked to develop a model of territorial governance that would allow them to break free from external control over their territory and natural resources. The Monkoxi call their own form of government *Nuxiaká Uxia Nosibóriki*.

This case study discusses this alternative political project, and the strategies used by the Monkoxi over the last four decades to advance this dream. It also analyses the transformations that have taken place in the making of this political project and the challenges still ahead.

We draw on results from a long-term collaboration between Universidad Nur, the University of East Anglia and CICOL (the Indigenous Union of Lomerío) who, since 2013, have used action-research to help strengthen the Monkoxi Indigenous governance of Lomerío. In 2017, as part of the ACKnowl-EJ project, we carried out a participatory assessment of strategies used by the Monkoxi over the last four decades to consolidate their territorial control and political autonomy in Lomerío. We used the Conflict Transformation Framework developed by Grupo Confluencias[2] (discussed in Chapter 2 of this book) to systematize strategies used by the Monkoxi in their struggle for 'freedom' and to trace the changes brought about over time in terms of cultural revitalization, political agency, local governance, control of the means of production and environmental integrity (Rodríguez and Inturias 2018; Rodríguez et al. 2019). We also used the Three Horizons methodology mentioned in Chapter 1 to envision the future challenges of an autonomous Indigenous government in Lomerío and filmed a series of video testimonies about autonomy, in which twenty Monkoxi leaders tell their individual and collective stories of transformation in the struggle.[3] In this chapter we present the results of this joint reflection process and discuss key actions that the Monkoxi began putting into practice immediately afterwards to deal with certain challenges identified.

We divide this chapter into six sections. In the first part we introduce recent changes made in the Bolivian plurinational nation-state to open the way for Indigenous autonomy. In section two we present how the Monkoxi have fought for and defined their dream for *Nuxiaká Uxia Nosibóriki* (their own form of government). In section three and four we present the results of a joint evaluation of Monkoxi strategies for liberation and the transformations that have taken place as a result. In section five, we discuss some new strategies the Monkoxi have been developing to deal with key challenges for self-government. We close in section six with some final reflections about the main achievements and challenges ahead for the Monkoxi to fulfil their *Nuxiaká Uxia Nosibóriki.*

The Plurinational Nation-state and Indigenous Autonomy in Bolivia

The 2009 Bolivian political reforms are considered the most transformative in Latin America in terms of recognizing Indigenous peoples' rights (Lupien 2011). In contrast to similar reforms such as those that occurred in Venezuela in 1999 and Ecuador in 2006, one of the most important changes brought about by the 2009 Bolivian constitution was a new regime of territorial organization of the state. This new territorial organization recognizes departmental, municipal, regional and Indigenous autonomies as new plural forms of political organization that seek to decentralize decision-making power and the management of public funds away from the central government.

Departmental, municipal and regional autonomies can be recognized within the pre-2009 territorial division of the state, as they can simply be superimposed over existing departments, municipalities or regions. However, Indigenous autonomies are more challenging to develop as they often overlap with more than one municipality or department and as a result demand greater institutional and legal changes.

The Indigenous autonomy model acknowledges the rights of Indigenous peoples to their self-determination, self-government and culture, and to the consolidation of their territorial entities within the framework of the unity of the state, opening the way for the conformation of Original Indigenous Peasant Autonomies (AIOCs). Gaining state recognition of their right to political autonomy is key for Indigenous peoples' ability to exert control over their natural resources and territories according to their own governance rules and procedures.

However, the process for gaining Indigenous autonomy is a lengthy one due to a wide array of legal requirements demanded from Indigenous nations. To date, only three claims out of more than thirty initiated have managed to concretize the constitution of their Autonomous Indigenous Peasant Governments (GAIOCs). One of the demands still pending is that of the Monkoxi Indigenous peoples of Lomerío, from the Bolivian lowlands.

The Monkoxi Dream for *Nuxiaká Uxia Nosibóriki* (Their Own Form of Government)

The communal indigenous territory (TCO) of Lomerío is an area of 256,000 hectares dominated by Chiquitano dry forests (Figure 9.1), located in the department of Santa Cruz in the Bolivian lowlands (Figure 9.2). Lomerío is home to around 7,000 Monkoxi settled in twenty-nine communities, which range in size from 100 to 1,500 inhabitants.

The Monkoxi peoples define Lomerío as a refuge, an area to which their ancestors escaped during colonial times to be free from Jesuit missions. Yet, much to their

Figure 9.1 Chiquitano dry forests, Lomerío (Photo by Iokiñe Rodríguez)

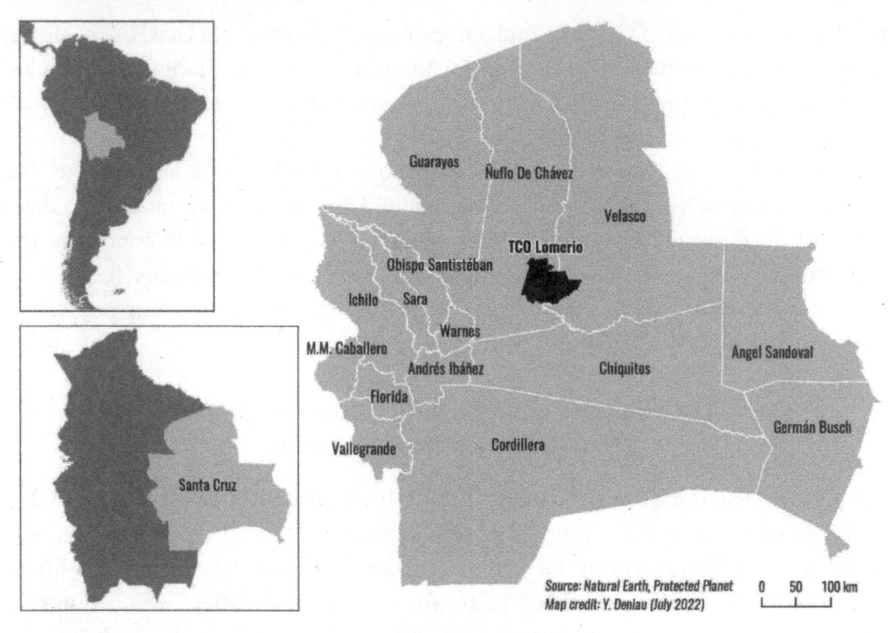

Figure 9.2 Location of the Indigenous Territory of Lomerío

disadvantage, shortly after the missionaries were expelled in 1776, Bolivian mestizo and white landowners took over their lands for agriculture and husbandry, forcefully enslaving the Monkoxi on rubber plantations. They were brutally exploited well into the late 1900s, thus turning their refuge into a prison for more than a century.

Despite this oppressive past, or perhaps because of it, the Monkoxi are one of the most emblematic Indigenous nations of the Bolivian lowlands in terms of political strength and organization. They have a long history of resistance to colonial rule and the patronage land tenure system. In 1964, in the Republican era, they began to engage in organized resistance through the formation of the Agrarian Peasant Union. In 1982 they were instrumental in the creation of CIDOB (the Confederation of Indigenous Peoples of Bolivia), and later, in 1983, formed the Indigenous Organization of Native Communities of Lomerio (CICOL). In the late 1980s, they were the first Indigenous nation in Bolivia to develop community forestry as a form of territorial control, and in 2006, after a long struggle, they succeeded in gaining legal rights over their communal Indigenous territory, officially denominated the Territory of Indigenous Communities (TCO) of Lomerío, which CICOL has the legal mandate to safeguard.

As seen from the statement below, territorial self-governance is considered the final step in freeing themselves from oppression:

> Our grandmothers and grandfathers gave their lives to give us a territory where we can be free, where we can make the dream of having our own form of government real, and thus turn our refuge into our road to freedom and our desire to live well. This is what we call: *Nuxiaká Uxia Nosibóriki*. (Masay and Chore 2018)

Thus, in 2008, the Monkoxi were the first Indigenous nation in Bolivia to use the United Nations Declaration of Indigenous Peoples Rights to give legal backing to their public proclamation as the first autonomous Indigenous territory of Bolivia. In 2009, they wrote and validated their autonomy statutes in a General Assembly with participation from the twenty-nine communities and initiated their legal claim for autonomy rights. Parallel to this, Monkoxi leaders actively took part in the 2008 constitutional reforms to ensure that Indigenous autonomy rights were adequately accounted for in the new plurinational nation-state framework.

The most salient features of the Monkoxi autonomy statutes are the following:

- The definition of the ancestral territory of Lomerío as the geographical limit of their government.
- The defence of communal democracy as the main form of collective decision-making.
- The emphasis given to principles, values and norms of communal and territorial life, such as freedom, sharing (*minga* or *bobikix*), equity, reciprocity, redistribution and solidarity.
- The establishment of Besiro as the official language and Spanish as the second one.
- The designation of the General Assembly, with representation from the twenty-nine communities, as the maximum decision-making authority.
- The importance given to customary rules and norms and Indigenous justice in regulating day-to-day communal life.
- The definition of communal economy as the desired form of development, aimed at achieving the *Uxia Nosibóriki* (*Vivir Bien*) of the Monkoxi nation, respecting Mother Earth, the spirits of the forest (*jichis*) and life in harmony with nature (Flores 2018).

In May 2018 the Constitutional Tribunal sanctioned in favour of the Monkoxi statutes, but they must fulfil a series of additional requirements before they are finally granted autonomy.

CICOL's Board of Directors is well aware that being granted the right to autonomy will not resolve all their problems. They know that despite having property rights over their territories and forests, they face many challenges for managing their territories sustainably, which their new form of self-government will have to deal with if and when it comes into power. Hence the interest of CICOL's Board of Directors in carrying out a joint assessment of their forty years of struggle that would help them collectively devise strategies to deal with the challenges ahead.

Assessing the Strategies in the Monkoxi Struggle for Liberation

In April 2017 we carried out a three-day workshop with eighty community leaders from the Lomerío Indigenous Territory to assess the Monkoxi's strategies in their struggle for liberation (see Figure 9.3).

Figure 9.3 Participation in conflict transformation workshop, Lomerío (Photo by Miguel Aragon)

The workshop participants grouped the strategies in their 'fight for freedom' into six platforms of struggle: community forestry, territorial management, political autonomy, education, health and gender, with most actions concentrating in the first three.

Each platform of struggle involved developing strategies to challenge hegemonic power in different ways, in some cases confronting and managing to change dominant structures in Bolivian society, while in others relying more on the development of alliances and networks to advance in their objectives. Strategies to impact cultural power were the least prominent in all platforms of struggle. In fact, strategies to impact cultural power only started to flourish in a more systematic way after carrying out this assessment.

Community forestry

The majority of Bolivia's forests are located in the lowlands and contain a high diversity of species of commercial value. Land use, and in particular forest management, has undergone different stages, from rubber production in the late nineteenth century to a boom of private logging concessions issued by the state in the 1970s. These were tumultuous times for Indigenous communities, including the Monkoxi, whose lands were confiscated and who often ended up (forcibly) working on rubber plantations as slaves.

This explains why the process to regain control over Monkoxi territory began in the forestry sector. One of their first strategies was to develop *alliances* with a number of national and international organizations. The aim was to strengthen the organizational capacity of CICOL in forestry management and gain access to economic resources to develop a communal forestry plan. CICOL received training from a variety of organizations, including APCOB (Apoyo Para el Campesino-Indígena del Oriente Boliviano), Oxfam America and BOLFOR (a sustainable forest management programme jointly run by USAID and the Bolivian government), on how to develop communal forestry plans. The first stage of this process involved proving that CICOL could add value to timber through a local sawmill. This not only led to eventually being granted forestry rights, but also to successfully expelling timber companies from their territories:

> Having a sawmill was a condition to access a forest concession. Thus, in 1988 we set up a sawmill with a social rather than business objective. Concretely, the sawmill contributed to the consolidation of the territorial demand, the improvement of livelihoods and the creation of working opportunities; and in 1992 and 1993 we expelled the private timber corporations, which were carrying out the illegal extraction of selective wood of commercially valuable species. (Peña 2020)

The other strategy involved seeking a competitive advantage in the international markets through forest certification, which they received in 1996.

The different alliances also resulted in CICOL's participation in national platforms fighting for normative and legal changes to the Forest Law to allow for participation of communities in commercial forestry activities. This was achieved in 1996 with the sanctioning of a new Forestry Law (No. 1700). By the end of the 1990s, Lomerío had established a Community Forest Management Plan that enabled all communities in Lomerío to benefit directly from forestry. Currently, twenty-three (of twenty-nine) communities in Lomerío have set aside areas for community-based commercial forestry, amounting to over a third of the TCO. Communities are allowed to retain 100% of the income from community forestry, though in theory 5% of the income should go to the legal authority over the territory, CICOL, to help cover administrative costs.

However, strategies to bring about change in the cultural sphere (narratives and knowledge) linked to forest management have been less pronounced. This is partly due to how ingrained epistemic violence is within the institutional culture of the state and in some of CICOL's allied organizations. For example, in the case of forestry management plans, a greater participation of Indigenous people in forestry activities has not equally led to recognition and revitalization of local forms of knowledge and forest management governance. In fact, despite their right to carry out community forestry and holding legal ownership of their forests, the Monkoxi have had limited control over forestry activity. Community forestry management must comply with regulations from the 1996 Forestry Law, including applying for a harvesting permit and following a set of specifications for tree harvesting. Furthermore, they remain heavily dependent on (external) timber companies for forest exploitation expertise, which often leads to unfavourable negotiations with these companies in the selling of timber.

In recent times, CICOL has begun taking actions to revitalize local knowledge of forestry management by seeking to add value to the forests through the creation of traditional seed banks and nurseries. They have also initiated discussions with officials from the forestry service to make changes to the 1996 Forestry Law in order to increase recognition of their governance structures and justice systems in forestry activities.

Territorial ownership

Parallel to community forestry, in the 1990s CICOL set in motion a variety of strategies aimed at gaining ownership rights of the Monkoxi territory. Some of them involved pressuring for changes to normative, legal and institutional frameworks. In 1990, representatives of CICOL participated in the first National Indigenous March, organized by the Indigenous Confederation of Bolivia, CIDOB, which set off from Trinidad city in the Beni department on 15 August. This was known as the 'March for Territory and Dignity'. In this march, CIDOB proposed to the then-government a project for an Indigenous law, aimed at achieving rights to self-determination for Bolivia's Indigenous people. The government split the proposal and issued three laws: the Popular Participation Law, the Educational Reform and Forestry Law 1700 (mentioned above).

This was followed by a second Indigenous march in 1996, 'For Territory, the Land, Political Rights, and Development', which successfully led to approval of Law 1715 or INRA Land Law, recognizing the right of Indigenous people to their Communal Lands of Origin (TCO, Tierras Comunitarias de Origen).

This new institutional framework allowed CICOL to start filing a demand for territorial property before the state in 1996.

In 2002 CICOL participated in the third Indigenous National March, 'For Popular Sovereignty, Territory and Natural Resources', which triggered a constituent assembly to rewrite the national constitution. CICOL achieved representation by electing Nelida Faldin Chuve (the first female chief general) as a constituent assembly member.

In this period, CICOL also established key alliances with international cooperation agencies to help advance their strategy for territorial ownership. For instance, the Danish Development Corporation (DANIDA) played a key role in financing the land sanitation process in Lomerío through a bilateral agreement with the Bolivian government.

In 2006, the National Executive declared Lomerío a TCO, thus officially granting the Monkoxi collective property rights over 259,188.7205 hectares, and the board of directors of CICOL became its legal steward.

Political autonomy (self-government)

Strategies developed by the Monkoxi to advance their claim for autonomy can be grouped in two categories: those developed in response to administrative requirements of the Bolivian government and those developed to unify interests within the territory, such as informative meetings and assemblies.

As mentioned before, the Monkoxi initiated their claim for political autonomy in 2009, self-proclaiming Lomerío as an autonomous territory and approving the autonomy statutes in a General Assembly. The fact that thirteen years down the line the Monkoxi still have not been granted the right to self-government speaks to the institutional structures and dynamics continuing to reproduce the logic of the monocultural state in Bolivia.

To make way for the autonomy claim, the Monkoxi had to adapt their autonomy statutes and structures to the new regulatory frameworks, including the 2010 Autonomy and Decentralization Law (No. 031/10), which makes the procedure a very complex and cumbersome one. The National Executive, Constitutional Tribunal, Legislative Assembly and Supreme Electoral Tribunal each demand a different set of requirements.[4] In addition, many of the public officers involved in the granting of autonomy rights continue to think and act in accordance with monocultural regulatory frameworks, which in practice has created obstacles for Indigenous forms of autonomy, pressuring them to opt for a short route to self-government: a conversion to municipal or departmental autonomy (Avila 2018).

To navigate these lengthy procedures, CICOL's establishment of sustained partnerships has proven key to help strengthen the technical and financial aspects of their claim. Of particular importance has been their continued work with CEJIS (Centro de Estudios Juridicos y Sociales), their main legal advisor for filing the claim. In recent times, the creation of strategic links with key central government players (such as the Minister and Vice-Minister of Autonomy) to jointly evaluate challenges and opportunities presented by Monkoxi Indigenous autonomy (Inturias et al. 2016) has helped make visible the demand nationally and internationally (see Figure 9.4), speeding up some of the procedures. Likewise, opportunities to share experiences with other Indigenous nations also in the process of claiming autonomy (Figure 9.5) has been instrumental for developing joint strategies to pressure the government to clear administrative hurdles obstructing the final approval of their demands (Inturias et al. 2019).

Education

The Monkoxi began developing strategies to gain access to formal education in Lomerío after the first major education reform in the 1950s, which made public education a right for all citizens. Although access to education has been slow, within the last four decades the Monkoxi have managed to install four educational centres in their territory and twenty-nine primary schools. However, the arrival of formal education in Lomerío came at a high cost. The accelerated process of *castellanizacion*

Figure 9.4 Presentation of the *Justicia Ambiental y Autonomía Indígena de Base Territorial en Bolivia. Un dialogo político desde el Pueblo Monkox de Lomerio* book in the United Nations Forum for Indigenous Affairs, New York 2016. Left to right: Henry Balderomar, Advisor to the Minister of Autonomy; Mirna Inturias, Nur University; Anacleto Peña, Chief General of CICOL; Hugo Siles, Minister of Autonomy; Iokiñe Rodríguez, University of East Anglia (Photo by Miguel Aragon)

Figure 9.5 Indigenous autonomy workshop, Santa Cruz de la Sierra, December 2018 (Photo by Iokiñe Rodríguez)

means that Besiro is now taught as a second language. High school graduates migrate to the city of Santa Cruz with the aim of continuing their studies and to seek economic alternatives. As a consequence, the territory is mostly inhabited by the elderly and children.

Health

Like the education sector, actions aimed at gaining access to public health services in Lomerío over recent decades have resulted in the presence of Western medicine in the territory, leading to the fragmentation of knowledge and the weakening of traditional medicine.

Gender

Lomerío is not free from patriarchal norms, which are present in social structures at different levels, including that of the family, the community, organizations and in the sphere of dominant knowledge. The participation of women in political spaces has increased little by little over recent decades. At the community level, women show high levels of participation in community life, but at the regional or national level, their participation is lower. One exception is Nelida Faldin, CICOL's first female chief general, who later represented Lomerío and the lowland Indigenous peoples in the constituent assembly.[5]

Assessing Transformations: Gains and Losses in the Struggle for Freedom

After carrying out an analysis of the conflict transformation strategies employed, workshop participants evaluated what they have achieved as a result, and concluded that just as there had been very important gains in the struggle, there had also been important losses. Using the five conflict transformation pillars – cultural revitalization, political agency, means of production, governance, and environmental integrity (discussed in Chapter 2 of this book) – we can get a clear sense of the Monkoxi's gains and losses in their struggle for freedom.

Diversity, recognition and cultural revitalization

In the last two decades important structural changes have been made within Bolivia's nation-state that have led to the recognition of Indigenous rights, greater participation in national politics, and an opening of a normative framework, giving the Monkoxi the possibility to start considering claiming their right to autonomy. As we have shown, the Monkoxi played an important role in making this possible. However, our joint assessment showed that after forty years of struggle for freedom, changes achieved within this conflict transformation pillar are less significant than the others, making cultural revitalization their most vulnerable area.

Although many of the Monkoxi's strategies aimed to achieve the right to cultural difference within the Bolivian state, comparatively they carried out significantly fewer strategies aimed at their own cultural revitalization. This has been problematic because current tensions between modern and more traditional Indigenous values pose important challenges for the process of Indigenous autonomy. Firstly, Monkoxi principles, values and norms of communal and territorial life, such as sharing, equity, reciprocity, redistribution and solidarity, are not necessarily the ones guiding community life. On the contrary, in many communities there is currently a dominance of individual interests and benefits regarding the use of many of the territories, such as the forests, which not only creates strong inter- and intra-community conflicts, but also threatens the integrity of the environment and territory. Secondly, in terms of cultural vitality, central aspects of the Monkoxi identity, like the Besiro language, are struggling to survive. Thirdly, and linked to the above, is the fact that younger Monkoxi generations are growing up with little awareness of their own history and cultural ties to the land due to the attraction of 'modern' or urban ways of life, which weaken their leadership capabilities to manage and safeguard their territory in the future.

Thus, the joint assessment clearly showed that working to strengthen the Monkoxi cultural integrity will be key for a successful self-government. In the final section of the chapter we will discuss some actions that CICOL has been developing since then to deal with this challenge.

Political agency

Although the long process of struggle has not substantially strengthened Monkoxi cultural integrity, it has strengthened the Monkoxi Nation's political agency. This is reflected both in a greater capacity of the Monkoxi to influence public policies and political and regulatory frameworks, and in the strengthening of leaders' and community members' individual capacities in different areas (education, health, forestry).

> I participated in mobilizations for the INRA, forestry and other laws, I have gained that, I have learned a lot in these activities, an experience that allows me to have a broad vision, a horizon of what we have pursued and continue to pursue as leaders. (Ignacio Soqueré Tomichá, Testimonies of Autonomy, April 2017)
>
> The greatest gain that I have been able to have in all this is the knowledge that I now have of Lomerío's history and of how CICOL and our grandparents have fought for our liberation. (Nelida Faldin, Testimonies of Autonomy, April 2017)
>
> I learned many things that I did not know, first, personal rights. Before they said that we did not have rights to natural resources; I learned in many training courses to see the situation and who I am. (Esteban Quiviquivi Parapaino, Testimonies of Autonomy, April 2017)

This has allowed them to be better prepared in their relationship with external actors (state, looters, etc.), as Anacleto Peña (chief general of CICOL) affirms:

> The Monkoxi now know the laws that support the rights of the Indigenous population. We can now do direct advocacy actions and we know how to defend ourselves from the

> external attacks from people that want to have access to our natural resources. We have not been able to eradicate them because it is a dynamic of society, but we have been able to advance in exercising our own rights, both political, economic and social and cultural as well. (Anacleto Peña, Testimonies of Autonomy, April 2017)

However, the workshop participants also shared the perception that strong tensions prevail in several communities due to the lack of some community leaders' interest in exercising their authority to implement collective resource management agreements before external actors who continue to loot the territory. This weakens the political strength of CICOL as a whole.

Control of the means of production

There is a strong feeling among Monkoxi leaders that as a result of the transformation strategies deployed, the Monkoxi have been able to secure their means of production. The combination of owning both their territories and forests gives them the possibility to control and define the forms of development desired and to derive a direct benefit from the use of natural resources. Thus, this pillar has been one of the most strengthened over the history of their struggle.

> We now have the legal security of our land; this legal security allows us to live in plenitude without the fear that tomorrow someone will come to displace you because you are settled in a territory. The security that all that is within our territory belongs to us, for example natural resources, fauna, everything that makes up biodiversity, we understand that our life is totally related to what nature is. Everything that nature has is for the benefit of the person, for example building a house: the materials come from nature, other things: handicrafts come from nature, hunting, fishing comes from nature. There is a set of elements that nature offers for the benefit of the person within its territory. (Pedro Ipamo Jimenez, Testimonies of Autonomy, April 2017)

However, there are still challenges to overcome in relation to market forces and the developmental logic of the state. The national development model continues to imprint a narrowly defined economic rationality regarding the use of natural resources. The state is far from putting the narrative of Living Well (*Buen Vivir*) into practice (despite such an approach forming part of the new constitution). In Lomerío, one example of this is the strong hold that market forces and bureaucratic procedures from the forestry department still have over dictating rules for community forestry, limiting the actual control the Monkoxi can have over this activity. Added to this is the fact that, although the Monkoxi have legal property rights for their territory, the subsoil remains state property. All mineral resources in Lomerío have been designated for mining concessions, without respecting prior informed consent procedures, making current and future mineral extraction a very contested issue.

Governance (institutions and forms of government)

Since the formation of the TCO, there have been important advances in collective governance. CICOL was created with the precise purpose of safeguarding Lomerío as a community commons. The organization has played a key role in different struggles and in the demand for autonomy before the state. In addition, in 1999 the Monkoxi managed to have Lomerío declared the first Indigenous municipality in Bolivia. This allowed them to gradually acquire experience in municipal management, which in turn serves as experience for a future self-government.

However, a future Monkoxi self-government faces an important internal challenge. There is strong resistance to the model of Indigenous autonomy by hegemonic Monkoxi families currently in charge of the municipal government, and questions regarding a communal democracy model. This group favours maintaining a representative democracy model and regards Indigenous customary decision-making procedures and justice systems as backward and primitive. Consequently, they have pushed for municipal autonomy as an alternative form of local government (one of the three options for self-government available in the new Bolivian constitution, as discussed in the first section of this chapter). Underlying this tension between different models of democracy are conflicting values and worldviews about what it means to be Indigenous, which is yet another way in which coloniality and modernity find expression in Lomerío.

Environmental integrity

Despite the Monkoxi owning their lands and forests, there is at the same time a marked sense of loss of control of their territory. In some communities, collective resources continue to be exploited individually to the detriment of the collective interests. There exist conflicting views regarding allowance of mineral extraction in the territory. Divisions caused by political parties, and leaders appointed by CICOL who leave the territory seeking job opportunities in the cities, also act as threats to the environmental integrity of the territory.

Reconnecting with the Roots

One key outcome of the joint assessment of Monkoxi strategies for liberation was the identification of CICOL's effectiveness in developing strategies to change the legal, institutional and political frameworks of the nation-state. However, it has been much less effective at confronting cultural power through strategies that could help challenge the coloniality of knowledge expressed in dominant discourses, knowledge and narratives of development.

The synergistic pressures of environmental, food and health crises generated by the 2019 and 2020 Amazonian and Bolivian lowland forest fires and the Covid-19 pandemic made it even more imperative to work with youth to help them reconnect with their identity and territory, and to recover the ancestral values, knowledge and visions held by their elders for Lomerío's future. As a result of the Covid crisis, many young people returned to Lomerío in search of refuge and security in these uncertain times, but found that they no longer had the relevant knowledge for many of the traditional livelihood practices needed to live in the territory for the long term. Thus, CICOL started putting into practice a series of activities to help youth reconnect with their roots.

Revitalizing the Besiro language

One of these strategies was to develop activities to revitalize the Besiro language. In 2019, CICOL signed a collaboration agreement with CEJIS, the European Union and the International Work Group for Indigenous Affairs (IWGIA) to develop a project in primary schools to help families and small children recover the use of Besiro in their households. This was followed by Besiro Language Fairs organized in schools to encourage the use of Besiro among young people.

Additionally, key publications produced by CICOL prior to 2017 about their struggle for freedom, such as the history of Lomerío published in 2016, have now been translated into Besiro and are being used in schools.

Researching traditional knowledge of fire

In 2019, Lomerío was heavily affected by forest fires (Rodríguez and Inturias 2020a), caused in part by extreme drought and severe frost conditions, occurring more frequently and acutely due to climate change. Studies also show a strong connection between the extent and damage of the fires and recent government policies and projects encouraging agricultural expansion (Devisscher et al. 2016). For example, the production of biofuels through soybean plantations and the expansion of livestock grazing for export fostered speculative clearing of new lands through the use of fire, both on private land (especially agro-industrial properties) and in adjacent areas on the agricultural frontier (Romero-Muñoz et al. 2019). In addition, since 2013, the government has sanctioned several regulations to allow for the burning and logging of forest areas up to 20 hectares. The result was an acceleration of land-grabbing and logging in the lowlands, evidenced by the concentration of the majority of fires in protected areas and titled Indigenous territories, affecting the Monkoxi people of Lomerío among other groups (Rodríguez and Inturias 2020b).

As a result, the Bolivian Forest Management Agency issued a resolution prohibiting the authorization of burning for agricultural use in the Chaco region. A total prohibition of the use of fire in Indigenous territories threatens the cultural and physical survival of its inhabitants. To ensure their food sovereignty, Indigenous peoples who depend on forests for their subsistence activities, such as agriculture, hunting and fishing, must be able to appropriately use fire as a way to manage their territories (Inturias et al. 2019; McDaniel et al. 2005).

In 2020, CICOL initiated a series of activities to ensure control and management of fire use in order to adapt to the increased risk of climate-induced fires in their territory. This included the establishment of community forest firefighters, the drafting of a burning protocol, and participatory research conducted by young Monkoxi researchers to recover ancestral knowledge about the use of fire.[6] The research seeks to recover local knowledge about the cultural significance of fire for the Monkoxi, its different uses, standards for its appropriate use in agriculture and current applications, perceptions of the impact of the 2019 fires, and visions for appropriate fire use in the territory. The aim is for the Monkoxi to reach a stronger position from which to dialogue with national authorities on appropriate regulations for the use of fire in their territory based on Indigenous knowledge systems, and to jointly agree on appropriate protocols.

Youth reconnection through photography

Another cultural revitalization activity led by CICOL in conjunction with Nur University, the University of East Anglia and National Geographic in the last two years, as part of the INDIS (Indigenous International Interactions for Sustainable Development) project, has been a PhotoVoice project to help young Monkoxi reconnect with their territory through the use of cameras.[7] The PhotoVoice project acted to bring young Monkoxi closer to the forest, the elders and their stories. For four months, eight young people were trained in participatory photography and documented their culture, way of life and challenges (Figure 9.6). The trained

Figure 9.6 Young Monkoxi receiving their PhotoVoice training certificates

photographers represented youth voices in documenting Monkoxi wisdom and knowledge, making new intergenerational and intercultural dialogues possible. Themes they chose to document included: medicinal plants to treat Covid-19; political issues in the territory; and the imminent impact of climate change on their communities (see Figures 9.7–9.10).

The PhotoVoice book was published in May 2022 (Chuvirú García et al. 2022) and presented in a community general assembly and a national book fair in Santa Cruz de la Sierra in June 2022.[8]

Figure 9.7 Photo taken by the young Monkoxi photographer Eliana Pena Chore

Figure 9.8 Photo taken by the young Monkoxi photographer Victor Hugo Garcia

Figure 9.9 Photo taken by the young Monkoxi photographer Johan Pedriel

Figure 9.10 Photo taken by the young Monkoxi photographer Maria Isabelle Garcia Parapaino

Designing a Monkoxi education plan

In 2018, in collaboration with researchers from the INDIS project, Monkoxi elders and leaders began a circle of reflection to analyse the state of Monkoxi knowledge systems. Thus, a core group made up of the council of elders and historical leaders (men and women) was formed, who gradually began to include school directors. Together they decided to draft a new education plan to strengthen the Monkoxi knowledge systems in the territory. The group proposed six axes of work: territorial development, network of traditional knowledge, educational spaces, educational subjects, education approach and strengthening of the Besiro language. A new Monkoxi education plan was published in May 2022 (CICOL 2022) and presented in the territory to all authorities and 155 teachers from Lomerío, as well as broadcast at the 2022 National Book Fair and in different media such as national newspapers and radios.

COVID and revitalizing traditional medicine

The Covid-19 pandemic generated a movement to renew traditional medicine in the territory, recovering ancestral plants as cures. In this context, Lomerío has become a laboratory for experimentation and resilience where knowledge is porous, dynamic and in constant transformation. CICOL has begun promoting family gardens with medicinal plants. Some of the PhotoVoice stories mentioned above document this traditional knowledge. The educational plan (CICOL 2022) considers health a relevant area of knowledge and thus new educational strategies are being designed to strengthen Monkoxi knowledge of traditional medicine. CICOL also became involved in international dialogues to discuss Covid's impact in Lomerío and to promote traditional medicine. For example, a webinar organized by the Global Tapestry of Alternatives took place in October 2020, titled: 'Self-determination and COVID in the Indigenous Territory of Lomerío'.[9]

Final Reflections

Over the last forty years the Monkoxi have made important progress in their struggle for liberation and the development of an alternative political model for their nation. This is reflected not only in their success in gaining ownership rights over their territory and forests but also in the systematic progression of their claim for political autonomy. The elements that have made possible this progression towards the consolidation of a self-government model in Lomerío include:

- *An organization that assumes its actions organically:* There is clarity, firmness and a commitment by CICOL's communal leaders regarding freedom and autonomy as a legitimate need of the people. One characteristic of this process has been the importance that CICOL gives to the systematic training of leaders and to developing their capacity for collective territorial management. Undoubtedly, the narrative built around autonomy exists in the Monkoxi's collective imaginary, in the same way that territorial consolidation was a dominant narrative of the struggle in the 1980s and 1990s. These are elements that generate cohesion and transformation. The role of leaders as transformative actors makes it possible for this narrative to remain alive thanks to ongoing actions and practices to achieve this end.
- *The establishment of sustained partnerships with different support organizations to strengthen technical and financial aspects of the strategies:* The achievements of the Bolivian Indigenous movement in terms of the recognition of collective rights (Indigenous territories and autonomies) did not happen behind closed doors or in isolation. Partnerships have been crucial. Obtaining territorial property rights necessitated working at different levels: local, national and international. The Monkoxi Nation as part of the Indigenous movement was part of different platforms. The international context had a decisive impact, and so have alliances between the key state actors and the Monkoxi, which have helped to develop public policies that managed to transform land tenure in Bolivia.
- *The use of a wide variety of strategies of struggle:* In the forty years of struggle, the Monkoxi have used a wide range of strategies: social mobilization, lobbying, strategic negotiations, participation in national decision-making institutions, demands for rights before the state, participation in the elaboration

> of constitutional reforms, participation in the formulation of new laws (e.g. Forestry Laws), training in various issues for the strengthening of political and social organization, and obtaining funds for territorial management. The organizational capacity to influence different spheres of power through the design of multiple and creative strategies has been key in their progression towards self-government and self-determination.

At the same time, important challenges remain in the making of the Monkixi *Nuxiaká Uxia Nosibóriki* (their own form of government).

One of these relates to the difficulty of sustaining alliances between Indigenous movements and external actors. These external actors can help develop counter-strategies to confront different forms of hegemonic power that are still present in the Bolivian nation-state, and which have significantly slowed the progress of claims to autonomy. Over the last decade conditions for Indigenous peoples in Bolivia to develop and maintain alliances have changed considerably. The Indigenous movement has become fragmented, with international cooperation less and less present, thus reducing the availability of international funds to help sustain CICOL's territorial management mandate in Lomerío.

Coupled to this are the contradictory public policies in the forestry, mining, Indigenous autonomy and education sectors which impede the Monkoxi's ability to effectively advance a coherent Indigenous self-government. Due to this unfavourable context and a weakened Indigenous movement, the Monkoxi run the risk of not being able to continue transforming or impacting the national frameworks that are limiting the advance of their alternative political model.

The Monkoxi also face a challenge to diversify their strategies for freedom and liberation even further. So far, CICOL has been very effective at developing strategies to change the legal, institutional and political frameworks of the nation-state, but has been much less effective at confronting cultural power, through strategies that can help challenge the coloniality of knowledge expressed in dominant discourses, knowledge and narratives of development. In recent times, CICOL has been much more actively engaged in developing strategies to strengthen its cultural power, but for the Monkoxi self-government to be effective in the long run, it will be necessary to continue devising strategies for the Monkoxi (particularly the younger generations) to stay connected with their identity, culture and territory.

Notes

1. The Monkoxi form of government.
2. Grupo Confluencias is a Latin American network of practitioners, researchers and institutions who have been working since 2005 on the development of a platform for deliberation, joint research and training in environmental conflict transformation.
3. Link to video series here: www.youtube.com/playlist?list=PL2tuXXfyo5WQix9C9MO-Z7OSOSiOUXPhR0.

4 Indigenous people must liaise with the National Executive to request initial certificates of public administration and self-government. The Plurinational Constitutional Tribunal has to approve the autonomy statutes and ensure that these are subject to a local referendum. The Plurinational Legislative Assembly must foresee and approve the necessary changes in legislation and sanction in favour of the creation of new territorial units. The Supreme Electoral Tribunal is in charge of coordinating consultations for the final approval of the claim to autonomy.
5 Nelida Faldin's video Testimony of Autonomy: https://youtu.be/apf0oipNEj4?list=PL2tuXXfyo5WQix9C9MOZ7OSOSiOUXPhR0.
6 Link to video: El fuego en la Nación Monkoxi, https://youtu.be/pL9D22qGGZQ.
7 This PhotoVoice process was led by National Geographic's young explorer Markus Martinez Burman.
8 See References for a download link. To access the book presentation see here: www.bivica.org/file/view/id/6140.
9 The full webinar can be accessed here: https://archive.org/details/gta_webinar_20201023.

References

Avila, H. (2018) Prologo. In Flores, E. *Sueños de Libertad. Proceso autonómico de la Nacional Monkoxɨ de Lomerio*. Santa Cruz de la Sierra: CICOL-CEJIS.

Chuvirú García, B., Peña Chore, E., Ipamo Ipi, G.I., Rodriguez Cesarí, J.P., Guizada Palachay, J.A., García Chuvirú, J.S., García Parapaino, M.I., García, V.H. (2022) *FOTOVOZ reconexión Monkoxi Un tejido de historias de autonomía, identidad y acción climática por jóvenes Monkoxi* (ed.) Inturias, M., Rodríguez, I., Martinez, M., Wershoven, J. Bolivia: Editorial Nur, Universidad East Anglia, Cooperaciòn Alemana ZFD-GIZ. https://ueaeprints.uea.ac.uk/id/eprint/82987/1/Fotovoz_Reconexion_Monkoxi.pdf.

CICOL (2022) *Plan educativo Monkoxi (2022–2026)*, ed. Inturias, M., Rodríguez, I., Gonzales, W., Peña, A. Bolivia: Editorial NUR, Universidad East Anglia, Cooperaciòn Alemana ZFD-GIZ. https://drive.google.com/file/d/1q9jsO_YAANv4f6ywcn4D1DGUJse-6a7b/view.

Devisscher, T., Anderson, L.O., Aragão, L.E.O.C., Galván, L., Malhi, Y. (2016) Increased Wildfire Risk Driven by Climate and Development Interactions in the Bolivian Chiquitania, Southern Amazonia. *PLOS ONE*, 11(9), e0161323. https://doi.org/10.1371/journal.pone.0161323.

Flores, E. (2018) *Sueños de Libertad. Proceso autonómico de la Nacional Monkoxɨ de Lomerio*. Santa Cruz de la Sierra: CICOL-CEJIS.

Inturias, M., Rodríguez, I., Valdelomar, H., Peña, A. (2016). *Justicia Ambiental y Autonomía Indígena de Base Territorial en Bolivia. Un dialogo político desde el Pueblo Monkoxi de Lomerio*. University of East Anglia, UK, Universidad Nur, Bolivia, Ministerio de Autonomia, Bolivia.

Inturias, M., Vargas, G., Rodríguez, I., Garcia, A., Von Stosch, K., Masay, E. (2019) *Territorios, justicias y autonomías*. Santa Cruz: Universidad Nur-Universitat de East Anglia-ZFD/GIZ-PNUD-CICOL.

Lupien, P. (2011) The Incorporation of Indigenous Concepts of Plurinationality into the New Constitutions of Ecuador and Bolivia. *Democratization*, 18(3): 774–96. https://doi.org/10.1080/13510347.2011.563116.

Masay. E., Chore, M. (2018) Presentación. In Flores, E. (ed.) *Sueños de Libertad. Proceso autonómico de la Nacional Monkoxi de Lomerio*. Santa Cruz de la Sierra: CICOL-CEJIS.

McDaniel, J., Kennard, D., Fuentes, A. (2005) Smokey the Tapir: Traditional Fire Knowledge and Fire Prevention Campaigns in Lowland Bolivia. *Society and Natural Resources*, 18(10): 921–31. https://doi.org/10.1080/08941920500248921.

Peña, A. (2020) *On the Road to Freedom: The History of the Monkoxi from* Lomerio. University of East Anglia, Nur University, CICOL. https://indisproject.org/wp-content/uploads/2020/09/SIMPLE-DOC-LOMERIO-ENGLISH.pdf.

Rodríguez, I., Inturias, M. (2018) Conflict Transformation in Indigenous Peoples' Territories: Doing Environmental Justice with a 'Decolonial Turn'. *Development Studies Research*, 5(1): 90–105. https://doi.org/10.1080/21665095.2018.1486220.

—— (2020a) Challenges to Intercultural Democracy in Plurinational State Bolivia: Case Study of the Monkoxi Peoples of Lomerío. *Beyond Development*. https://beyonddevelopment.net/challenges-to-intercultural-democracy-in-the-plurinational-state-of-bolivia-case-study-of-the-monkoxi-peoples-of-lomerio/.

—— (2020b). Bolivia: Contribution of Indigenous People to Fighting Climate Change is Hanging by a Thread. *The Conversation*, 11 February. https://theconversation.com/bolivia-contribution-of-indigenous-people-to-fighting-climate-change-is-hanging-by-a-thread-129399.

Rodríguez, I., Inturias, M., Frank, V., Robledo, J., Sarti, C., Borel, R. (2019) *Conflictividad socioambiental en Latinoamérica: Aportes de la transformación de conflictos socioambientales a la transformación ecológica* (Cuaderno de la transformación, 3). Ciudad de México: Friedrich-Ebert-Stiftung.

Romero-Muñoz, A., Jansen, M., Nuñez, A.M., Toledo, M., Almonacid, R.V., Kuemmerle, T. (2019) Fires Scorching Bolivia's Chiquitano Forest. *Science*, 366(6469): 1082. https://doi.org/10.1126/science.aaz7264.

10
On the Cusp: Reframing Democracy and Well-being in Korchi[1]

Neema Pathak Broome, Shrishtee Bajpai and Mukesh Shende

Introduction

Mainstream governance and development models – characterized by seemingly democratic but inherently centralized and top-down governance and extractive, commercially motivated, capitalist economic policies – have failed to achieve minimum levels of well-being for a very large part of humanity and non-human species. They have in fact caused large-scale human and environmental injustice. However, there are also counter-trends in either resisting current models or towards developing and defending alternative forms of governance and well-being (Singh et al. 2018). In this chapter we explore and discuss the emergence of one such process towards direct democracy and well-being in Korchi *taluka* in the Gadchiroli district of Maharashtra state in India. We use Zografos's definition of direct democracy as a 'form of popular self-rule where citizens participate directly, continuously, and without mediation in the tasks of government' (Zografos 2019).

India has a federal democratic system that is decentralized in form but retains strong political and administrative centralization in its spirit and functioning. As per the latest government report, India has a forest cover (very dense, moderately dense, and open forest) of 21.71% (Forest Survey of India 2021: 28–9). As per the 2011 census, 8.6% of India's population is classified as *adivasi* or tribal (scheduled tribes) – roughly 10.42 million people (MoTA, 2013). About 200 million people – *adivasi* and other traditional forest dwellers (OTFD) – spread over 170,000 villages across India, are estimated to be directly dependent on these forests for their subsistence, livelihoods, and cultural and spiritual needs (CFR-LA 2017). Yet since the colonial times of the nineteenth century, they have had little control, use, access rights or decision-making powers over their surrounding forests and customary territories. These communities have resisted their systemic alienation from use, access, governance and management of their surrounding forests by colonial and post-colonial governments through strong grassroots movements and uprisings. In 2006, submitting to this long-standing grassroots struggle, the Parliament of India enacted a landmark legislation: The Scheduled Tribes and Other Traditional

Forest Dwellers (Recognition of Rights) Act 2006, also called the Forest Rights Act of India (herein referred to as FRA). The FRA provides for recording fourteen kinds of pre-existing but legally unrecognized customary forest rights to scheduled tribes and other traditional forest dwellers. Most important among these are the *gram sabhas* (village assemblies) to claim rights to use, manage and conserve their traditional forests and protect them from internal and external threats – called the Community Forest Resource (CFR) Right. The Act also requires free, prior and informed consent of the *gram sabhas* before their customary forests are diverted for non-forestry purposes. Among the many radical provisions of this law, and the most significant, is the powerful envisioning of the basic unit of governance as the *gram sabha* – to be self-determined by a group of people residing in a settlement, which may or may not have been described as a village in the government records. This Act, along with another radical law for the tribal areas, the 1996 Panchayat (Extension) to the Scheduled Areas (PESA), has paved the way for transformative democratic processes to take shape for *adivasi* and other forest-dwelling communities in India.

This chapter attempts to understand and analyse how these laws were used by the *gram sabhas* in Korchi *taluka* to move towards direct democracy and greater economic, social, ecological and political well-being. We describe and discuss the alternative democratic processes in Korchi and what they hope to achieve; factors that led to the emergence of such processes; and the constraints and hurdles that they face. This description and analysis will help foster a greater understanding of the causes and conditions that often lead to the emergence and sustainability of transformative alternative initiatives.

Background

Korchi *taluka* is one of the twelve sub-districts in the Gadchiroli district of the western Indian state of Maharashtra. Gadchiroli covers a geographical area of 14412 km^2 and most of its land is under forests (~76%, GoM 2022), with nearly 39 % of its population comprising of *adivasi*.[2] Korchi's population is 42,844 (Census 2011) of which almost 75% are *adivasi* belonging to the Gond and Kanwar tribes. Almost the entire population of Korchi depends heavily on the forest for a cash-based and also subsistence livelihood. Besides forests being important for local economies and livelihoods, they are an integral part of the *adivasi* socio-cultural practices and political identity. Like in the rest of India, until recently, local communities had restricted and limited forest rights, leading to oppression, atrocities and a culture of bribes for accessing forests.

Forests in Gadchiroli, as in other parts of India, have also been important for the state. Commercial extraction of timber and other non-timber products has conventionally been carried out by the forest department through leases given to contractors and paper and pulp companies and, in more recent times, to mining companies. Between 1990 and 2017, twenty-four mining leases were sanctioned or proposed in the district, collectively impacting approximately 15,000 hectares of dense forest directly, and over 16,000 hectares indirectly. In Korchi *taluka* alone, around twelve mining leases are proposed despite strong local opposition, impacting over 1,032.66 hectares of forest (Pathak Broome and Raut 2017).

Severely affected by the centralized, top-down and oppressive forest policies and practices, the Gadchiroli district has seen a number of resistance movements demanding self-determination and village self-rule. It has also seen strong women's mobilization and grassroots movement both within their communities against external imperial forces and against the systems of social discrimination within their own patriarchal societies. The most significant of these was the 'save people and save forests movement' of the mid-1980s, which demanded greater tribal autonomy, control over decision-making, and rights over forests and natural resources. Slogans like *Mawa mate mawa sarkar* ('We are the government in our village') and 'Our representatives govern from Delhi and Mumbai but we are the government in our village' emerged from Gadchiroli, intensifying the self-rule movement in the district. Villages such as Mendha-Lekha declared de facto village self-rule, inspiring many others to follow suit (Pathak Broome 2018). Despite these resistance movements, control over forests has remained in the hands of the forest department and forest leases continue to be issued for commercial extraction (Ali 2016; Pinjarkar 2013), including for mining (Newsclick 2018; Pathak Broom and Raut 2017), in a clear violation of the country's legal provisions related to FPIC (free, prior and informed consent) provided under the FRA 2006 and PESA 1996.

Towards Transformative Alternatives Through Direct Democracy in Korchi

Forest and village governance

The FRA was enacted in 2006 and PESA Rules for the state of Maharashtra were drafted in 2014. However, considering the potential of these two laws to strengthen self-governance and FPIC over forests (Padel 2014), they have faced stiff opposition from existing power centres including the forest department. Consequently, by 2016 (over a decade after its enactment) only about 3% of the FRA's minimum potential had been unlocked throughout the country (CFR-LA 2016). Due to a range of factors, mainly the people's movement, Gadchiroli has fared much better, having achieved over 60% of the FRA's potential and bringing around 38% of forests in the district under the control of local *gram sabhas* (CFR-LA 2017).

Within Gadchiroli district, the village of Mendha-Lekha was the first to file a CFR claim over the forests of which it had de facto taken charge. It became one of the first villages to receive a legal title and started to sustainably manage, conserve and earn revenue from forest produce (Das 2011). Many villages both within and outside the district went to Mendha-Lekha to learn from them, including the local leaders from Korchi *taluka*. A sustained and effective campaign by a collective of non-government agencies, local community leaders and government officials led to most villages across Gadchiroli filing claims and receiving titles. By 2012, a total

87 of the 133 village *gram sabhas* in Korchi had claimed and received CFR Right titles over their traditional forests. The 2014 PESA Rules for Maharashtra also strengthened village self-rule.

Although officially administered by the Gadchiroli District Administration and elected *panchayats* (executive committees of one or more villages elected to be the first unit of governance in India's Panchayati Raj System, or local self-government), people in Korchi *taluka* continue to have their informal traditional village-level governance systems and a *taluka*-level collective of village elders called the *ilakas* (territories constituting of multiple villages) *sabha* (assembly). With few political and economic powers however, these informal institutions largely focused on socio-cultural activities or conflict resolutions.

In 2016, some local community leaders made efforts to understand how laws like the FRA could strengthen and empower *gram sabhas*, and also to mobilize local people towards understanding how legally empowered *gram sabhas* could work towards self-determination and self-governance, including asserting greater but equitable control over forests and local economies. This led to *taluka*-level discussions on the concept of *gram sabhas* and the implications of their empowerment, the role of the FRA and PESA in strengthening *gram sabhas*, mining as a means of development and the idea of development itself, among other things. The discussions on development were triggered by multiple proposals to begin mining operations within the customary forest boundaries of some of the villages. Over a period of time, multiple open and transparent public debates and discussions, including during cultural ceremonies and gatherings, influenced villages in Korchi. By 2017, eighty-seven villages in Korchi *taluka* were moving towards reconstituting and strengthening their village *gram sabhas* to be inclusive decision-making bodies, transparent and accountable at the village level. Each *gram sabha* opened a bank account and appointed a secretary and a president, who were in charge of village communication with the government officials and other outside actors. Consequently, some *gram sabhas* began to gain empowerment and recognition; however, it was important for them to get collectively stronger in order to support other *gram sabhas* which were just beginning to re-organize themselves, as well as for the required mutual learning and support. Under the FRA and PESA, ownership of non-timber forest produce (NTFP) and the right to harvest and sell was now with the *gram sabhas*. Sustainably harvesting commercially important NTFP and selling them required skills, knowledge and strength to deal with the market forces. Understanding and addressing the divisive strategies of the mining companies required collective action. The intense *taluka*-level debates and discussions in 2016 led to the realization that individual *gram sabhas* by themselves were not strong enough to prevent exploitation by the market forces as they ventured into the collection and trade of forest produce. It was also recognized that the traditional *ilaka sabhas* had their limitations in being able to address the above-mentioned issues. In 2017, therefore, a decision was taken to establish a federation of all ninety *gram sabhas*, called the Maha Gramsabha (MGS), which would be more inclusive, fair and transparent than any of the existing traditional *taluka*-level bodies or formal institutions.

Within a short period, the *gram sabhas* at village level and MGS at *taluka* level began to emerge as institutions of self-governance. Individual *gram sabhas* began organizing regular village-level meetings while the MGS would meet once a month in Korchi town. *Gram sabhas* formally wanting to join the MGS would pass a resolution to this effect after a detailed discussion within their village. As per the rules of the MGS they would then nominate two women and two men to represent them in the MGS general body and they would agree to pay an annual membership fee of 5,000 rupees to cover the MGS's operating costs (earned from the sale of NTFPs). A couple of years later, in order to facilitate greater interaction and more frequent meetings between neighbouring *gram sabhas*, seven clusters were formed with ten to twelve villages each. The MGS executive body comprises fifteen members, made up of one woman and one man from each of the seven clusters and one person with disabilities. The remaining fourteen members represent all social groups (caste, class and gender) in accordance with their demographic structure in Korchi *taluka*. The MGS has since evolved into a *taluka*-level pressure group for oversight on all issues related to forest management and conservation, forest produce sale and benefit-sharing mechanisms, and other aspects of local well-being.

Consequently, *gram sabhas* began to negotiate and market their NTFP. The profits were now entirely with the *gram sabhas*, unlike previously when the forest department had controlled the trade and retained the profits. The *gram sabhas* paid for the labour (all families in the village), retained some percentage for the village bank account and distributed the remaining as profit shares to the collectors. *Gram sabhas*, which until then were economically and legally disempowered, began to gain in both these areas. For example, from nearly zero income in 2014, the eighty-seven *gram sabhas* in 2017 had a total income of over 120 million rupees (about US$1,700,000) from the sale of NTFPs.

The role of women in institutions of direct democracy

In this predominantly patriarchal society, women had little say in traditional village and forest governance. Women also faced a number of social challenges, including domestic violence abetted by alcoholism, and lack of resources, property or decision-making rights. A change in these areas has been made possible because of a fairly long history of women's mobilization in Gadchiroli through empowering women's self-help groups (SHGs)[3] and their federations, the *mahila parisar sangh* (MPS) (women's collectives), in which Amhi Amchya Arogyasathi (AAA), a local NGO, has played a crucial and catalytic role. Over the decades, these MPS became a support group for women facing injustice, oppression, violence or any other issue within the family or in the larger society. As the awareness among the women increased and they found the confidence to voice their opinions, many women highlighted how their well-being and that of their families was integrally linked to the well-being of the forests. Hence, it was important for women to discuss issues of forest degradation and the rights to use and protect them. This became particularly critical for women in 2009, especially in villages which discovered that their traditional forests were being leased out for mining. Through their MPS, women became one of the formidable forces in the resistance against mining. Their physical opposition and vocal expression in various meetings against mining, including the state-sponsored public hearings, ensured that the mining leases have remained pending until this date in the Korchi *taluka*. Among themselves, through their MPS, women have had numerous discussions around the impact of mining on their lives, families and forests, and the need for protecting forests. The MPS have also been crucial in platforming women leaders to narrate their struggles and opinions, including their conception of well-being, which is deeply linked to healthy forests. During the resistance against mining in the Korchi *taluka*, women leaders in the MPS began to realize and discuss that while women were always at the forefront of the resistance, they had no space in traditional decision-making processes, about the village or the forests.

By 2016, the discussions on *gram sabhas* as units of self-governance were gaining ground, *taluka*-level meetings were being organized, and implementation of the FRA was being spoken about in various *taluka*- and *ilaka*-level meetings. However, none looked at the issues of women's participation, women's rights under the Acts or the economic empowerment of women from the forest produce. Some of the women leaders began participating in the *taluka*-level meetings. In one of the first meetings of the MGS, women leaders insisted that along with challenging the hegemonic and top-down bureaucracies, it was also important to challenge the established traditional structures that legitimized oppression on women and restricted women's role in decision-making, including decisions around forests. They ensured that the MGS included two women representatives along with two men from each *gram sabha*. The MPS also ensured that Korchi *taluka* was one of the few in the country where the rights of the women under the FRA were being focused on. The FRA provides for joint land titles for wife and husband. In many villages in Korchi, titles have been issued jointly, and in some in the name of women as first owners or women as exclusive landowners. Empowered through this process, women leaders began to question why other traditional systems like the *ilaka sabhas* also should not include women. This led to the inclusion of women in the *ilaka sabhas* as well, a unique feature of this process.

Distinguishing Features of the Process Towards Village and Forest Governance Through Direct Democracy in Korchi – Unique Features and Enablers

Mere enactment of radical laws such as the FRA and PESA is not enough to bring about transformative alternatives. These laws are applicable across India but only in a handful of places like in Gadchiroli have they led to radical changes on the ground. These changes are a consequence of well-considered processes adopted by the actors involved and multiple enabling factors unique to Gadchiroli and Korchi. In the following sections we explore the processes which distinguish this combination of formal and informal governance systems in Korchi from other processes including state institutions of decentralization, i.e the Panchayati Raj System (PRS), and the enablers which made this possible.

Panchayati Raj System and gram sabha *empowerment in Korchi*

The PRS in India has been heavily criticized in recent years for its failure to secure meaningful democracy. Continued colonial distrust of local institutions in independent India meant that there was limited devolution of power and responsibility to the *panchayats*. These were subsequently further curtailed due to the decline in their performance. Consequently, most of the development programmes, meant to be implemented by the *panchayats*, are administered directly by the parallel administrative bodies. The factors that led to the *panchayats*' decline include the following (Banerjee 2013):

1. The most important reason for their decline is attributed to the otherwise centralized tendencies of operation in the country's political and administrative system. So while implementation of predetermined programmes and schemes is decentralized to *panchayats*, the financial and legal powers are largely centred in the state/central state institutions.
2. *Panchayats* themselves are not seen as institutions of direct democracy, as the power of decision-making is in the hands of the elected representatives. *Panchayats* are often constituted at the level of a cluster of widely dispersed hamlets or villages, making it difficult for members of all constituent villages to participate in their general body meetings, which are held at least eight times a year.
3. The PRS has seats set aside for women and members of disprivileged castes as office bearers, but in practice the participation of women (except in a few cases) has been symbolic, with their husbands assuming the actual power. The environment of *panchayat* general body meetings has been unsupportive of women's participation, consequently limiting their involvement.
4. Unaccountability, lack of transparency, inefficiency, corruption, nepotism, favouritism, uncertainty and irregularity have been intricately linked with the functioning of the *panchayats* across the country.
5. *Panchayat* elections have not been held in many states, and where they are held they are increasingly influenced by the national and regional political parties. It has become common practice for these parties to establish roots in a village through candidates standing for *panchayat* elections. This has created political divisions and factions within the villages and *panchayats*, often leading to a murky politics of power rather than elections for this basic unit being based on issues of local significance.

Thus while intended as a means to achieve direct democracy, *panchayats* have been largely reduced to an extension of representative electoral democracy, playing as pawns for political parties, fuelled by nepotism and patriarchy in society and further encouraging these to enhance their own power and control. The envisioning of the *gram sabhas* being empowered as a first unit of decision-making in the FRA and PESA is centred on overcoming the above limitations, while also additionally empowering them to sustainably use, access, manage and govern forests within the traditional village boundaries, and be responsible for the conservation and protection of biodiversity and natural and cultural heritage. The community leaders saw a potential to achieve better village and forest governance through direct democracy and to overcome the limitations of the PRS. This potential was realized through the various processes explored below.

Coordination with panchayats *and other state and non-state actors*

In Korchi, the *panchayats* continue to perform all government administrative and political functions at their level. However, at the village level *gram sabhas* are the decision-making bodies. At the cluster and *taluka* level the federation of *gram sabhas* or the MGS (which also includes members from the *panchayat*) operates as a support, guidance, advisory and conflict redressal body to the *gram sabhas* on the one hand and a pressure, coordination and liaison body to hold *panchayat*, *taluka*, district, state and non-state agencies and actors accountable to the decisions of the *gram sabhas*, on the other. The MGS itself remains accountable to the *gram sabhas*. Figure 10.1 shows a flow chart to better understand the politico-administrative system of governance.

Better-informed and -trained gram sabhas

Lack of awareness, information and legal/administrative/financial training prevents *gram sabhas* from realizing the true potential of the empowering provisions of the law. The community leaders have therefore placed great importance on ensuring that *gram sabhas* are empowered to make informed decisions. This is attempted through discussions in the MGS and regular training programmes organized and conducted by the MGS members. Another important component of this is peer-to-peer learning. MGS and *gram sabha* members visit other *talukas* where similar processes are unfolding. They stay connected with each other through social media and use local media to spread awareness. Traditional religious and cultural ceremonies are also used for self-empowerment and knowledge-sharing.

Ensuring transparent and open functioning

Many of the weaknesses in conventional governance systems, including *panchayats*, emanate from a lack of openness in approach and transparency in functioning. Having worked with the *panchayats*, community leaders were well aware of this and

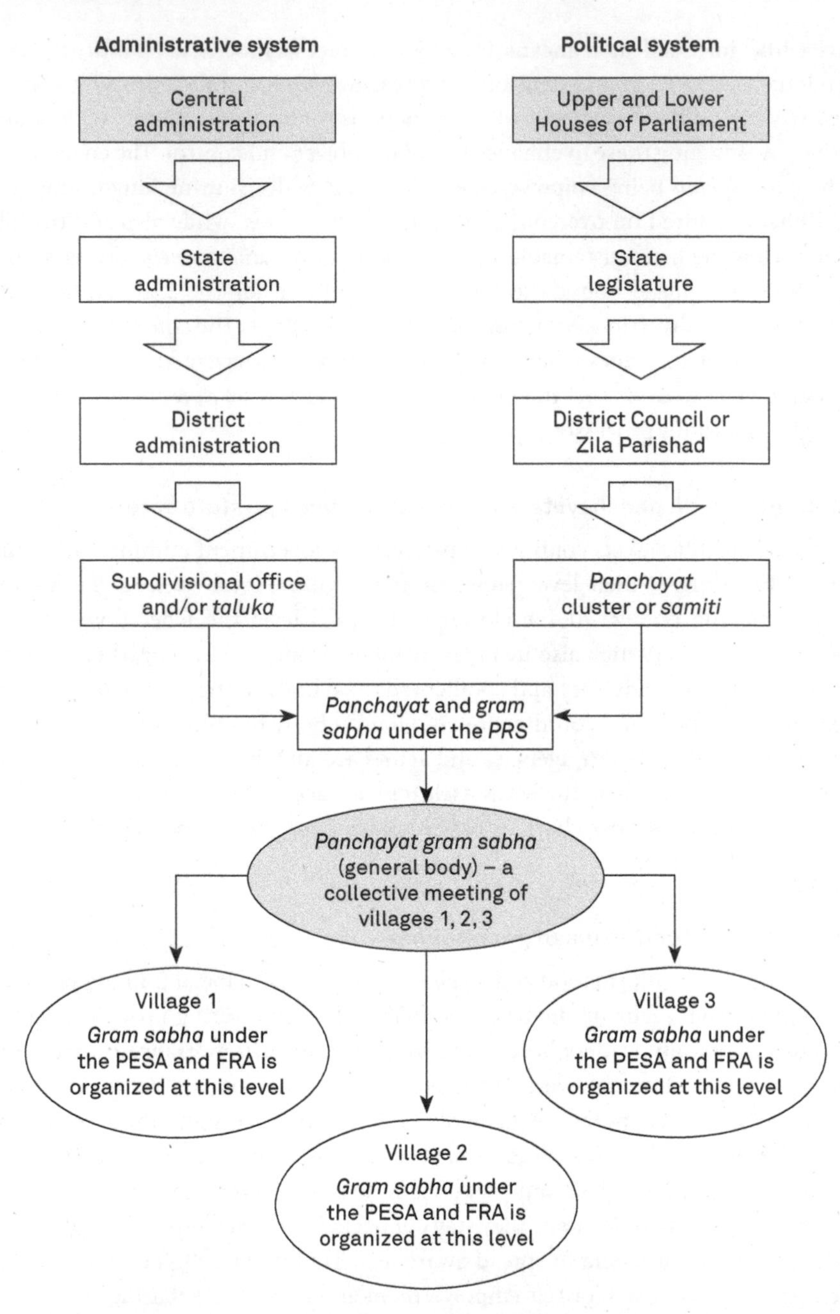

Figure 10.1 Political and administrative structure in India under the PRS and under the FRA and PESA

within the MGS openness and transparency was attempted right from the outset. It was ensured that the MGS meets monthly, meetings are held regularly, they are open to all members, and they are held at a widely known date and time.

Meetings include updates of all activities undertaken, new emerging issues, concerns of the *gram sabhas*, expenditure details of the previous month, planning for the next month, and any other issues raised by the members. *Gram sabha* members attending the meetings are expected to take the discussions to their own *gram sabhas*. No decisions are taken at the MGS meeting; respective *gram sabha* delegates take any proposals for decision back to their own *gram sabhas* and then share the eventual decision with the MGS.

Addressing power imbalances without destabilizing the process

During the evolution of the processes towards *gram sabha* empowerment and the constitution of the MGS, the local leaders demonstrated maturity and adaptability by transforming a potentially damaging conflict situation between different ethnic groups (*adivasis* and non-*adivasis*) into an opportunity for creating dialogue towards a more open and inclusive institutional arrangement. They did this by taking into account the concerns of the minority non-tribal groups while also addressing the unexpressed fears around falling into insignificance of the traditional leaders of majority tribal groups. Ensuring a balance in power and privilege was important. The wisdom lay in doing away with the power imbalances within the traditional and non-traditional existing institutions, without fostering fears and ill will. It was therefore ensured that traditional *adivasi* leaders would be granted the traditional respect and be included in various capacities, like advisory elders, but would not be the only voice of or for their community.[4] Similarly, it was important to continue involving the official members of the *panchayats* as their skills and resourcefulness would be useful for the process, while their antagonism could destabilize it. This was a delicate balance between challenging traditional or conventional hierarchies and power relations within and between the communities while minimizing the isolation, exclusion and antagonism of those who had been in power. The MGS has been successful in achieving this balance thus far, acknowledging that this is a continuous process and challenges will need to be addressed as they arise.

Gender-inclusive decision-making

The inclusion of women and their concerns in the processes has likewise been a unique feature of the democratic process in Korchi. The *mahila parisar sangh* works constantly to ensure that women do find space in all decision-making processes, including *gram sabhas* and the MGS, and receive equitable benefits. Since the *gram sabhas* now serve as the decision-making bodies, not the *panchayats* (far away from the village), there are many more opportunities for women's participation. Some *gram sabhas* have also made special efforts to ensure that meetings are held at times when women are able to attend. However, few *gram sabhas*, if any, have female office bearers; women are continuing their efforts to change this. The MPS members have played an important role in training *gram sabhas* in bookkeeping and accounting procedures thanks to the leaders' extensive experience in SHG accounting. Korchi *gram sabhas* are currently handling millions of rupees and maintaining the most transparent accounts (as also acknowledged by the local government agencies). In fact, one of the leaders, Kumari Bai, has also been appointed an advisor and financial consultant to the MGS. This is in addition to other SHG members holding executive positions within the MGS. Encouraged by the MGS, many *gram sabhas* have also taken the decision that women will get the daily wage labour as well as the profits from the sale of NTFPs directly in their own accounts, instead of their husbands' (a conventional practice). In fact, one village, Sahle, has decided that the entire profit earned by the family from forest produce will go only to the accounts of the women of the family – a powerful and unique decision.

Youth-inclusive decision-making

With party politics exerting ever greater influence, many *panchayats* in India are increasingly drawing in young people to engage in divisive party politics, which focuses on individuals and their amassing of power. In Korchi, however, the *gram sabha* processes have inspired many youths who are engaging in the harvesting of forest produce, forest management and conservation, resistance against mining, and the administrative activities of the *gram sabhas* (which require skills in account keeping, record maintenance, networking and alliance-building, among others). This is not a universal phenomenon, however, and there are many young people who are caught in a tussle between these unfolding local processes and the *adivasi* way of being on the one hand, and the lure of the market, glamour of the dominant outside society, and pull of right-wing religious elements[5] on the other. Many of them also support development processes like mining, and work to influence the *gram sabhas* in their favour. This is more common among those who have been through higher education outside the villages. The MGS is constantly thinking of ways to include such youths in local processes, by engaging in cultural activities, monitoring education institutions and establishing a library, among other methods.

Control over the local forest economy, strengthening local livelihoods and increasing financial stability for gram sabhas

The process of direct democracy in Korchi is closely linked to the local forest-based economy and gaining control over the means of production. In many ways the success of the governance process in Korchi is dependent on the ability of the *gram sabhas* and the MGS to help sustain forest-based livelihoods and economies. The *gram sabhas* began collecting and selling *tendu patta* (leaves of *Diospyros melanoxylon*) and bamboo, two important forms of commercial forest produce in the region, in 2017 (this had previously been exclusively controlled by the forest department). By 2019, the *gram sabhas* collectively received about 160 million rupees (US$2 million) from these forest products in addition to the daily wages paid to the collecting families. Different *gram sabhas* have retained differing shares of this total (ranging from 5% to 20%) to cover their administrative overheads while sharing the remaining amounts equally with all families who participated in the collection, including women.

Our analysis also showed that currently about 70% of the population is directly engaged with forest economy, 50% of whom are women. Our calculations also indicate that taking over the sustainable harvest and sale of these forest products has brought about a 70–80% increase in family incomes and, for the first time, has generated income for the *gram sabhas* (which up until then had no income or funds), empowering them financially to undertake activities for village well-being. In some villages, women are now their family's breadwinner, traditionally the role of the male head of the family. Directly participating in activities related to trading, marketing, record maintenance and other associated activities also means increased awareness and skill enhancement among the *gram sabha* members (including women). The overall revival and localization has reduced outmigration, which was rampant just a few years ago. Although outmigration continues, it is rarely as much of a compulsion as it was before.

Greater control over the forest-based economy has also helped the MGS demystify the employment and development promises being made by the mining companies. With generally declining employment rates in the country, the local leaders' calculations have indicated that the current combination of options open to villagers best protect local livelihoods and well-being. Agriculture and the forests provide food, while the trade of forest produce and other associated activities provide cash, leaving villagers with ample time to participate in community and collective cultural and political activities. They claim that standing forests provide more for longer and without the destruction that mining would cause. Mining companies would employ a handful of local people, mainly men and largely in unskilled work, while destroying the forests and forest-based income, affecting agriculture, causing water and air pollution, and cultivating an insecure and unsafe environment for women and children while taking away their income.

The Covid-19 pandemic and gram sabhas *in Korchi*

During the first phase of the pandemic when India saw an unprecedented reverse migration of migrant labourers in 2020, Korchi remained unaffected as few people had migrated out. *Gram sabhas* in Korchi also stepped up to provide basic relief for families who were affected by the nationwide lockdown from the funds they had collected. This was even before the government and NGO welfare systems could react. This set an example for many village communities across the country and was followed by many other offerings, particularly from those who had financial capabilities.

Engaging with and addressing party politics

In 2017, the MGS discussed the issue of *panchayat* representatives and felt that they had failed the community in their struggles, including those against mining, and were instead representing the corporate-political interests in the region's economy. The local *gram sabhas* therefore decided to participate in *panchayat samiti* and *zila panchayat* (district- and subdistrict-level) elections to help them gain political control over all three tiers of the PRS. The *gram sabhas* fielded candidates under an oath to follow ethical principles accepted by the *gram sabhas*, but lost the elections. The results of the election and events during the election period were discussed, analysed and found to be divisive, corrupting and to have taken a heavy toll on the unity of the collective. The *gram sabhas* felt that it may be better for the MGS to work as a pressure group from outside rather than trying to engage with electoral politics. An assessment of the historical events in the district also showed that the local leaders who engaged with electoral politics were co-opted and unable to achieve the objectives for which they engaged with this system in the first place. Within the *gram sabhas* and MGS people from all ideologies and political affiliations are welcome; however there is a clear understanding that within these two institutions ideological differences and party affiliations and interests are kept aside. theThe leaders constantly attempt to ensure that this is adhered to, although it is not always an easy balance to achieve.

Towards ecological wisdom, integrity and resilience

The recognition of rights has revived a sense of belonging over the forests that had eroded over generations because of alienating colonial policies. Since forest-based livelihoods are now locally controlled, ensuring the ecological sustainability of the forests is also seen as a local responsibility. These once rich forests, which have deteriorated over the years because of unregulated overuse and the encroachment of individual plots of land, are now being viewed differently. After receiving rights under the FRA, many *gram sabhas* have started devising rules and regulations regarding the management and protection of forests, including a system of regular forest patrols. Such protection and conservation systems are encouraged by the MGS. Controlling forest fires has resulted in greater regeneration and richness in forest biodiversity. The FRA requires all *gram sabhas* to formulate management plans and strategies, including for sustainable harvesting and sale of the commercially important NTFPs. Using funds from the Tribal Development Department and in some cases supported by outside agencies such as AAA, some *gram sabhas* have begun drafting formal management plans. With or without management plans, however, many villages have carried out successful plantations of a diversity of local species. In almost all cases, extraction of NTFPs is carried out on rotation (ensuring that all parts of the forest are not extracted in one go). Using the FRA's FPIC clause, villagers have already registered their rejection of the mining proposals, although the threat posed by mining is far from over and keeps resurfacing every now and then.

Enablers of Resistance and Transformative Processes

In this section we describe some of the actors and factors behind the transformative alternatives in Korchi.

Culture of community life and leadership

A communal way of being, setting aside time for the commons and collective action including community celebrations, festivities and community welfare activities is integral to tribal cultures. Regular community gatherings and celebrations (e.g. *yatras*) have been crucial to discuss and develop collective strategies. Community leaders built upon these existing traditions and systems and used them for socio-political discourses and discussions which build the foundation for the process. The existing culture of respect for elders combined with the presence of unique social leaders has played a critical role in this movement. Such leaders and elders have guided the processes and movements, often investing personal time and resources without expecting much in return.

Long history of struggles and movements, and confluence and dialectics of different socio-political ideologies

The *adivasi* regions like Gadchiroli have a long history of oppression and resultant resistant movements through pre-colonial, colonial and post-colonial times. The independence of India in 1947 did not mean greater autonomy or freedom of decision-making for the *adivasis* and other traditional local communities. The country continued to follow extractivism and capitalism-based development; centralized and top-down forest governance; and exclusive and representative electoral democracy. All of the above stand in direct contradiction with the worldview and socio-political organization of the *adivasis*. These contradictions have been historically and are currently the underlying causes of a continuous environment of conflict, within which some events, policies and actions trigger stronger episodes of resistance. The environment of resistance has kept political awareness and consciousness alive, leading to the emergence of collectives like the *mahila parisar sangh* and the MGS. Such autonomous, organic, inclusive and discussion-based collectives are crucial in creating an inherent understanding that resistance alone is not enough to challenge the root causes of injustice. The local processes towards strengthening self-rule are critical to impact the political economy of the region.

District-level study circle, and peer-learning and support processes

One of the key factors of the effective and successful implementation of the FRA in the district as a whole has been the district-level study circles initiated by some civil society actors to strengthen *gram sabhas*. Study circles provide a forum to understand local contexts, learn from each other and deliberate upon issues. They helped create a district-wide campaign calling for the implementation of the FRA as soon as it was introduced and led to Gadchiroli becoming the only district in the country where over 60% of the potential of the FRA had been realized by 2016. In addition to the district study circle, *gram sabhas* have also created means of exchange and learning among themselves across the district, as mentioned above.

The role of Amhi Amchya Arogyasathi (AAA)

Amhi Amchya Arogyasathi (AAA), a local NGO, has been active in Korchi for several decades and has worked towards improving health, forest management and women's empowerment. The AAA has also supported local social leaders, including women as *karyakarta* (village activists), in a range of projects and has provided them with opportunities to interact with actors at district, state and national level and be part of various discussions and debates. This has helped them to enhance their existing levels of awareness, information and leadership skills, and gain respect and acceptance within the larger community. The AAA has also provided timely help in accessing information and building capacity through various training programmes. This NGO has played a unique supportive role by implementing projects but not imposing and taking control of the local processes.

Jeevanshalas: *schools with a difference*

A unique education programme called the *jeevanshalas* ('schools of life') was implemented for three years in some of the villages where AAA works. The concept of *jeevanshalas* was based on the *Nai Talim* (Gandhi 1962) system of education. This was particularly important for the tribal children who, tuned into their forests, often found the classroom- and alphabet-based education system of regular government schools constraining and uninspiring, resulting in huge numbers of school dropouts. As two of the local leaders said:

> we were able to be what we are because we didn't go to the formal school after [the] initial few years. School was oppressive, difficult to understand and nothing much to learn. On the contrary when we roamed the forests we learned so much more. We also had time to be part of the collective community activities.

Jeevanshalas envisioned education differently, where learning from the local surroundings and ecosystem was key. Those influenced by its philosophy are among the main leaders of both resistance against mining and the movement in support of transformation processes.

Characteristics of Transformative Processes in Korchi

The analysis below is specific to the context of Korchi in particular and Gadchiroli district in general. Resistance and transformation are contextual and dynamic processes and what is relevant in one context may or may not apply to others. Yet there are commonalities and potential for learning across situations.

Continuous yet episodic and spiral – keeping political consciousness active

The alternative transformative processes in Korchi indicate that resistance and transformation are continuous and yet episodic. Political economy-related conflicts have been an integral part of the process of transformation. While some past resistance movements and transformative actions have continued to sustain and inspire new ones, others have been co-opted and undermined by the established power structures. So, while the progression may appear to be circular as similar events arise over a period of time, coming apart and reinventing themselves, they are not exactly the same in their reinvented forms.

Scalar, temporal and evolutionary

Transformation is an evolutionary process, having both scalar and temporal dimensions. Multiple factors emerging at different times in history can be transformative at different levels – for example sometimes transforming individuals, sometimes individual villages or an entire *taluka*, and at other times an entire district. These transformations are subjective and do not impact the society uniformly. Instead, they contribute to the overall evolution of the transformative processes, particularly when these differently transforming processes and actors come together. Thus, transformative process is a result of evolution over a period of time and also a coalition or friction between various individuals, ideologies, civil society groups and deliberative processes at different scales at any given point in time. District-level study circles and their influence in the process in Korchi, *taluka*-level federations and their interactions, individual villages like Mendha-Lekha located in other *talukas* but influencing processes in Korchi (and vice versa) – these are the scalar dimensions of learning and evolution within the transformation process in Korchi.

Locally rooted but also addressing traditional and customary discriminatory practices

The transformation process in Korchi is definitely embedded in local socio-cultural and political values, conceptions of well-being, principles and histories. Simultaneously, these processes and practices have also incorporated many modern and contemporary ideas of political economy, human ecology, equity and social justice. For example, while the principles of consensus-based, inclusive decision-making and collective community action are integral to the *adivasi* traditions, greater emphasis on gender participation in decision-making, women being equal or primary beneficiaries of local economic activities, and inclusion of non-*adivasis* (particularly scheduled castes) in decision-making bodies are newer inclusions.

Conceptions of well-being can be internally diverse and conflicting

The conceptions of well-being and transformation are not universally accepted conceptions in all ninety villages. There are several diverse and internally contradictory views influenced by different actors and factors. The capitalist and extractive economy and its propaganda machinery have been effective in influencing a large part of the population, particularly the youth. The existing state education system further alienates them from their own culture and creates consumptive and career-based aspirations, motivating them to support mining and its promised jobs. The right-wing religious groups have also influenced a large part of the population and their agenda aligns more closely with the growth-based model of development. Considering the presence of multiple ideologies and conceptions of well-being, which ideology influences the dominant processes at any point in time depends on multiple causes and conditions. In many ways, the transformative processes in Korchi are continuously impacted and evolve because of the dialectics of these multiple conceptions of well-being.

Non-static – constantly adapting but in keeping with core values

All processes are dynamic and non-static, continuously changing and evolving towards the larger goal of greater local autonomy and greater systemic accountability to be able to achieve equity, justice and well-being. *Gram sabha* members, *parisar sangh* members and MGS members have to continuously deal with new challenges and opportunities, and constantly learn and adapt – and while doing so keep to the values of transparency, inclusiveness, openness, mutual respect, and collective actions and benefits.

Located in and dependent on inherently contradictory contexts

Within the processes of resistance and transformation, there are many inherent internal contradictions. Among the most significant is heavy dependence on the state and its adopted exploitative capitalistic model of economy and representative electoral democracy. *Gram sabhas*, the institutions of direct democracy, remain dependent on state institutions which remain centralized in their spirit and disconnected from local issues. Similarly, NTFP trade such as *tendu* leaves and bamboo, which are the mainstay of the people in the region and have been crucial in causing a radical shift in the local economy, are themselves dependent on the external capitalist markets. Market fluctuations and vagaries have serious impacts on their sustainability.

Conclusion

The *gram sabhas* in Korchi are at different stages of empowerment. While some *gram sabhas* have established systems of equitable, transparent and inclusive decision-making and benefit-sharing, others are striving to reach that stage. The MGS is also continuously evolving in its structure and operation. *Gram sabhas* and the MGS face numerous internal and external challenges, the most significant among them being existence within the nation-state and its adopted exploitative capitalist model of economy and representative electoral democracy. Party politics, having entered all other levels of governance, now strives to control the *gram sabhas*. The PRS institutions at all levels are beginning to feel threatened by the emerging power of the *gram sabhas*, creating friction with the MGS. Religious right-wing (Hindu in this case) and cultural right-wing tribal outfits are using identity politics for political gain; some of these are supported by the mining companies and often create hurdles for the MGS and *gram sabhas* opposed to mining. Many local activists, including one of the core team members of this study, have been imprisoned under the Unlawful Activities Prevention Act (UAPA), which gives the state draconian powers to arrest without a warrant or evidence and keep people in police custody without bail for a certain period of time.

Despite these challenges, focusing on strengthening the smallest unit of direct decision-making and ensuring that these are inclusive, transparent, financially strong and fair structures has influenced nearly all spheres of social organization, including economic, political, ecological, cultural and social elements in Korchi. The government's decentralization efforts are different from the people's movement towards self-rule and direct democracy in that the former remains fixated on the external structure rulebooks at the cost of the spirit of decentralization, while the latter focuses on that spirit by constantly adapting and evolving strategies, structures, rules and operations to address the opportunities and challenges encountered, all while ensuring that the core principles of transparent dialogue, consensus-based decision-making and equity are not compromised. As a Gondi proverb says, '*Changla jeevan jage mayan saathi sapalorukoon apu apuna jababdarita jaaniv ata pahe*' ('To achieve well-being, everyone needs to know what their responsibility is'). The MGS members believe that to be more effective politically, different *taluka*-level collectives need to come together to form a district-level federation and must also have their delegates in the state legislature, which is yet to be achieved. They hope to slowly move in that direction.

Abbreviations

CFR: Community Forest Resource Rights, or the right to use, conserve and sustainably manage forests over which rights were granted under the FRA 2006
FPIC: Free, Prior and Informed Consent
FRA: Forest Rights Act, also called the Scheduled Tribes and Other Traditional Forest Dwellers (Recognition of Forest Rights) Act 2006
MGS: Maha Gramsabha, or federation of *gram sabhas* in Korchi
NTFP: Non-Timber Forest Product
PESA: Panchayat (Extension to Scheduled Areas) Act 1996
PRS: Panchayati Raj System
SHG: self-help group

Glossary

gram panchayat: The elected village executive committee forming the smallest unit of decision-making within India's PRS; a *panchayat* could cover one or more villages
gram swaraj: Village self-rule (or village republic)

panchayat samiti/mandal parishad/block samiti: The PRS has three levels – *gram panchayat* at village level, with *panchayat samiti/mandal parishad/block samiti* at the higher level called the *mandal/taluka/block*, which constitutes a cluster of villages
Panchayati Raj System: System of governance adopted by India in which the *gram panchayats* are the basic unit of local administration and governance
sarpanch: Elected head of a *panchayat*
taluka: An administrative unit at the level of multiple villages
zila parishad: The third tier of the PRS, covering a district, which constitutes multiple *talukas/blocks*; multiple districts constitute the state

Acknowledgements

This chapter builds upon a report of a study carried out by Kalpavriksh, with Amhi Amchya Arogyasathi (AAA) and members of the Korchi Maha Gramsabha as part of an ACKnowl-EJ (Academic-Activist Co-produced Knowledge for Environmental Justice) project. ACKnowl-EJ is a network of scholars and activists engaged in action and collaborative research that aims to analyse the transformative potential of community responses to extractivism and alternatives emerging from resistance.

The authors would like to thank all *gram sabha*, *maha gramsabha* and *mahila parisar sangh* members from Korchi, in particular G. Kumaribai Jamkatan, Ijamsai Katenge, Zhaduram Salame, Siyaram Halami, Govind Hodi, Sheetal Netam, Nand Kishore Varagade, Hirabhau Raut, Bharitola, Lalita Katenge, Suresh Madavi, Dashrath Madavi, Sundar Bai, Indirabai, Kamala Bai, Manbai, Dev Sai, Deepak Madavi, Sumaro Kallo, Sunul Hodi, Narobai Hodi, Amita Madavi, Ramdas Kallo, Makau Hodi, Rameshwari Bai and Babita Bai. Zendepar, Salhe, Bodena, Phulgondi, Padyal Job, Kodgul and Tipagarh village *gram sabhas* for their kind hospitality and conversations. Subhadha Deshmukh and Satish Gogulwar from Amhi Amchya Arogyasathi for support and guidance during the study. Ashish Kothari, Mariana Walter, Iokiñe Rodríguez, Jérôme Pelenc, Madhu Ramnath and Suraj Jacob for their valuable comments on the original report. Special thanks to Mahesh Raut, for whom we eagerly await justice to be done.

Notes

1 This chapter is based on a larger report: Pathak Broome et al. 2022.
2 *Adivasi* – a Hindi word meaning 'original inhabitant' – is a common term used to describe all Indigenous groups in India, of which there are over 700. We stick to using this term throughout the chapter, as that is how the Korchi residents self-identified themselves in interviews.
3 Small committees of ten to twenty women and men initially set up in rural India for providing financial support to women, which have in many cases evolved to become agents for change through the general empowerment of women.
4 At the same time, the traditional *jat panchayats* also continue to exist. They have not remained unscathed by the ongoing debates and discussions. Many of the local social leaders are also members of the *jat panchayats*. *Jat panchayats* have therefore made some significant changes in their oppressive, discriminatory socio-cultural practices. For example, women are now part of the decision-making process even here.

5 The more right-wing Hindutva outfits have long striven to have *adivasis* considered Hindus. With the right-leaning party in power in the country, such efforts have increased in recent times.

References

Ahmad, T. (2017) *National Parliaments: India*. February 2017. The Law Library of Congress, Global Legal Research Center. https://www.loc.gov/item/2016478967/.

Ali, M. (2016) Chanda Villagers Refuse to Withdraw Chipko Movement. *Times of India*, 2 June. http://timesofindia.indiatimes.com/city/nagpur/Chanda-villagers-refuse-to-withdraw-Chipko-movement/articleshow/52544294.cms. Accessed 14 December 2019.

Banerjee, R. (2013) What Ails Panchayati Raj? *Economic and Political Weekly*, 48(30).

Census (2011) Gadchiroli Population, Maharashtra. Gadchiroli: Government of India.

CFR-LA (2016) *Promise and Performance: Ten Years of the Forest Rights Act in India. Citizens' Report on Promise and Performance of The Scheduled Tribes and Other Traditional Forest Dwellers (Recognition of Forest Rights) Act, 2006, After 10 Years of Its Enactment.* India: Community Forest Resource Rights Learning and Advocacy Process (CFR-LA).

—— (2017) *Promise and Performance: Ten Years of the Forest Rights Act in Maharashtra. Citizens' Report on Promise and Performance of the Scheduled Tribes and Other Traditional Forest Dwellers (Recognition of Forest Rights) Act, 2006.* India: Community Forest Resource Rights Learning and Advocacy Process Maharashtra.

Das, D. (2011) Mendha Lekha is First Village to Exercise Right to Harvest Bamboo. *Times of India*, 24 April.

Gandhi, M. (1962) *Village Swaraj*. Ahmedabad 380014: Navjivan Mudralaya.

GoM (2022) District at a glance. District Gadchiroli. https://gadchiroli.gov.in/district-at-a-glance/. Accessed 1 July 2023.

ISFR (2021) Forest Survey of India. Ministry of Environment Forest and Climate Change, Uttarakhand, India.

MoTA (2013) *Statistical Profile of Scheduled Tribes in India*. Government of India, Ministry of Tribal Affairs.

NewsClick (2018) Mining Operations in Gadchiroli Face Stiff Resistance from Villagers. 28 August. www.newsclick.in/mining-operations-gadchiroli-face-stiff-resistance-villagers. Accessed 14 December 2019.

Padel, F. (2014). The Niyamgiri Movement as a Landmark of Democratic Process. Vikalp Sangam, 24 July. http://vikalpsangam.org/article/the-niyamgiri-movement-as-a-landmark-of-democratic-process/. Accessed 14 December 2019.

Pathak Broome, N., Raut, M. (2017) Mining in Gadchiroli – Building a Castle of Injustices. *Countercurrents*, 17 June. www.countercurrents.org/2017/06/17/mining-in-gadchiroli-building-a-castle-of-injustices/. Accessed 14 December 2019.

Pathak Broome, N. (2018) Mendha-Lekha – Forest Rights and Self-Empowerment. In Lang, M., Konig, C., Regelmann, A. (eds) *Alternatives in a World of Crisis*. Brussels and Ecuador: Global Working Group Beyond Development. Rosa Luxemburg Stiftung, Brussels Office and Universidad Andina Simon Bolivar, Ecuador.

Pathak Broome, N., Bajpai, S., Shende, M., Raut, M. with Jamkatan, G.K., Katenge, I., Salame, Z., Halami, S., Deshmukh, S., Gogulwar, S. (2022) *The Forest Resource Rights, Gram Sabha Empowerment and Alternative Transformations in Korchi*. Pune, India:

Kalpavriksh, Pune with Maha Gramsabha Korchi and Amhi Amchya Arogyasathi, Kurkheda.

Pinjarkar, V. (2013). Forest Development Corporation of Maharashtra seeks 630 sq km New Forest Area for Operations. *Times of India*, 21 August. http://timesofindia.indiatimes.com/city/nagpur/Forest-Development-Corporation-of-Maharashtra-seeks-630-sq-km-new-forest-area-for-operations/articleshow/21945126.cms. Accessed 14 December 2019.

Singh, N., Kulkarni, S., Pathak Broome, N. (eds) (2018) *Ecologies of Hope and Transformation: Post-development Alternatives from India*. Pune, India: Kalpavriksh and SOPECOM.

Zografos, C. (2019) Direct Democracy. In Kothari, A., Salleh, A., Escobar, A., Demaria, F., Acosta, A. (eds) *Pluriverse: A Post-Development Dictionary*. Delhi, India: Tulika Books.

11

Transformative Strategies Forged on the Frontlines of Environmental Justice and Indigenous Land Defence Struggles in So-called Canada

Jen Gobby and Leah Temper

Introduction

Canada's natural resource-based economy, driven by extractive industries such as forestry, mining, and oil and gas, has been wreaking havoc on social and ecological systems across the country and throughout its history. The path of destruction has been met with strong resistance from communities and movements; their efforts have led to delaying, altering and halting projects as well as to legislative victories, new legal precedents, evolving relationships between settlers and Indigenous peoples in Canada, and innovation of transformative strategies. These resistance efforts have been led largely by Indigenous people, sometimes with support from allies and environmental groups. People have set up temporary blockades and long-term protest camps, launched international boycotts, manually shut down pipelines, taken industry to court and hit the street en masse to draw attention to injustice and to amplify the voices and demands of those resisting.

This chapter presents a quantitative and qualitative analysis of fifty-seven cases of environmental conflict on the EJAtlas Canada map. Through this analysis we identified six *transformative strategies* that have been forged on the frontlines of environmental justice and Indigenous land defence struggles in Canada. The strategies include the physical disruption of resource flows, international boycotts, assertion of Indigenous sovereignty, placing the alternatives in the pathways of unwanted development, and others. We discuss these six strategies in dialogue with the Conflict Transformation Framework (Rodriguez and Inturias 2018) – specifically the three forms of power described by the framework – to explore what has been transformed, and how, through these environmental conflicts.

Unsustainable extraction of natural resources is driving global environmental crises which threaten much of life on Earth, including human life (Field et al. 2014; Ripple et al. 2017). The local impacts of this extraction – as well as the impacts of the resulting environmental and climate crises – are being borne disproportionately by Indigenous, racialized and poor communities (Martinez-Alier et al. 2016; Parks and Roberts 2006; Taylor 2014; Waldron 2018). Governments around the world are failing again and again to make socially just and sustainable decisions about natural resources use, enabling the continuation of extractivism.

Amid this failure from above, communities on the frontlines of climate change and extractivism, supported by social movements, have been resisting unwanted mines, dams, oil and gas pipelines, and other projects. These resistance efforts have been influencing outcomes, powerfully transforming institutions and relations and developing place-based alternatives to extractivism (Salick and Byg 2007; Scheidel et al. 2020; Temper et al. 2018).

These dynamics can be clearly witnessed in Canada – where a natural resource-based economy, driven by extractive industries such as forestry, oil, gas and mining, has been wreaking havoc on social and ecological systems across the country and throughout Canada's history (Chodos 1973). This has been ongoing from as far back as the fur trade, right up to the current moment of tar sands expansion and pipeline development.

The path of destruction has been met with strong resistance from communities and movements; their efforts have led to delaying, altering and halting projects as well as to legislative victories, new legal precedents and evolving relationships between peoples, and between people and resources in Canada (Black et al. 2014; Gosine and Teelucksingh 2008).

These resistance efforts have been led largely by Indigenous people, sometimes with support from allies and environmental groups. People have set up temporary blockades and long-term protest camps, launched international boycotts, manually shut down pipelines, taken industry to court and taken to the street en masse to draw attention to injustice and to amplify the voices and demands of those resisting (EJAtlas 2020).

In this chapter we argue that while the governments in Canada have been failing to transition the energy and economic systems away from fossil fuels and extractivism in general, frontline communities and social movements are powerfully shaping outcomes in Canada through the development and practice of *transformative strategies*. Our analysis of fifty-seven cases of environmental conflict in Canada, accessed through the EJAtlas, provides concrete examples of these bottom-up transformative strategies being forged on the frontlines of extractivism. We present the results of our analysis in the form of six transformative strategies and explain how these strategies have been and are being deployed by communities and social movements in Canada to drive much more just and sustainable outcomes, while transforming systems and relations in the process.

Both of us come to the writing of this chapter as activists and as researchers. We are both settlers who are continually learning about the disastrous impacts that settler colonialism has brought to these lands, knowing that we still benefit from the extractive and exploitative status quo in deeply problematic ways. We are both involved in many collaborative projects with Indigenous and non-Indigenous environmental and climate justice activists in Canada. Leah Temper is currently an environmental justice campaigner with the Canadian Association of Physicians for the Environment. Jen Gobby has organized for years with Climate Justice Montreal, collaborates on research with Indigenous Climate Action and is currently Director of Research for Front Lines, a network that actively supports research led by Indigenous and other communities on the frontlines of the fight for environmental and climate justice across Canada. We are both continually navigating the dual roles of activist and researcher and working to find ways to use our research skills to serve and support grassroots environmental justice struggles. We are interested in practising research as a form of allyship and in employing the tools of research to contribute to the transformative power of radical movements.

We've conducted this analysis with the hope that by identifying transformative strategies that are having a powerful impact in this country, we can offer something useful to the movements of which we are a part. Although we conducted the analysis of the data ourselves, the data itself (cases on the EJAtlas) was co-produced in collaboration with other activists and organizers involved in the many struggles across Canada. We each led the creation of many different cases on the Canada map of the EJAtlas prior to this analysis. This process included preliminary online research to create a draft description of the case. We then contacted those directly involved with the resistance effort to seek their feedback, corrections and additions, to ensure that the story in the EJAtlas reflects their understandings of the people directly involved. In several cases, we added cases that we had some involvement with as activists and were able to draw on our own knowledge and experiences to help tell the story.

In the following section we present our theoretical framework, explaining our understanding of *transformation, transformative power and transformative strategies*. In the third section we offer an overview of the methods used in identifying and analysing the transformative strategies being used by communities and movements in Canada to shape outcomes and transform systems and relations. The fourth section offers the results of the analysis in the form of six transformative strategies, bringing them into dialogue with the three forms of transformative power. The final section brings findings together to discuss and propose the implications of theorizing and practising *strategies as forms of transformative power*.

Theoretical Framework

There is increasing acknowledgement from both academic and activist communities that in order to address the climate, environmental and social crises, there needs to be a profound transformation in the economic, political and thought systems driving them (Beddoe et al. 2009, IPCC 2018, Moore et al. 2014), a 'fundamental restructuring of the way modern societies operate' (Scheidel et al. 2017: 11). We define transformative change as intentional change that confronts the root causes of social injustice and environmental unsustainability, including unequal power relations, and alters the overall composition and behaviour of the system in ways that drive desirable change across temporal and spatial scales, towards increased social well-being, equality and ecological sustainability (Gobby 2020; Temper et al. 2018).

Social movements are one of the 'social forms through which collectives give voice to their grievances and concerns about the rights, welfare, and wellbeing of themselves and others' (Snow et al. 2008: 3). Social movements – ordinary people coming together, engaging in collective action to push for change – are crucial for bringing about transformative change (Carroll and Sarker 2016; Choudry 2015, Kothari et al. 2014; Scheidel et al. 2017; Solnit 2016; Temper and Del Bene 2016). Though there are other ways by which social change is driven, such as through legislation and court proceedings, educational systems and electoral outcomes, social movements provide regular people a means by which to combine forces to influence change without needing to hold certain specialized or elite roles in society (Glasberg and Shannon 2010).

McBay (2019) further distinguishes between social movements and resistance movements, examining how and when resistance movements can become effective. According to him, a resistance movement is a 'social movement that contends existing power structures in society are unjust and takes action to disrupt and dismantle those power structures'. McBay argues that transformative change happens when those in power have no choice but to change their actions due to the political and economic force that those resisting have exerted upon them through the effective deployment of the right tactics and strategies.

Our interest here is to identify the tactics and strategies resistance movements against extractivism in Canada are deploying that are effectively contributing to transformative change. We consider that they do this through the use of strategies to contest and transform power relations. As explained in Chapter 2, this may include institutional power, relational power and discursive power. The Conflict Transformation Framework helps us with the knotty task of identifying when transformation is occurring and in which of these spheres, and what gets transformed as a result.

Institutional power

Institutional power is transformed when existing legal and economic frameworks are made to acknowledge new rights and forms of difference, leading to new forms of production and social relations that are more equitable and just. This can happen through lobbying, engaging in plebiscites, intervening in public consultations for environmental assessments and other decision-making.

Institutional power can also be transformed through enacting alternative institutions outside the formal existing systems, through autonomous governments and forms of territorial control; and through recapturing democratic processes, such as holding local referenda to decide on territorial governance questions, the holding of assemblies, and the creation of local management plans.

Resistance movements can work to transform institutional power by engaging with existing economic and political institutions in order to influence more just and sustainable decision-making. Or they can build institutional counter-power by innovating and invigorating alternative, autonomous institutions.

We see this dual form of institutional power in Indigenous resistance and resurgence. Taiaiake Alfred, whose early work on resurgence influenced the current generation of Indigenous scholars, wrote: 'Many of my own generation of scholars and activists hold on to ways of thinking and acting that are wrapped up in old theories of revolution. Those theories centre on convincing the settler society to change their ways and restructure their society through the use of persuasion or force.' He asks: 'What if settlers never choose to change their ways?' (2008: 11). Resurgence 'refocuses our work from trying to transform the colonial outside into a flourishment of the Indigenous inside' (Simpson 2011: 17). Drawing on anti-colonial political philosopher Frantz Fanon, who urged those struggling against colonialism to turn away from the colonial state and 'find in their own decolonial praxis the source of their liberation', Glen Coulthard discerns and advocates an Indigenous resurgence directed not towards recognition by the state but 'toward our own on-the-ground struggles of freedom' (2014: 48). Rather than pathways for change that are contingent on changes in settler society, resurgence calls for a 'turning inward to focus on resurgence of an authentic Indigenous existence and recapturing the physical, political, and psychic spaces of freedom'.

Relational power

Relational power is built when dialogue, deliberation and collaboration between actors is increased (Rodríguez and Inturias 2018). Intersectional feminist scholars argue that power is inherently relational; that 'power is better conceptualized as a relationship than as a static entity ... Power constitutes a relationship' (Collins and Bilge 2016: 28). Foucault also conceived of power in relational terms as well, moving through networks, flowing or shared between institutions or people (1971). Rowlands (1997) provides another framework for understanding different forms of power: power within, power with, power to and power over. Where power over is the power of domination and hierarchy, power with – the power of collective action, of organizing together, of acting in solidarity with others – is how people combine individual power such that they begin to grow their shared capacity to influence social outcomes (power to) and to resist and oppose power over.

Building people power through relations and collaborations is how we build a counter-force to the hegemony of those who are working hard to maintain the destructive and unjust status quo. Building a counter-hegemonic force requires overcoming the divisions and fragmentations of the left, created by the divide and rule politics of the elite, and forging a strong alliance politics of the left (Harvey 2014). Unity across struggles can be forged, forming coalitions that make possible 'a coherent, counter-hegemonic alternative to the dominant order' (Epstein 1990: 51).

Relational power is forged through alliances *across difference*. Dominika and Piotrowski (2015) argue that cooperation across difference is more transformative than among so-called 'natural allies' that are easy to align with. They suggest that 'productive frictions' between actors and the resultant conflicts and need for negotiation leads to transformations of those involved. It is also important to note that as relations of domination – sexism, racism, classism – play out in movement spaces, this weakens the relations power of these movements to transform systems (Gobby 2020). As such, part of building relational power is building strong, just relations based on equality and reciprocity within and across movements.

Discursive power

Transformative moments can also occur through symbolic actions, which upset and challenge dominant cultural codes. By offering new interpretive frames, or ways of comprehending the world, the way we understand knowledge and power can be transformed. This can be achieved through popular education, artistic activism, media campaigns and the development of frames which contest dominant meanings.

Building popular support for resistance and for transformative alternatives, bringing new people to the movements and forging links between different movements all require compelling narratives – narratives that help people see what is at stake and what else is possible, that help people see the structural links between climate change, racism, colonialism, violence against women and other forms of injustice. We need the kind of narratives that inspire people to see how these diverse social and environmental justice struggles are inextricable, that make clear that we need to fight together.

The resurgence of diverse Indigenous cultures, languages, economies, governance systems and land-based practices can all be considered forms of cultural power transformation. As Corntassel (2012: 88) writes:

> Being Indigenous today means struggling to reclaim and regenerate one's relational, place-based existence by challenging the ongoing, destructive forces of colonization. Whether through ceremony or through other ways that Indigenous peoples (re)connect to the natural world, processes of resurgence are often contentious and reflect the spiritual, cultural, economic, social and political scope of the struggle.

Transforming systems towards justice, equity and sustainability requires the transformation of power – institutional, relational and discursive power. That said, it's very important to acknowledge that transformative narratives and visions need to be supported materially. It's not enough to offer visions and narratives of a better world; we need to be 'changing on the ground practicalities' (Gobby 2020). This brings us to the final form of power – *physical power.*

Physical power

Understanding resistance to extractivism in Canada, as elsewhere, requires attention to physical power as well. We build on Rodríguez and Inturias's (2018) three forms of power by adding this fourth. This is the realm of physical resistance to extractivism, the use of physical means to stopping unwanted resource extraction, of refusing entry to territories, of direct action and shutting down infrastructure, of putting bodies between industry and the resources they seek to exploit. Physical power can be exerted in resistance efforts, as in blockades and land occupations, and it can also be exerted in the form of alternatives to extractivism, as in permaculture gardens, social housing, or community-owned and operative renewable energy installations. All this generates a counter-force to ongoing state power exerted physically through force and direct violence.

In this chapter, we engage with this theoretical framework to explore the strategies being used by communities and movements to resist extractivism in Canada, framing these strategies as forms of transformative power. We explore what is being transformed, through which forms of power and the deployment of these strategies. We offer this analysis in an effort to contribute to our collective understanding of what is working, where and why, to stop unwanted extractive projects while forging more just and sustainable systems and relations.

Methods Used

The data

We conducted qualitative and quantitative analysis of the fifty-seven cases of environmental conflict that make up the Canada map on the EJAtlas, an online database and interactive map (Figure 11.0). Each case on the EJAtlas includes the following kinds of data: general characteristics (location, relevant background information, type of project or commodity being contested); project details; companies, finance institutions and government actors involved; social and environmental impacts; actors mobilizing; forms of mobilizations used; conflict outcomes; and references to relevant legislation, academic research, videos and other media (Scheidel et al. 2020).

The database aims to develop a system whereby environmental conflicts can be described, analysed, compared and interpreted, where quantitative data from the activity at the source of the discontent can be gathered, and where patterns of mobilization and the rates of success in stopping extractive projects or introducing new regulations can be discerned and productive lessons can be learned (Temper et al. 2015).

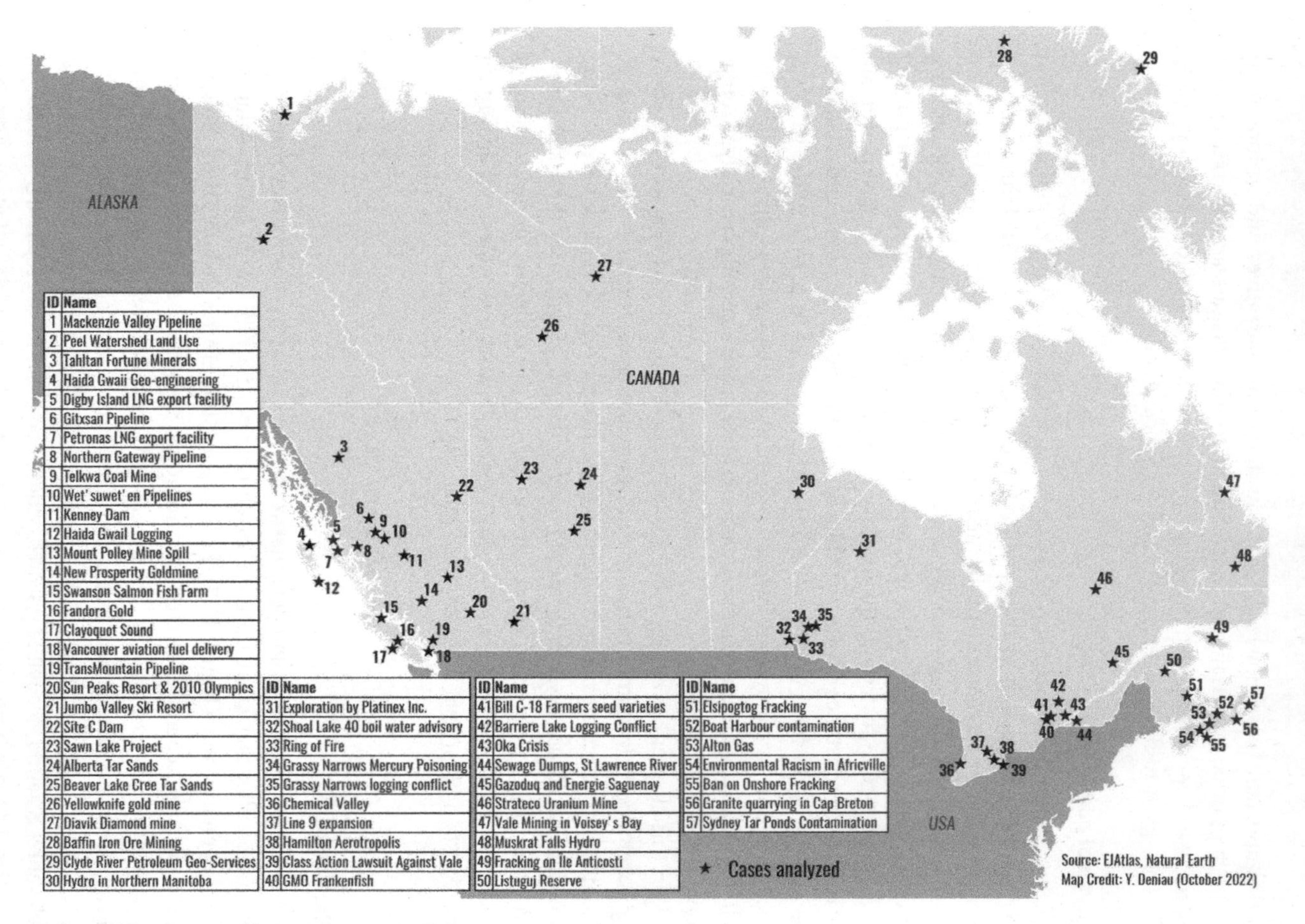

ID	Name
1	Mackenzie Valley Pipeline
2	Peel Watershed Land Use
3	Tahltan Fortune Minerals
4	Haida Gwaii Geo-engineering
5	Digby Island LNG export facility
6	Gitxsan Pipeline
7	Petronas LNG export facility
8	Northern Gateway Pipeline
9	Telkwa Coal Mine
10	Wet' suwet' en Pipelines
11	Kenney Dam
12	Haida Gwaii Logging
13	Mount Polley Mine Spill
14	New Prosperity Goldmine
15	Swanson Salmon Fish Farm
16	Fandora Gold
17	Clayoquot Sound
18	Vancouver aviation fuel delivery
19	TransMountain Pipeline
20	Sun Peaks Resort & 2010 Olympics
21	Jumbo Valley Ski Resort
22	Site C Dam
23	Sawn Lake Project
24	Alberta Tar Sands
25	Beaver Lake Cree Tar Sands
26	Yellowknife gold mine
27	Diavik Diamond mine
28	Baffin Iron Ore Mining
29	Clyde River Petroleum Geo-Services
30	Hydro in Northern Manitoba
31	Exploration by Platinex Inc.
32	Shoal Lake 40 boil water advisory
33	Ring of Fire
34	Grassy Narrows Mercury Poisoning
35	Grassy Narrows logging conflict
36	Chemical Valley
37	Line 9 expansion
38	Hamilton Aerotropolis
39	Class Action Lawsuit Against Vale
40	GMO Frankenfish
41	Bill C-18 Farmers seed varieties
42	Barriere Lake Logging Conflict
43	Oka Crisis
44	Sewage Dumps, St Lawrence River
45	Gazoduq and Energie Saguenay
46	Strateco Uranium Mine
47	Vale Mining in Voisey' s Bay
48	Muskrat Falls Hydro
49	Fracking on Ile Anticosti
50	Listuguj Reserve
51	Elsipogtog Fracking
52	Boat Harbour contamination
53	Alton Gas
54	Environmental Racism in Africville
55	Ban on Onshore Fracking
56	Granite quarrying in Cap Breton
57	Sydney Tar Ponds Contamination

Figure 11.0 Canada Map and case studies

Data analysis

The fifty-seven cases on the Canada map of the EJAtlas were analysed in three stages. In the first stage, we identified cases that were particularly successful in achieving movement goals. This was determined through an amalgamation of several variables into one code, which served as an indicator of 'success'. The variables used to construct this dummy variable included the following kinds of conflict outcomes: project stopped, project temporarily suspended, court decision (victory for environmental justice), new legislation, application of existing regulations, moratorium, withdrawal of company investment, institutional change, and the level of success attributed to the case by the person who entered it into the EJAtlas. Each variable counted as 1, and with this each case was attributed a score of 0–11. Scores of 3 or above were categorized as successful (n=18), whereas scores below 3 were considered as not successful (n=39).

We conducted an exploratory statistical analysis using Fisher's exact test at 5% significance level, to find associations between outcome (success/not success) and 1) different forms of resistance (e.g. protests, blockades, boycotts, lawsuits, etc.) and 2) different actors involved (e.g. Indigenous groups, environmental organizations, farmers, etc.). We also explored the association between outcome and the median number of strategies used in any given case (Wilcoxon rank-sum test at 5% significance).

In order to further investigate which resistance strategies are correlated with successful outcomes we computed, for each form of resistance, the number of successful cases involving that form and compared that to the number of unsuccessful cases to have employed that form.

We then conducted a qualitative analysis of the *case descriptions texts* for each of the eighteen 'successful' cases to better understand the cases and how the resistance strategies played out, what worked and what didn't, to identify situations and conditions under which a given strategy is most effective and most transformative. Put together, our analysis methods were employed to identify and understand how different strategies are 1) impacting decision-making outcomes about extractive projects, and 2) helping transform Canada's social, economic and political systems. Through these three stages of qualitative and quantitative analysis, six transformative strategies were identified as central to the more successful cases. Each of these are presented in turn in the following section.

Findings

Describing the fifty-seven cases of environmental conflict on the Canada map

The EJAtlas's map of Canada had fifty-seven cases of environmental conflict at the time of analysis, spanning temporally from 1919 to the present, all involving some form of resistance. Most of the conflict in Canada has been over fossil fuels or mining, with many cases also triggered by extraction of biomass such as timber and fish, and 75% of the conflicts taking place in rural areas. In terms of geographical distribution, conflicts have been spread across the country, though there is a high concentration of conflicts in British Columbia (BC) and the other coastal areas. In 89% of cases, Indigenous people were involved with or led the resistance. In most cases resistance was mobilized in order to *prevent*, though in some cases, resistance was organized *in response to* project implementation or to *seek reparations after* the impacts of the project were already being experienced. In 21% of cases the project was successfully stopped. In 17% of cases the conflict resulted in new legislation, and an equal percentage resulted in the application of existing regulations. In 15 cases (26%), new environmental assessment processes were triggered and five cases (9%) resulted in moratoriums. This overview of the data suggests that resistance is indeed affecting decision-making outcomes around natural resource extraction in Canada, and it supports our contention that understanding and transforming systems requires the theorization and practise of *strategies as forms of transformative power*. We now turn to the kinds of strategies that are shaping outcomes and transforming systems.

Strategy 1 – Physical disruption of resource flows: occupations and blockades

Our analysis found a significant positive association between successful outcomes and the use of blockades ($p=0.048$), such that 61% of successful cases involved blockades whereas only 35% of unsuccessful ones did. We also found a trend towards significance between successful outcomes and land occupations ($p=0.067$). Where 50% of successful cases involved land occupation, only 28% of unsuccessful ones did. Half of the cases where Indigenous peoples were involved employed blockades and land occupation as a strategy, most of these involving disputes over unceded territories.

Blockades and land occupations were extensively employed in one of the most successful cases of resistance to extractives in recent history in Canada. Beginning in the 1970s, the Haida people and their supporters contested industrial logging on their island territories. This sustained resistance successfully stopped logging operations and led to victorious court decisions, the application of existing regulations, the withdrawal of financial investment, institutional changes and a localized moratorium on industrial logging. Haida resistance has protected old-growth forests, helped bring about greater acknowledgement of treaties and other Indigenous rights, and helped innovate new forms of co-management. The conflict led to the creation of the 1,495 square kilometre Gwaii Haanas National Park Reserve and Haida Heritage Site. Beyond just stopping an unwanted project, this Haida resistance transformed governance and government–First Nations relations in Canada. Indeed, 'the 1985 protests marked the end of one epoch and the beginning of another, an end of the time when governments could ignore First Nations' (Pynn 2010: n.p.).

As a more recent example, the Unist'ot'en resistance camp was set up in BC by members of the Wet'suwet'en Nation re-occupying their traditional territory to block the construction of multiple oil and gas pipelines over the course of almost a decade. These proposed pipelines are part of an 'energy corridor' aiming to connect the Alberta tar sands and shale gas extraction projects with ports and LNG-processing terminals in Kitimat and Prince Rupert on BC's west coast. The camp has contributed to companies withdrawing investment and multiple pipeline projects being diverted or cancelled. In January 2020 the BC supreme court approved an injunction to remove the blockade to make way for the construction of the Coastal Gas Link pipeline, and the RCMP (Royal Canadian Mounted Police) were sent in to violently remove the land defenders; solidarity blockades by other First Nations located in key arteries across the country were erected. The rail shutdown, under the banner of #shutdowncanada, stranded hundreds of millions of dollars in goods and stalled some cross-border trade. These solidarity blockades provoked a political crisis, placing tremendous pressure on the federal and provincial governments to negotiate with the Wet'suwet'en. It also served to demonstrate the power of grass-roots resistance efforts to impact on economic and resource flows, and the power of multiple nations working in concert, and it provoked an unprecedented nationwide discussion about Indigenous governance and the band council system as well as natural resource extraction and its costs.

With the blockades and occupations and other such strategies used to physically disrupt resource flows, we see the exertion of *physical power*. These are communities of people using built structures, their lands and their bodies to physically hinder the ability of extractive companies to access sites, to extract, remove and/or process resources.

This physical power exerted through blockades and land occupation serves several purposes. On one hand they serve to interrupt the contentious economic activity that is often being forced through territories without consent. However, the transformative impact of blockades is largely due to the ways they create a space for the control and practise of Indigenous economic and political authority in the face of the cultural and economic dislocation forced upon them. Within these spaces, they may enact their aspirations and identity and assert their sovereignty over their lands.

According to our analysis, blockades and land occupations are most potent in affecting outcomes and most transformative to economic and political systems 1) when they are led by Indigenous people, bringing to the struggle their unique and powerful set of rights, responsibilities, worldviews and connections with their lands and waters, 2) when they are disruptive of resources that are highly valued and are centrally located, such that the blockade means business cannot continue as usual, and 3) when they are employed in concert with other strategies such as lawsuits, media campaigns and coalition-building.

Strategy 2 – Boycotts and other financial pressure

Our analysis found a significant positive association between successful conflict outcomes and the use of shareholder and financial activism ($p=0.024$), such that 27% of successful cases involved financial activism whereas only 0.5% of unsuccessful ones did. Though our analysis did not find a significant association between successful outcomes and boycotts, it is interesting to note that 11% of successful cases involved boycotts of company's products, while only 2.5% of unsuccessful ones did.

Where blockades and land occupations *physically* hinder the flows of resources and capital, boycotts and other forms of financial activism can serve to block such flows in economic and indirect ways. Boycotts are targeted at either state or corporate actors and aim to draw attention to practices and products that violate environmental, human and Indigenous rights.

One of the most emblematic cases of effective boycotts in Canada was in the Clayoquot Sound protests in the 1980s and 1990s against clear-cut logging on Vancouver Island, BC in the unceded territory of the Nuu-chah-nulth First Nations. The protests culminated in 1993, in what was the largest act of civil disobedience in Canadian history, when over 900 people were arrested (Grant 2010).

The resistance against clear-cut logging involved a large-scale international boycott of timber products – from lumber to toilet paper to books. During the extended conflict, activists switched strategies from trying to influence the BC government to putting direct economic pressure on forest companies, notably MacMillan Bloedel (MB) through retail boycotts of the MB's industrial customers for pulp. They did this through internationalizing the struggle, working with activists in Europe and the US. Throughout the 1990s, various companies divested their holdings in the area, cancelling millions of dollars' worth of contracts for Clayoquot Sound wood products (Clapperton 2019).

A significant outcome of this campaign was the tarnishing of the BC forestry sector's image in Europe, with Canada's ambassador bemoaning that up to $3 billion worth of annual forest-product sales to Europe was threatened as a result (Stanbury and Vertinsky 1997).

That said, MB quickly found new suppliers at a higher cost, so the actual economic damage inflicted was minimal (Stanbury and Vertinsky 1997). While the boycott did not lead to material impacts on MB, it managed to spur BC to implement its environmental agenda, including the Forest Practices Code and a suite of environmental regulations of the industry. These provisions led to reduced availability of crown timber and higher harvesting costs. Average logging costs increased from $62 per cubic metre in 1992 to $106 in 1996 (Stanbury and Vertinsky 1997). Furthermore, it was due to the Clayoquot protests that forestry companies became more willing to cooperate with the environmental movement (Tindall 2013).

The protests of 1993 also prompted a transformation in public perception about clear-cut logging (Vanchieri 2011) and contributed to the stigmatization of forestry (Tindall 2013). While clear-cut logging lost broad social acceptability, a significant area was protected from logging, eco-based planning was implemented and Indigenous control over resources and decision-making was increased. In the end, government and industry bowed to public pressure to change forest management standards and limit clear-cuts (Clapperton 2019).

Although there are cases in which boycotts were less successful, under certain conditions boycotts can generate economic pressure that helps strengthen resistance efforts. They tend to be most effective when product substitutes are easily available, and when boycotts are backed up by blockades, protests and strong media campaigns as well as the support of big ENGOs with substantial budgets for said media campaigns. It also seems to be important that the site of extraction is broadly valued recreationally or aesthetically by the public to generate broad support.

With this strategy, of launching boycotts and the mounting of other financial pressure tactics, we see the exertion of *institutional power* which aims at impacting on legal, economic and political frameworks and institutions. To counter the disproportional power that extractive industry tends to have on government decisions, communities and movements deploy economic pressure tactics to amplify their resistance to projects and to drive more just and sustainable outcomes. Institutional power can be exerted in an outwards way whereby those resisting engage in existing institutions and decision-making structures in order to have influence. In the case of boycotts, they are engaging in the financial markets, creating financial pressure that in turn led to transforming public opinion about clear-cut forestry as well as forestry practices and regulation.

However, given that many existing institutions are inherently predisposed to maintaining the unjust and unsustainable status quo, communities and movements may opt instead (or as well) to exert institutional power by creating new institutions, enacting alternative forms decision-making and resource management and enforcing their own policies and decisions. It is these forms of institutional power we turn to next.

Strategy 3 – Enacting Indigenous sovereignty, law and governance (institutional power)

Though it was not statistically significant, it is interesting to note that 100% of successful cases involve Indigenous peoples, whereas only 82% of less successful cases do. And many of the most successful cases of resistance led by Indigenous people involved Indigenous communities defending their lands and waters by evoking their rights and enacting their own legal systems, forms of governance and sovereignty.

The Unist'ot'en camp is a prime example of this. They govern themselves by a 'system of natural Laws that are based on their Indigenous Laws or Responsibilities, which they refer to as Anuk Nu'at'en (Wet'suwet'en law)', which includes – as their website states – 'Responsibilities to ourselves, our families, and the lands and waters we have relationships with'. Espousing their message 'Heal the People, Heal the Land', their Healing Lodge offers experiences of healing to their community members 'to live on the land and have a connection with the natural world and our teachings … It is a chance to return to some of our traditional teachings and land-based wellness practices of our ancestors' (Unist'ot'en Camp n.d.).

Another important example is the Listuguj First Nation conflict over fishing and logging rights, which led to the successful assertion of Mi'gmaq governance over fisheries on the Restigouche River. Listuguj First Nation successfully defended their rights to fish and manage fisheries according to their own laws and governance system. They took over the management of the salmon fishery without being under a contract with provincial or federal authorities; in fact, the province of Quebec opposed them. Nor did they do it by asking permission or receiving a request from some other government. Nor did they do it by force. They did it by passing, implementing and enforcing a law (National Centre for First Nations Governance 2010). 'Ultimately, the Listuguj Mi'gmaq fought back with the tools of governance: by making credible law – Mi'gmaq law – and then backing it up with competent management and enforcement' (National Centre for First Nations Governance 2011).

In 2013, during the Tahltan Nation v. Fortune Minerals conflict over mining, the Tahltan occupied a mining camp near the community of Iskut in the Skeena Mountains and issued an eviction notice to the mining company. In March 2019, Mi'gmaq matriarchs served an eviction notice to Alton Gas, enacting Mi'gmaq sovereignty in the face of state and corporate violation of their rights. These evictions upend the colonial assumption of jurisdiction and control over resources and lands and are powerful acts of reclamation of Indigenous self-determination and governance over territories.

Here *institutional power* is being exercised outside the formal institutions of the Canadian state while having an impact on those decisions being made through those formal systems. Through these enactments of Indigenous sovereignty and the raising up of Indigenous governance systems, more just and sustainable alternatives are being lived, and the decision-making authority and legitimacy of the colonial state is being brought into question and weakened, thus transforming political and economic systems in so-called Canada.

Strategy 4 – Winning the battle of ideas: media, communications and new imaginaries

As has been mentioned in relation to several strategies above, resistance efforts can be rendered more effective when supported by strong media campaigns to garner wide support. Though we didn't find evidence of statistical significance, it's worth noting that 66% of successful cases employed media-based activism, whereas only 51% of less successful cases did. And 72% of successful cases involved a public campaign, while only 58% of unsuccessful cases did. The effect of media- and communications-based strategies are evidenced in the Canada map of the EJAtlas both in the presence of media and public campaigns in the more successful cases, and in the marked lack of coverage of those conflicts that had less successful outcomes.

One of the widest-spread and longest-standing environmental conflicts in Canada is also one of the least well known. Massive historical and ongoing hydro development in Northern Manitoba has led to ecological and social devastation in Cree territories. Benefits are accruing to southern Manitobans while the costs are being born by the Cree communities in the north. Although the communities' opposition efforts were sufficiently united and gained enough public support within the province to force a modern treaty on Manitoba Hydro, the terms of agreement were systematically violated by the province over the next decade. This was facilitated by the low national profile of these projects (kulchyski 2012). The Cree have been enormously creative in their political resistance, developing their own governance system and generally making life difficult for the utility by trying to force it to live up to its promises. 'Whether they manage to make any gains will depend in part on their story getting a wider hearing in Canada and internationally than it has so far' (kulchyski 2012: n.p.). Unlike the rallying around tar sands and oil and gas resistance, hydro conflicts have not been taken up by the environmental movements in Canada and massive social and environmental injustice ensue away from the public gaze.

The 1990 Siege at Kanehsatake – commonly referred to as the Oka Crisis – was an eleven-week armed stand-off between Mohawk Warriors, Quebec provincial police and the Canadian forces on Mohawk territories over the expansion of a golf course. The siege sparked Indigenous-led blockades on railways, highways and bridges across the country in solidarity. This was the first well publicized conflict between First Nations and the Canadian government in the late twentieth century (kulchyski 2003: n.p.). In the end the golf course was not expanded. The siege at Oka captured international media attention and galvanized a remarkable wave of civil dissent in Indigenous country across Canada. This conflict made headlines across Canada for months and through this coverage, 'many of Canada's open wounds were exposed to the world: racism towards aboriginal people at Mercier bridge; failure of institutional vehicles like land claims process to deal with legitimate aboriginal land rights; and an emergent new militancy in Indian country were all made more apparent' (Kulchyski 2003: n.p.).

This high-profile and high-intensity conflict generated huge political pressure which led directly to the creation of a Royal Commission on Aboriginal Peoples, a major institutional process that dominated the early half of the 1990s and resulted in a massive report recommending a drastic change in direction for aboriginal policy (kulchyski 2003). That said, the land claim has still not been resolved and unwanted development is still going on to this day in this contested territory.

In some cases, such as that of the Unist'ot'en, a long-term and widespread media campaign was extraordinarily helpful in garnering public support to help drive successful and transformative outcomes, as well as ensuring that the world was watching and responding when the camp was violently invaded by the police. Other cases, such as conflicts over hydro development in Manitoba, show that some conflicts, regardless of sustained and creative efforts to garner media attention, remain out of the national and international public eye, rendering support for the resistance limited. In yet other cases, such as the Siege at Oka, the intensity of the conflict generates such attention that no intentional media campaign is required. In all cases, media and communications appear critical for shaping the outcomes of resistance efforts.

These cases emphasize the importance of *discursive power* as a crucial element of transformative strategies. To shape outcomes and transform institutions and relations, the work of unmasking unjust systems while fostering new meanings, discourses, narratives, norms, values and worldviews is critical for both delegitimizing the current systems and informing alternative ones. Discursive power, exerted through media campaigns and other strategies, is also critical for raising awareness about ongoing environmental injustice and garnering widespread support for those resisting extractivism.

As new understandings, values and norms are shared and taken up over time, 'systemic changes in cultural power can take place' (Rodríguez and Inturias 2018: 23). Cultural power then can have influence on decision-making outcomes, leading to outcomes based on local communities' needs, interests and worldviews and grounded in their own conceptions of the environment, the land and development (Rodríguez and Inturias 2018).

Strategy 5 – Transformative alliances: building support across cultures, sectors, movements and regions

Our analysis found that the greater the diversity of actors involved in a conflict, the greater the likelihood of successful outcomes. For successful cases, the median number of kinds of actors involved was seven, and only five in unsuccessful cases. This association trends towards significance ($p=0.083$). We found a statistically significant association between successful outcomes and the involvement of women ($p=0.042$) as well as with the involvement of national and international NGOs ($p=0.022$). The involvement of Indigenous groups or traditional communities ($p=0.058$), social movements ($p=0.088$) and fisher people ($p=0.090$) trend towards significance. Another finding of interest is that among the unsuccessful cases, only 30% involved the development of network/collective action as a mobilizing strategy, versus 78% of successful cases. We interpret this set of findings to emphasize the importance of developing broad alliances of involvement, in engaging with many kinds of communities and groups.

In both the Oka and Unist'ot'en cases, when the local community was faced with violent police and military incursion, solidarity blockades emerged rapidly across the country. These relations of solidarity are critical and are based on the slow and long work of relationship-building. This kind of relationship-building between Indigenous nations across Canada has been strengthening and new alliances have been forming in the context of resistance struggles. For example, the leadership at Unist'ot'en helped forge the Sacred Fire Network to coordinate and share information and resources between many Indigenous frontlines in BC, and the fight against the Energy East pipeline led to the creation of the Treaty Alliance Against Tar Sands Expansion. Similarly, the fight against the Northern Gateway pipeline was successful in no small part because of the Yinka Dene Alliance which constituted an unprecedented coming together of many different First Nations from across BC.

Industry and governments know that the power of individual First Nations is strengthened when they work in alliance, and they strategically use divide and conquer tactics with Indigenous communities, to ensure that their power remains manageable (Aorta Coop 2017; Russell 2018). Several cases in the EJAtlas made this clear. The Ring of Fire is a 5,000 square kilometre massive chromite deposit in the James Bay Lowlands in Ontario on Ojibwe and Cree territory. In this case, the Ontario government has been using divide and conquer tactics in attempts to weaken opposition by First Nations. In the conflict over hydro development in Manitoba, the public utility has been dealing with First Nations separately, intentionally weakening opposition. 'Among the reasons for Hydro's continued colonial success is that it now deals with communities one at a time, so opposition is fragmented' (Kulchyski 2012: n.p.). Resisting such divisive tactics by industry and government is crucial to building transformative relational power.

Another form of strategic alliance forged in environmental conflicts in Canada is between Indigenous communities and environmental groups. While alliances between environmentalists and Indigenous people have been powerful and important in achieving important wins over the last fifty years, they have also been fraught due to conflicting motivations and goals, disagreements over strategy and unequal access to power and resources (Gobby 2020). There has been an ongoing problem of environmental groups supporting Indigenous rights when those rights are useful for the environmental cause and not actively supporting Indigenous rights when they are not. In the Clayoquot Sound conflict, 'environmentalists fought for an end to this logging practice, [and] much of their campaign hinged on recognition of the local Nuu-chah-nulth First Nations' Aboriginal rights to their traditional territories' but 'environmentalist support for First Nations was actually ambivalent and sought to erase Indigenous peoples' presence from the land (Clapperton 2019: 182).

These dynamics have been slowly improving and settlers are learning better ways to be in alliance (Gobby 2020). In the recent fight against the Transmountain pipeline, environmental groups made more concerted attempts to put Indigenous rights at the centre of their campaigns and to play a supportive role, leaving the strategic planning and messaging to the Indigenous people of the territories.

This transformative strategy – of building broad alliances across difference – is the forging of *relational power*. These alliances are exceedingly important. Canadian journalist Naomi Klein has pointed out that there is a long legacy in Canada of movements working in silos on 'separate issues' and failing to see the crucial overlap in their visions. She contends that moving past this siloed approach and creating a 'movement of movements' is necessary for creating a strong enough force to shift the trajectory of the Canadian energy and economic systems (2014).

Others have also argued that, in Canada, social movements' efforts have been fragmented, often working at cross purposes and unwilling to collaborate, which renders them unable to build a 'counter-hegemonic' political force (Findlay 2002). Tensions over conflicting end goals and theories of change, as well as over preferred tactics, are driving the fragmentation. But ongoing racism, sexism, classism and other forms of domination playing out within and across movements is also blunting the transformative power of these movements (Gobby 2020).

These tensions 'exist across the broader left, where sectarianism has been a disastrous and weakening force' (Dixon 2014: 233). They hinder the creation of widespread and strong collaboration, and inhibit our ability to envision strategies and futures together. As Naomi Klein put it: 'The intellectual fencing has constrained the progressive imagination for so long it's lying twisted on the ground' (2017: 263). In these and other social movements, deep ideological and other rifts are 'rendering much of our work useless' (Brown 2017: 62).

As such, the alliances and collaborations that have been forged in the cases outlined in this chapter hold much promise for cultivating the kind of relational power required to transform Canada.

Strategy 6 – Multi-pronged approaches: building transformative power by combining different strategies

Our analysis found that successful cases of resistance tended to employ a wider range of strategies. The median number employed in successful cases is nine different strategies, whereas the median number employed in less successful cases is six. We found these differences of medians statistically significant, at 5% significance level according to a Wilcoxon rank-sum test.

These results suggest that using multiple strategies concurrently can lead to successful and transformative outcomes. For example, the successful fight against the Northern Gateway oil pipeline in British Columbia had a strong three-pronged strategy – legal, direct action and media. The pipeline was cancelled in 2016. In the case of Haida Gwaii as well, a multi-pronged strategy was successful:

> From at least the mid-1970's, the Haida pursued a number of strategies, from participation in cooperative management processes to illegal roadblocks, in order to reclaim control over the forests and waters of Haida Gwaii ... By pursuing both the official processes of consultation and government planning, as well as direct civil disobedience the Haida were able [to] assert their interests in determining the structure of forest policy on Haida Gwaii. (Dean 2009: 46)

We also found a significant association at 5% between successful outcomes and cases that employed lawsuits, court cases and/or judicial activism (p=0.013) along with other strategies such as blockades, media activism and protests. Fighting for Indigenous rights and against destructive projects is happening more and more in the courts of law. Legal battles have proven to be powerful when they win, but they are extremely expensive and risky strategies for transformation. For example, Sydney Tar Ponds in Nova Scotia is North America's largest toxic waste site, a result of toxic runoff from steel industry coke ovens. In 2018, Canada's Supreme Court rejected a class action lawsuit on behalf of Nova Scotians suffering from negative health impacts linked to exposure to the ponds. The claimants were hit with almost CA$1 million in legal fees. This case has potential to now serve as precedence for the future rejection of environmental class action cases in Nova Scotia. This example serves as a warning that although multi-pronged strategies involving lawsuits can be powerful, they are risky.

Another form of combining different strategies has been emerging over the last few years in Indigenous communities; they are building solutions in the pathway of the problem. From the Healing Lodge and permaculture gardens at the Unist'ot'en camp and the Treaty Truck House at the Mi'gmaq protest camp against Alton Gas in Nova Scotia, to the Tiny House Warriors fighting the Kinder Morgan pipeline in Secwepemc territory and the Watch House on Burnaby Mountain, Indigenous people are building low-carbon, beautiful, culturally grounded alternatives and placing these alternatives strategically to block the way of unwanted oil and gas projects being pushed into their territories. These alternatives are offering inspiration by making clear that there are other ways to build economies. At the same time, they are enacting Indigenous sovereignty and lifeways. This new strategy of placing the alternatives in the pathway of the problem, or direct-action solutions (Gobby 2020), has been changing what environmental justice organizing and protest looks like in Canada.

In these innovative combinations of different strategies, we see a different form of *relational power* contributing to transformative change. In 'Strategy 5 Transformative alliances' we discussed examples of working across difference (different movements, different sectors, different communities) to build relational power. Here we see relational power being built through working across different strategies. Where movement efforts can be rendered more transformative by breaking out of single-issue silos and forging unlikely alliances, perhaps they are also made more transformative by thinking and organizing across different strategies, and combining them in novel and effective ways.

As promising as this relational power built through multi-pronged strategy is, it is not immune to repression from industry and government. In April 2019, the Alberta fossil fuel corporation AltaGas brought in a bulldozer and tore down the Treaty Camp Strawbale House that had been built by water protectors. AltaGas pulled it down under the protection of for-profit security and the colonial police. In January 2019 and again in January 2020, the RCMP, enforcing an injunction filed by Coastal Gas Link pipeline company, violently invaded the blockades on Wet'suwet'en territory, including the Unist'ot'en camp. Matriarchs were arrested while in ceremony and apprehended. While the strategy of direct-action solutions is at the forefront of Indigenous resistance and resurgence in Canada and is proving to be powerful – in its very ability to inspire support across the country while blocking unwanted projects – it is also being subject to violent repression.

Discussion and Conclusion

This analysis has explored the strategies being deployed and innovated on the frontlines of resistance to extractivism in Canada. We've described the kind of strategies that are succeeding in influencing outcomes and transforming systems, and have done so through the lens of four forms of power – physical, relational, discursive and institutional – in an effort to better understand how transformative strategies are being used to impact on existing power structures and relations and generate counter-power.

These six transformative strategies have shed light on several key things. One is the centrality of Indigenous-led resistance and resurgence to the fight for more sustainable and just economic and political systems in Canada. Another is the important role resistance plays in opening space and opportunities for innovating new and alternative modes of living, managing resources and governance in general. Also emphasized is the importance of relations of solidarity for bringing resistance efforts to successful outcomes.

The six strategies also illustrate the ways and the extent to which resistance is shaping outcomes and transforming how such decisions are made. From the triggering of the application of existing regulations to the development of new legislation and new environmental assessment processes, to the cancellation of proposed projects and the creation of moratoriums, we see the many ways that resistance is regularly influencing outcomes. Furthermore, these resistance efforts are also impacting on 'how power and responsibilities over natural resources are exercised' (IUCN n.d.) – for example Indigenous nations serving eviction notices to industry, and court rulings that acknowledge and uphold the rights of Indigenous nations to self-determination. These forms of resistance are upending the assumption of settler colonial governments' authority to make these decisions on unceded territory. Blockades then provide material, physical force behind these more symbolic acts of Indigenous sovereignty.

If we seek to understand and transform economic and political systems in Canada, it is not enough to look at formal, top-down forms of governance. Doing so ignores critically important dynamics and risks glossing over the most transformative forces for justice and sustainability at play in Canada. This is important to note, and our data analysis found evidence suggesting that cases where communities engaged in formal consultation processes are less associated with successful outcomes. We found a statistically significant association between boycotts of official procedures/non-participation in official processes and successful outcomes ($p=0.047$), at 5% significance level. In other words, participating in formal governance processes can reduce the likelihood of community resistance being successful.

We also found evidence to suggest that some forms of seeking to influence outcomes – namely the creation of alternative reports/knowledge – is negatively associated with success (p=0.033). It appears that the more outside of the official process, and the more conflictual and material the forms of resistance, the more likely they are to influence decision-making outcomes. This lends force to our contention that resistance needs to be central to the theorizing and practise of social transformation.

Resistance movements are not only helping drive more just and sustainable decision-making; they are contributing vital forces for *transforming* economic, political and social systems in Canada, compensating for the failure from the top to do so. 'Struggle and resistance are not simply sources of failure of governance, but contribute to the formation and distribution of power, knowledge, norms and renegotiation of scale, opening up possibilities for new forms of governance' (O'Malley 1996: 310).

For example, the many instances of resurgence of Indigenous governance and self-determination discussed in this chapter are more than transformative strategies – they are transformative acts in and of themselves, altering rules and practices such as laws, procedures and customs (Moore et al. 2014), shifting legal, economic and political frameworks (Temper et al. 2018). It is in this way that the cases where Indigenous communities have resisted unwanted projects by enacting their own governance on their own lands have been arguably the most transformative. They are living and defending just and sustainable lifeways and relations, here and now.

Resistance efforts in recent Canadian history have altered public perceptions, delegitimized whole industries and called into question who has authority over decision-making and who does not. As discussed in 'Strategy 2 – Boycotts and other financial pressure', the Clayoquot protests transformed public perception about clear-cut logging (Vanchieri 2011), contributed to the stigmatization of forestry and changed the ways in which many settler Canadians saw forests – including increasing the perceived importance of ecology and biodiversity (Tindal 2013). The case of Haida Gwaii resistance to logging ended an era where governments could ignore First Nations (Pynn 2010). Further transformative impacts have come from military stand-offs in the Oka Crisis in 1990 and the recent police raid of the Unist'ot'en camp. Both cases demonstrate the extent – force, violence, criminalization – to which the state will go to secure its control over land and resources and decision-making power. One can hope that this use of force against people defending lands and waters will contribute to the delegitimization of the state and open up more possibilities for the dismantling of oppressive institutions and practices that hold the status quo in place, thus transforming systems.

Though we've been arguing the crucial role that resistance plays in transformative change, it's important to note that it is not sufficient on its own. Nor should we rest assured that because communities are resisting extractivism in Canada, all will be well. Indeed, even in the most successful cases, communities continue to face extractive pressure as companies violate agreements and as governments continue to pursue profit. Even some of the most successful cases have had mixed, incomplete outcomes. And furthermore, communities – especially Indigenous communities defending their lands and waters – are facing violent repression, criminalization and surveillance (see Monaghan and Walby 2017; Nikiforuk 2019). In the 2018 book *Policing Indigenous Movements*, Crosby and Monaghan investigate how policing and other security agencies have been working to surveil and silence Indigenous land defenders and other opponents of extractive capitalism. They make the case that the expansion of the security state and the criminalization of Indigenous land defence have been allowed through the norms of settler colonialism.

Our analysis found a significant association between successful outcomes and the violent targeting of activists (p=0.028), at 5% significance level. This suggests that a consequence of successfully resisting extractivism in Canada is violence and criminalization. As such, a critical part of theorizing transformative strategies is acknowledging the huge risks and costs presently borne by Indigenous communities on the frontlines, and developing ways to broadly share and redistribute those costs and burdens.

In looking at the cases on the Canada map of the EJAtlas, we see that Indigenous communities are doing most of the heavy lifting when it comes to resisting injustice and unsustainability and working to transform conditions and structures in Canada. As the researcher who contributed the Lelu Island natural gas conflict into the EJAtlas wrote, 'it is a crime that Gitwilgyoots and activists had to waste so much resources to prevent the project, but they were successful in preventing it. In a more just world Petronas would have to reimburse the community for forcing them to waste so much of their lives in resistance' (EJAtlas 2017: n.p.).

Indigenous people have been bearing the brunt of transforming Canada despite their communities already facing devastating levels of poverty, suicide and health crises. To transform Canada, settlers need to do more of the heavy lifting, taking up more of the share of risk, burden and cost that comes with resisting the state and industry. This also means that governance scholars and researchers can and should do their work in ways that raise up the voices of frontline Indigenous communities, and actively support their work on the ground, including funnelling resources to these communities.

Researchers interested in systems transformation are well advised to pay attention to, and actively support, Indigenous resistance to extractivism as well as the vital alternatives to extractivism offered by Indigenous governance and economic systems. Indeed, these many stories of Indigenous resistance offer hope for different forms of relations among people and nature. But in order to take inspiration and guidance from these in ways that are not replicating colonial relations, we must recognize that 'alternate ways of being have always already been there' (Rutherford 2011: 197), and that for these alternatives to once again flourish there needs to be concerted and collective recognition of Indigenous sovereignty as an essential condition for generating better, more just and sustainable systems of governance (Manuel and Grand Chief Derrickson 2017; Ostrom 1990).

For non-Indigenous peoples, Indigenous resistance and alternative models hold the possibility for opening different ways of seeing nature and natural resources, which can inform how the environment is understood, defended and governed (Rutherford 2011). These models can be most effectively supported by non-Indigenous peoples through decolonization as a process of restoring governance and self-determination – and land – back to Indigenous peoples (Doxtater 2011).

Acknowledgements

Additionally to thanking (ISSC) and the Swedish International Development Cooperation Agency (Sida), Leah Temper would like to acknowledge that her work was funded by the Leadership for the Ecozoic project, and Jen Gobby's postdoctoral fellowship was funded through the Government of Canada's Social Sciences and Humanities Research Council.

References

Alfred, T. (2008) Opening Words. In Simpson, L. (ed.) *Lighting the Eighth Fire: The Liberation, Resurgence, and Protection of Indigenous Nations*. Winnipeg, MB: Arbeiter Ring.

Aorta Coop (2017) Understanding and Resisting Divide. https://www.powershift.org/sites/default/files/resources/files/Divide-and-Conquer-1.pdf.

Beddoe, R., Costanza, R., Farley, J. et al. (2009) Overcoming Systemic Roadblocks to Sustainability: The Evolutionary Redesign of Worldviews, Institutions, and Technologies. *Proceedings of the National Academy of Sciences*, 106(8): 2483–9.

Black, T., D'Arcy, S., Weis, T. (eds) (2014) *A Line in the Tar Sands: Struggles for Environmental Justice*. Oakland, CA: PM Press.

Brown, A.M. (2017) *Emergent Strategy: Shaping Change, Changing Worlds*. Chico, CA: AK Press.

Carroll, W., Sarker, K. (2016) *A World to Win: Contemporary Social Movements and Counter-Hegemony*. Winnipeg: ARP Books.

Chodos, R. (1973) *The CPR: A Century of Corporate Welfare*. Toronto: James Lorimer & Company.

Choudry, A. (2015) *Learning Activism: The Intellectual Life of Contemporary Social Movements*. North York: University of Toronto Press.

Clapperton, J. (2019) Environmental Activism as Anti-Conquest: The Nuu-chah-nulth and Environmentalists in the Contact Zone of Clayoquot Sound. https://prism.ucalgary.ca/bitstream/handle/1880/109482/9781773850054_chapter08.pdf?sequence=10&isAllowed=y.

Collins, P.H., Bilge, S. (2016) *Intersectionality*. Hoboken, NJ: Wiley & Sons.

Coulthard, G.S. (2014) *Red Skin, White Masks: Rejecting the Colonial Politics of Recognition*. Minneapolis: University of Minnesota Press.

Corntassel, J. (2012) Re-envisioning Resurgence: Indigenous Pathways to Decolonization and Sustainable Self-determination. *Decolonization: Indigeneity, Education and Society*, 1(1).

Crosby, A., Monaghan, J. (2018) *Policing Indigenous Movements: Dissent and the Security State*. Halifax, NS: Fernwood Publishing.

Dean, M. (2009) 'What they are doing to the land they are doing to us': Environmental Politics on Haida Gwaii. Doctoral dissertation, University of British Columbia.

Dixon, C. (2014) *Another Politics: Talking Across Today's Transformative Movements*. Oakland, CA: University of California Press.

Dominika V.P., Piotrowski, G. (2015) The Transformative Power of Cooperation Between Social Movements: Squatting and Tenants' Movements in Poland. *City*, 19(2–3): 274–96. https://doi.org/10.1080/13604813.2015.1015267.

Doxtater, T.M. (2011) Putting the Theory of Kanataron:non into Practice: Teaching Indigenous Governance. *Action Research*, 9(4): 385–404. http://doi.org/10.1177/1476750311409766.

Epstein, B. (1990) 'Rethinking Social-Movement Theory.' *Socialist Review*, 20(1): 35–65.

Field, C.B., Barros, V.R., Mastrandrea, M.D. et al. (2014) Summary for Policymakers. In *Climate Change 2014: Impacts, Adaptation, and Vulnerability. Part A: Global and Sectoral Aspects. Contribution of Working Group II to the Fifth Assessment Report of the Intergovernmental Panel on Climate Change*. Cambridge: Cambridge University Press.

Findlay, P. (2002) Conscientization and Social Movements in Canada. In Lankshear, C., McLaren, P. (eds) *The Politics of Liberation: Paths from Freire*. London: Routledge.

Foucault, M. (1971) Orders of Discourse. *Social Science Information*, 10(2): 7–30.

Glasberg, D.S., Shannon, D. (2010) *Political Sociology: Oppression, Resistance, and the State*. Thousand Oaks, CA: Sage Publications.

Gobby, J. (2020) *More Powerful Together: Conversations with Climate Activists and Land Defenders*. Halifax, NS: Fernwood Publishing.

Gosine, A., Teelucksingh, C. (2008) *Environmental Justice and Racism in Canada: An Introduction*. Toronto: Emond Montgomery.

Grant, P. (2010) Clayoquot Sound. *Canadian Encyclopedia*. www.thecanadianencyclopedia.ca/en/article/clayoquot-sound.

Harvey, D. (2014) *Seventeen Contradictions and the End of Capitalism*. Oxford: Oxford University Press.

IPCC (2018) Summary for Policymakers. In Masson-Delmotte, V., Zhai, P., Pörtner, H. et al. (eds) *Global Warming of 1.5°C. An IPCC Special Report on the impacts of global warming of 1.5°C above pre-industrial levels and related global greenhouse gas emission pathways, in the context of strengthening the global response to the threat of climate change, sustainable development, and efforts to eradicate poverty*. Geneva: World Meteorological Organization.

IUCN (n.d.) Natural Resource Governance Framework. www.iucn.org/commissions/commission-environmental-economic-and-social-policy/our-work/knowledge-baskets/natural-resource-governance.

Klein, N. (2014) *This Changes Everything: Capitalism vs. the Climate*. New York: Simon and Schuster.

—— (2017) *No Is Not Enough: Resisting the New Shock Politics and Winning the World We Need*. New York: Knopf Canada.

Kothari, A. (2014) Radical Ecological Democracy: A Path Forward for India and Beyond. *Development*, 57(1): 36–45.

Kulchyski, P. (2003) 40 Years in Indian Country. *Canadian Dimension*, 2 November. https://canadiandimension.com/articles/view/40-years-in-indian-country-peter-kulchyski.

—— (2012) Flooded and Forgotten – Hydro Development Makes a Battleground of Northern Manitoba. *Briarpatch*, 28 February. https://briarpatchmagazine.com/articles/view/flooded-and-forgotten.

Manuel, A., Grand Chief Derrickson (2017) *Reconciliation Manifesto: Recovering the Land, Rebuilding the Economy.* Toronto: James Lorimer.

Martinez-Alier, J., Temper, L., Del Bene, D., Scheidel, A. (2016) Is There a Global Environmental Justice Movement? *Journal of Peasant Studies*, 43(3): 731–55.

McBay, A. (2019) *Full Spectrum Resistance, Volume One: Building Movements and Fighting to Win.* New York: Seven Stories Press.

Monaghan, J., Walby, K. (2017) Surveillance of Environmental Movements in Canada: Critical Infrastructure Protection and the Petro-Security Apparatus. *Contemporary Justice Review*, 20(1): 51–70.

Moore, M.L., Tjornbo, O., Enfors, E. et al. (2014) Studying the Complexity of Change: Toward an Analytical Framework for Understanding Deliberate Social-Ecological Transformations. *Ecology and Society*, 19(4): 54.

National Centre for First Nations Governance (2010) *Making First Nation Law: The Listuguj Mi'gmaq Fishery.* http://fngovernance.org/publication_docs/Listuguj_Mi-gmaq_Fishery_FINAL_Dec.15.pdf.

—— (2011) Making First Nation Law: The Listuguj Mi'Gmaq Fishery. https://caid.ca/MiqFisLaw2010.pdf.

Nikiforuk, A. (2019) When Indigenous Assert Rights, Canada Sends Militarized Police. *The Tyee*, 17 January. https://thetyee.ca/Analysis/2019/01/17/Indigenous-Rights-Canada-Militarized-Police/.

O'Malley, P. (1996) Indigenous Governance. *Economy and Society*, 25(3): 310–26. http://doi.org/10.1080/03085149600000017.

Ostrom, E. (1990) *Governing the Commons: The Evolution of Institutions for Collective Action.* Cambridge: Cambridge University Press.

Parks, B.C., Roberts, J.T. (2006) Globalization, Vulnerability to Climate Change, and Perceived Injustice. *Society and Natural Resources*, 19(4): 337–55.

Pynn, L. (2010) Lyell Island: 25 Years Later. Wilderness Committee, 17 November. www.wildernesscommittee.org/news/lyell-island-25-years-later.

Ripple, W.J., Wolf, C., Newsome, T.M., et al. (2017) World Scientists' Warning to Humanity: A Second Notice. *BioScience*, 67(12): 1026–8.

Rodríguez, I., Inturias, M. (2018) Conflict Transformation in Indigenous Peoples' Territories: Doing Environmental Justice with a 'Decolonial Turn'. *Development Studies Research*, 5(1): 90-105. https://doi.org/10.1080/21665095.2018.1486220.

Rowlands, J. (1997) *Questioning Empowerment*. Oxford: Oxfam.

Russell, E.D. (2018) Resisting Divide and Conquer: Worker/Environmental Alliances and the Problem of Economic Growth. *Capitalism Nature Socialism*, 29(4): 109–28.

Rutherford, S. (2011) Governing the Wild: Ecotours of Power. Minneapolis: University of Minnesota Press.

Salick, J., Byg, A. (2007) *Indigenous Peoples and Climate Change*. Norwich: Tyndall Centre Publications.

Scheidel, A., Temper, T., Demaria, F. Martinez-Alier, J. (2017) Ecological Distribution Conflicts as Forces for Sustainability: An Overview and Conceptual Framework. *Sustainability Science*, 13: 585–98.

Scheidel, A., Del Bene, D., Liu, J., Navas, G., Mingorría, S., Demaria, F., Avila, S., Roy, B., Ertör, I., Temper, L. Martinez-Alier, J. (2020) Environmental Conflicts and Defenders: A Global Overview. *Global Environmental Change*, 63: 102–4.

Simpson, L.B. (2011) *Dancing on Our Turtle's Back: Stories of Nishnaabeg Re-Creation, Resurgence and a New Emergence*. Winnipeg, MB: Arbeiter Ring.

Snow, D.A., Soule, S.A., Kriesi, H. (eds) (2008) *The Blackwell Companion to Social Movements*. Hoboken: John Wiley & Sons.

Solnit, R. (2016) *Hope in the Dark: Untold Histories, Wild Possibilities*. Chicago: Haymarket Books.

Stanbury, W.T. Vertinsky, I.B. (1997) Boycotts in Conflicts over Forestry Issues: The Case of Clayoquot Sound. *Commonwealth Forestry Review*, 76(1): 18–24.

Taylor, D. (2014) *Toxic Communities: Environmental Racism, Industrial Pollution, and Residential Mobility*. New York: New York University Press.

Temper, L., Del Bene, D. (2016) Transforming Knowledge Creation for Environmental and Epistemic Justice. *Current Opinion in Environmental Sustainability*, 20: 41–9.

Temper, L., Del Bene, D., Martinez-Alier, J. (2015) Mapping the Frontiers and Front Lines of Global Environmental Justice: The EJAtlas. *Journal of Political Ecology*, 22(1): 255–78.

Temper, L., Walter, M., Rodríguez, I., et al. (2018) A Perspective on Radical Transformations to Sustainability: Resistances, Movements, and Alternatives. *Sustainability Science*. https://ddd.uab.cat/record/271379.

Tindall, D. (2013) Twenty Years After the Protest, What We Learned from Clayoquot Sound. *Globe and Mail*, 12 August. www.theglobeandmail.com/opinion/twenty-years-after-the-protest-what-we-learned-from-clayoquot-sound/article13709014/.

Unist'ot'en Camp (n.d.) https://unistoten.camp/.

Vanchieri, N. (2011) Environmentalists Defend Old Forest in Clayoquot Sound, B.C., Canada, 1993. Global Nonviolent Action Database. https://nvdatabase.swarthmore.edu/content/environmentalists-defend-old-forest-clayoquot-sound-bc-canada-1993.

Waldron, I. (2018) *There's Something in the Water: Environmental Racism in Indigenous and Black Communities*. Halifax, NS: Fernwood Publishing.

12

Sandhani: Transformation Among Handloom Weavers of Kachchh, India[1]

Kalpavriksh and Khamir[2]

> *My loom is my computer; I have to continuously think, innovate, it is not only mechanical, and I have to use my head, my hands, and my heart to make a product that you will like.*
>
> – Prakash Naran Vankar, Bhujodi village, Kachchh, India

Introduction

This is a study of the multiple dimensions of transformation taking place in the livelihoods of the *vankar* (weaver) community of Kachchh, Gujarat, India resulting from a revival of the handloom weaving craft (*vanaat*) from a time when it was in sharp decline. It examines whether the changes taking place can be said to be in the direction of a systemic or structural change towards justice and sustainability. It looks at how well-being has increased, along with positive transformations around caste, gender and generational relationships and a flowering of innovation and creativity, hybrid knowledge and learning systems – but also how there are some regressive trends in issues of class and ecological sustainability. This chapter examines how the transformation has been enabled by the agency of the weavers, their adaptability, resilience and innovativeness, as well as the facilitation of institutions enabling innovations in production and marketing. It also looks at external factors (in economy and society) that contribute directly or indirectly to transformation, such as new consumer tastes and markets, new techniques and technologies, and, in this case, a massive earthquake that was a crisis turned into opportunity. The narrative suggests that weaving, while linked inextricably to the market, has escaped some of the detrimental effects of commodification as envisaged by Karl Marx, as the craftsperson's labour has not been alienated from them, and the means of production – especially the loom – remains in the ownership or control of the producer family. Perhaps most importantly, this enables creativity and an embedded worldview that sees weaving as not only a commercial activity but also a cultural one, with important emotional, psychological and affective aspects, as well as the dignity of fulfilling labour, as stressed by Mahatma Gandhi.

Methodology

The study involved three main actors: Kalpavriksh, a forty-year-old environmental action group based in Pune/Delhi; Khamir, (www.khamir.org), an organization based near Bhuj supporting crafts that had a significant role in the revival of *vanaat* in Kachchh; and the *vankar* community. A number of consultants and advisors were part of the study, including experts in Indian crafts, an ecologist and an economist. The study was initially motivated by the interest of Khamir, which sought to understand the ecological footprint of *vanaat* in order to determine its sustainability compared to the powerloom and industrial cloth production, and/or consider interventions to improve its sustainability. Khamir's interest later expanded to understand the holistic transformations that have taken place in the sector – across social, economic, cultural, environmental and political dimensions – in order to aid the organization to reflect on its own work. There was also interest among the weaver community, as expressed by some of its elders, for documentation of its history and changes in the last few decades, to help transmit these developments to new generations.

The study also stems from Kalpavriksh's interest in using the Alternatives Transformation Format (ATF, described in Chapter 2) to study not only the ecological dimension but also other dimensions of transformation. It is the first ever study in India to look at a craft from multiple dimensions (economic, socio-cultural, political, ecological, ethical), and the first in the world to use the ATF. The *vankar* community was considered to be a good candidate because, prima facie, there appeared to be many dimensions to the transformation taking place in the lives of *vankars*, and different study partners were interested in different dimensions as well as understanding the broader dynamics at play.

Beyond these immediate impacts, the study also aimed to generate further discussions among all partners, identify interventions that could benefit *vankars* as well as the environment, and provide lessons for similar multidimensional studies of the handloom sector in other parts of India, or even studies of other craft sectors.[3]

Study methods included choosing a sample of fifteen villages[4] (out of a total of sixty-four) based on prior understanding of these settlements (while an associated baseline survey was also carried out during the process, to get a broad understanding of the entire *vankar* community as a context, data from this survey has not been included here); selecting a team of senior *vankars* and a group of *vankar* women for involvement with the study; participatory video documentation with youth *vankars* and production of six documentary films;[5] visits to each of the fifteen villages for group discussions, one-to-one interactions and observations; visits to other sites for the ecological footprint assessment (EFA), including farmers growing *kala* cotton, traders, spinning and ginning mills, and dye units; focused group discussions with women, youth and elders; perusal of secondary literature available from civil society groups; and discussions with some key people involved in interventions or studies with *vankars* in the last three decades. Discussions about key results were carried out with *vankars*, including the various teams mentioned above, and separately with women, with youth and in clusters of villages. The ecological footprint assessment (EFA) used a mix of quantitative and qualitative methodologies in an attempt to gain a broad understanding of the craft's impact, compare the impacts of organic (*kala*) cotton and genetically modified (Bt) cotton, and was developed with the goal of generating a template that could possibly be used for handlooms elsewhere and with modifications for other crafts.

Study limitations included the fact that aspects like the inter- and intra-community political dynamics could not be studied in any depth, and some cultural aspects that had initially been included for study were left out at the request of the community. The core team of weavers involved with the study consisted only of male weaver entrepreneurs whom Khamir was familiar with (though later on women and youth also were involved in some aspects). Language was occasionally a limiting factor when the community interaction teams did not contain Kachchhi- or Gujarati-speaking people. There were inadequate interactions with non-*vankar* communities, to get a better understanding of the changes in relations between them and the *vankars*.

Transformations: Findings of the Study

Background

The study was located in the district of Kachchh, in the Western Indian state of Gujarat. Kachchh is extremely dry, and ecologically distinct from the rest of India due to the extensive grassland and salt desert ecosystems that cover most of its area. It has a long history of nomadic and settled pastoralism, and other livelihoods based on its unique ecological features. Traditionally these livelihoods were also associated with the religious and caste identities, though this is now becoming less predominant, especially where urban and industrial economies have developed.

Vankars, part of the Marwada (and in very small numbers, the Maheshwari and Gurjar) subgroups of the Meghwal community, which originally came from neighbouring Rajasthan, are spread over much of the district. Traditionally, they have been involved both in weaving and in other occupations including farming, leatherwork and labour in construction sites, etc. More recently, many of them have become full-time weavers, and the widespread use of the identity '*vankar*' to refer to themselves as a community is only two to three decades old.

The 'conflict' that forms the focus of this study relates to the economic and social distress and discrimination that *vankars* as a whole have faced, particularly in the early years of the twenty-first century. A series of changes in the larger economic and social milieu of which Kachchh was a part, including the rise of industrial cloth production flooding the market with cheap products (e.g. shawls made on powerlooms in Ludhiana, imitating Kachchhi designs), alterations in habits and tastes among communities that *vankars* made cloth for (and their inability to buy the more expensive newer products), the reduction in availability of traditional yarns (especially sheep wool), and the intervention of government and civil society agencies that attempted to link weaving with external markets, were already changing the nature of how (and for what) *vankars* were using their craft in the last couple of decades of the twentieth century. A devastating earthquake in 2001 caused severe loss of productive capacity (literally, damaging looms in hundreds of households) and access to markets. At the same time, newer job opportunities arose, for instance in industries being set up by the government or private sector, or migrant labour in the Middle East. The *vankar* community faced difficulties recovering from these multiple drivers, especially as they were already on the economic margins of society and due to their marginalization as *Dalits* (the so-called 'outcastes' of Hindu society).[6]

A number of factors began to turn the tide from about 2005–6. This included the intervention of Khamir, an organization set up with the mandate of reviving and sustaining Kachchh's crafts, and re-establishing relevant economic value chains. The revival in production of the Indigenous (and organically cultivated) *kala* cotton was a key element of this and coincided with renewed consumer interest in handwoven cloth. This led to marketing opportunities in the Kachchh desert festival, exhibitions and markets in India's big cities and in Europe, interventions by handloom and design schools to help train young weavers in entrepreneurship or design innovation, the distribution of looms, other help by the government immediately after the earthquake, and other such factors. Handloom weaving has revived in a significant way in the last decade, transforming the lives of at least a part of the *vankar* community.

Amongst the fundamental structural aspects that lay at the base of the conflict described above are an economic system that has marginalized *vankars* (including their skills, knowledge and products) as producers, and a social system (the caste system) that has long relegated *vankars* to the lowest social status. Within the *vankars* too, there are structural issues like patriarchy and toxic masculinity that marginalize women, and generational inequities with youth having little say in community matters. There is finally the issue of the distance between weaving and nature. It was perhaps never a very strong *direct* relationship (as for instance would have been the case for pastoralists), but when wool from local sheep and camels, dependent on local ecosystems, was the main yarn, and there was some amount of use of natural dyes from local plants, the relationship was likely stronger.

Brief history

The *vankars* in Kachchh are mostly from the Marwada ethnic group (and some are Maheshwaris and Gurjars) from the Meghwal community of Rajasthan, which migrated to Kachchh about 500 to 600 years ago. These landless Marwadas lived with the local pastoral Rabari community and other communities such as Ahirs,

Darbars and Patels, with relations of economic exchange among them that were passed down the generations. While originally handloom weaving was one among many occupations, it became more prominent over time as other communities came to depend on them for clothing. In the several decades before India's independence (1947), however, cheap industrial mill-made cloth affected their markets, and interdependence among local communities diminished. Weavers were forced to service more urban customers and began experimenting with different raw materials and techniques. The introduction of *khadi* (handspun cloth, advocated by Mahatma Gandhi), and the setting up of cooperatives or *vankar mandalis* – aimed at safeguarding the economic, social and cultural interests of this community – aided the revival of the craft somewhat. At the time of independence, around 5,000 to 5,500 weavers came together as part of a weavers' *sammelan* (gathering). In the mid-twentieth century, and into the 1970s, a number of technological changes to production and yarn-making, and governmental attempts to promote weaving, also influenced *vanaat* in significant ways. Several thousand weavers found livelihoods through these processes until the late twentieth century.

In the 1990s, however, powerloom units in North India began imitating Kachchhi products, making them much cheaper. This, as well as a series of natural calamities in the late 1990s and early 2000s (including the 2001 earthquake mentioned above), nearly decimated the handloom sector in Kachchh. It was only in 2005–6, with the intervention of Khamir and some initiative by enterprising *vankars*, that a revival began, based on the reintroduction of Indigenous *kala* cotton into the production chain and its marketing to urban centres in India and abroad. Other innovations, like the use of plastic waste in weaving, were brought in by Khamir, and some design schools helped provide further training to young *vankars* in adapting designs and products for the external market. From the mid-2000s until the present, the revival has been sustained, with varied results that are examined below.

Transformations in different dimensions

The revival of *vanaat* has had ramifications in various spheres of *vankar* life. A simple overview of the transformations taking place is given in Figure 12.1; each of these is explained in the sections below.

Economic sphere

The revival of an economic livelihood that was in severe decline in the early part of this century, leading to a measure of livelihood security for a part of the *vankar* community, is the most obvious aspect of transformation. It is important to note that this revival builds on local traditional knowledge and locally (family-) owned means of production. This is in contrast to other communities in Kachchh or elsewhere in India whose traditional occupations and livelihoods have declined. Millions of farmers, fishers, craftspeople, forest-dwellers and others dependent

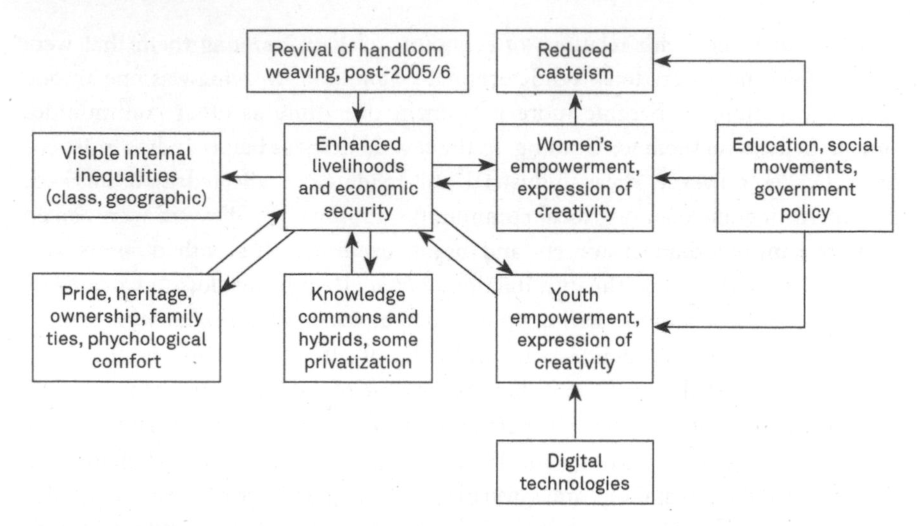

Figure 12.1 Multi-dimensional transformations in the lives of *vankars* as a result of the revival of *vanaat* (the arrows represent direction of influence)

on the 'primary' sector have been 'deskilled' through loss of productive natural resources (including land) or outright displacement, and have had to switch to insecure, lowly paid and alienating jobs in industry, construction, services or other parts of the 'modern' economy, where their own knowledge is considered useless, and where the means of production are wholly or predominantly owned by others (state or private) (Sadgopal 2016; Shrivastava and Kothari 2012).

The revival of handloom weaving has been accompanied by an increase in incomes for at least a part of the *vankar* community, as evidenced by changes in lifestyles, housing and other indicators. Several entrepreneur weavers (those who are able to sell their own products) have much greater access to outside markets in India's metropolitan cities and in European countries directly.

Other weavers also increased earnings. This is partly due to the overall increase in market opportunities for the sector, and partly due to the deliberate action of Khamir to provide higher wages for the weavers working for them, forcing the entrepreneurs to increase wages.

As the study found, the uplift in handloom weaving has also arrested or reversed occupational migration to some extent, with many young people staying in the region or coming back to it.[7] This is in contrast to a general trend of new generations across India leaving traditional occupations, and shows the potential of such livelihoods, when viable, to generate interest in youth. This runs counter to the narrative, pushed by government and corporations, that large-scale industrialization is the only pathway to economic prosperity and job creation in Kachchh and elsewhere. The search for such alternatives is urgent given that industrialization is already leading to serious ecological and social disruption in the region.

While the production of *vanaat* was previously based primarily on patronage, this has now changed, with entrepreneurial weavers hiring other weavers, and both groups working in part-time and full-time, and semi-skilled and skilled arrangements. Further, some use acrylic only and others use *kala* and sometimes mix it with other materials; some weavers supply mainly to external markets and others continue to produce for a (declining) local market. There are also new, younger weavers, many of them women, who are establishing their own market connections.

The *vankar* community, with the facilitation of Khamir, seems to have some control over defining the market, rather than purely responding to it. For instance, a market has been created for *kala* cotton cloth and clothes (somewhat similar to the market for organic food in parts of India, or globally). Entrepreneurs in Bhujodi, the most thriving of the *vankar* settlements, have created a demand for the 'Bhujodi sari'. When the demand for *dupattas* decreased, *vankars* came up with products like stoles and scarves. Costing of such products is also to some extent in the hands of the producers or intermediaries like Khamir, with many producers spoken to during the study expressing satisfaction at the prices they get. However, this does not change the fact that the revival is dependent primarily on external (including foreign) markets,[8] with their own issues of vulnerability,[9] absence of (or only weak) challenges to macro-economic and class inequalities both locally and in larger society, absence of challenges to capitalism as a system, and (as described below), some negative ecological ramifications.

Over the last few decades, weaving has transformed from being primarily confined to local economic exchange to being more integrated with global markets. For instance, raw material like wool used to come from local pastoralists, who then used the cloth woven by *vankars*; there were other similar localized processes. A major change occurred when acrylic produced outside the region was introduced as a yarn in the 1970s, enabling an expansion of weaving as it was easier to use, but with the produce being sold both locally and outside the region. Currently the situation is mixed; while *kala* cotton is locally produced, establishing or re-establishing a link with local farming communities, the sale of cloth made from it is primarily to external markets.

What has significantly increased is the link to rich consumers in Indian metropolizes and in foreign countries; this means that the local revival is based on elite consumers, an irony that cannot be ignored. The silver lining to this is that it is catering to that section of the global market that has at least some level of concern for ecological, health and social issues, assuming that they are buying these products with the environment, the producer, and their own and others' health in mind.[10]

The downside of increased prosperity in the community is what appears prima facie to be an increase in economic inequality among the *vankars* themselves. We do not have the baseline data for earlier periods to show this conclusively, but senior *vankars* agree that inequality has increased, at least compared to four to five decades back when most *vankars* were at similar levels of earning. This is especially so between the entrepreneur weavers (particularly those who have managed to significantly enlarge their market base and even employ a lot of workers) and the job workers. The study noticed some rather glaring discrepancies in the living conditions of the two in some villages, though we also came across several job workers who expressed significantly better conditions now than before. Among at least some of the job workers, there is also a sense of alienation, with weaving being seen as just physical labour (*majuri*) like any other, and for these the notion of alienated labour does seem to be appropriate.

Another noticeable phenomenon is the geographic inequality within the *vankar* community. Villages close to the district capital Bhuj, and/or where institutions like Khamir have been active, are clearly doing much better than others. The surveys referred to above (beyond the fifteen villages chosen for focused study) strongly indicate that distress continues among *vankars* of villages that do not have these advantages. In discussion with *vankar* elders, there appeared to be clear recognition of these inequities; one suggestion that came up was that entrepreneurs should focus more on giving *kala* cotton and other work to weavers in 'remote' areas, and also enable greater innovation/diversity in acrylic, to reduce at least some geographical inequities. They also pointed to the fact that settlements like Bhujodi had a greater critical mass of *vankars*, who were able to stand up to caste oppression and economic marginalization more than in villages where *vankar* families are a tiny minority.

Interestingly, while on social matters there is reported to be a high level of cohesion and inter-personal or intra-community support (e.g. donations or loans for weddings or other functions if someone cannot afford them), this does not translate into the economic domain. Little thought has gone into the economically better-off *vankars* helping the worse-off ones, though there have been occasional discussions (including some as part of this study). Yet we did not come across strong resentment or a sense of conflict due to the inequality. Senior *vankars* who were in the core study team mentioned the following factors behind this: economic inequalities are still not large enough to be perceived as exploitative, and there is no feeling that some weavers have deliberately created conditions for such inequalities; in some cases there is a lack of proactiveness to improve one's situation, or weaving capabilities are low, which is recognized; mutual aid (for social issues) and relations of care still work; and finally there remains a strong sense of social equality and common religious or spiritual identity between those who are economically unequal. And yet, it appears that commercialization and greater orientation towards the market may be reducing the hold of traditional principles, such as that of equality embedded in elders advising *vankars* to keep a limit to the profit margins they were setting, to help prevent a wider gap of inequality among the weavers, and because weaving was not just a profession but also their identity and spiritual responsibility.

It is also interesting that most *vankars*, especially the job workers, do not wear the products they are weaving, especially if it is *kala* cotton, as they can't afford it. This shows a certain economic marginalization. Some of the entrepreneurs have made it a point to wear *kala*, and there is discussion on how this needs to be increased among others (especially the youth), perhaps by enabling some extra yardage to be retained by weavers (particularly those who work for entrepreneur weavers) rather than having to sell off everything.

Socio-cultural sphere

Indian society in general has been characterized by moderate to severe inequities within and between communities. These relate to caste (prevalent especially in Hindu society but also communities practising other major religions), patriarchy and toxic masculinity (almost universal), ethnicities (between *adivasis* and non-*adivasis*, or different religious groups), ability (differently abled people being discriminated against), age (the domination of elders going beyond justified respect to a serious lack of space for the young to express themselves), and others. The study attempted to understand if and how transformations may have taken place in some of these aspects.

Caste: There is a clear and consistent reporting of reduction in casteism, especially its worst forms of untouchability, to which *vankars* were traditionally subjected. Several elder *vankars* talked about various practices of social stigma they had experienced, such as not being allowed to enter the houses of other castes that they had economic relations with, or their cloth not being accepted without sprinkling water on it to rid it of its 'impurity'. The worst forms of casteism such as these practices associated with untouchability are reported to be well on their way out, and relations with other communities are increasingly seen on a more equal footing, especially among the youth. But by no means is casteism eradicated; several *vankars* mentioned continued discrimination of various kinds, including restrictions on entry to some temples, or in subtler forms such as furtively keeping separate cups or glasses for use by *vankars* in some shops. It appears that where it has reduced, in form and intensity, it is at least partly due to the enhanced economic status of *vankars*; several *vankars* said their greater wealth status, their contacts with the outside world, greater confidence levels and other such aspects associated with the revival in *vanaat* were important factors. But many also talked about social reformers, education and law, and declining economic (and therefore social) relations with other communities as being other key factors, with the new generation being much less steeped in casteist traditions. The relative role of such 'background' factors or circumstances compared to that of transformation in *vanaat* is not possible to determine without deeper and wider study.

Gender: There appears to be a general increase in women's empowerment, with weaving revival being one factor. Indicators include: women doing actual weaving (traditionally virtually non-existent) including in some cases teaching their husbands, women's role in pre-weaving being explicitly recognized and valued, women becoming involved in processes in which they were not included earlier (e.g. dyeing, yarn treatment), young girls getting into the process and being more vocal, a greater ability of women to leave the home or village for weaving-related events, participation in events like Women's Day, and so on. Innovation in weaving waste plastic into products has empowered women in Avadhnagar village especially, by providing an independent source of income. However, gender discrimination continues, such as the generally greater ability of men and boys to relate to the outside world. One issue is that women who now take a much greater part in weaving saw an increase in the amount of work they do since they are doing both weaving and housework. In general, while women are taking on more of what the men used to do, very few men are doing the converse, taking on what have been typically women's household tasks.

By no means is the revival of weaving the only or perhaps even the main reason for women's empowerment. Official policies and programmes, including reservations for positions within village- and wider-level decision-making bodies, access of girls to education opportunities, civil society programmes, also have an impact.

Age: Youth are clearly quite active in the *vanaat* revival and are able to speak about it confidently and assertively, and are exercising a greater role in design, market linkages and networking among themselves (see Figure 12.2). Major contributing factors to this include the use of digital platforms for learning new designs, and marketing and training in artisanal craft, handloom and design schools, which allow young *vankars* to not only learn new skills, but also put a more 'modern' stamp on a traditional occupation. The return of youth to weaving is an important (even if for the moment small-scale) phenomenon. Further, youth expressed that beyond simply perceiving weaving as an economic activity, for them there are strong elements of culture, emotion, psychology and affective relations with the work. Interestingly, deference to elders in both economic and socio-cultural matters still appears to be very strong.

Ecological sphere

Handloom weaving, on its own, is considered to be an 'environmentally' sound manufacturing process, as compared with powerloom and mill-based weaving which typically consumes substantial power and water and generates polluting effluents. But even within the handloom sector, there are a number of changes that imply a very different set of ecological linkages and impacts. These include the switch from local wool (traditionally traded family to family between pastoral Rabari and the *vankars*) to externally produced acrylic or 'super' wool (merino, initially from Australia, now also produced in India), and, in the last couple of decades of the twentieth century, the revival of *kala* and other yarns, as well as Bt (or other hybrid) cotton. Dyes have also diversified, with much use of chemical dyes but a strong revival of natural dyes too. Further, there has been a shift in the scale of the value chain, as most sales today predominantly take place outside Kachchh.

Figure 12.2 Sheetal Hiteshbhai, part of youth helping revive handloom weaving in Kachchh, western India

Source: Ashish Kothari.

While the initial intent of this study was to assess the ecological footprint of the handloom weaving sector as a whole, this was found to be impossible, given the enormous range of factors involved, and the limited time and resources available for the study. While overall observations were obtained on several dimensions of the sector, a more detailed and quantitative assessment was conducted comparing the ecological footprint of the *kala* cotton value chain with other (mostly the genetically modified Bt) cotton.

The value chain of cotton consists of five critical stages: 1) cultivation of cotton (different cotton varieties and their input-production systems); 2) production of lint cotton through a mechanized ginning process; 3) production of cotton yarn through a mechanized spinning process; 4) production of fabrics/textiles by weavers, including yarn procurement, yarn dyeing and the weaving process; and 5) marketing of finished products. Each of these stages has an ecological footprint of a different type. While the use of chemicals, water, energy (electrical energy in ginning and spinning mills) and fuel (mainly for transportation) are key factors, a shift in the use of different biological resources during the entire value chain also contributes to the ecological footprint. It is important to mention here that the figures emerging from these case studies (or from key informants at each value chain) cannot be applied to 'represent' the entire cotton weaving sector across Kachchh. Also, it was not possible to calculate and include the biodiversity component at this stage; hence, in this case,

Table 12.1 Ecological (Carbon and Water) Footprint of *Kala* and Bt Cotton

Weaver/unit[11]	CO_2 eq per kg of cotton value chain attributed to weaver		Blue water* use per kg of cotton value chain attributed to weaver (litres)	
	Kala	Bt	***Kala***	Bt
Case 1	4.85	9.45	194	1,838
Case 2	5.42	10.15	168	1,884
Case 3	5.09	9.94	172	1,837
Case 4	4.63	10.30	163	1,884
Case 5	5.43	**N/A**	**146**	**N/A**
Average	**5.08**	**9.96**	**169**	**1,860**

*Blue water is the water that is drawn from rivers, wells and the ground for irrigation purposes, hence using energy.

the ecological footprint is a calculation of two key components, carbon and water. The results are presented in Table 12.1 above.

The study shows that the footprint of *kala* cotton is nearly half that of Bt, indicating that a locally produced organic cotton variety has the smaller environmental impact of a locally produced organic cotton variety, despite the expanded trading of raw material and final fabric from outside the region.

The study also found that if one looks at the footprint at each level of the value chain for *kala* and Bt, the dyeing of yarn, especially when using chemical dyes, contributes heavily to both the water and carbon footprint. In addition to this, transportation for sales across the national and global markets add considerably to the carbon footprint from transportation.

While the use of *kala* cotton in weaving is clearly ecologically less damaging than the use of Bt cotton, it should be a matter of concern that the shift away from localized production chains (e.g. local wool to local sale of wool-based clothes) to more global ones (yarns coming both from outside and within the region, and a lot of clothes being sold outside it) has increased the ecological footprint. This is not restricted to Kachchh handlooms, but appears to have been a trend in many parts of India where crafts have revived on the basis of national and global markets. There are a few areas of enquiry that were left out in this study, but which are considered vital to improve the overall result. This includes 1) the transportation of lint cotton to other states for spinning purposes and their reverse transportation to Gujarat with yarns of different counts, 2) the production and transportation of agriculture chemicals and fertilizers, 3) the production of natural and chemical dyes, and 4) other impacts like biodiversity and pollution. Also, while carbon and water impacts are crucial parts of the EFA, other factors like the impacts on biodiversity and pollution are also critical, and no attempt has been made to include them in this study.

There are other ecological aspects of the handloom weaving revival (not necessarily linked to the EFA per se) that are important, but which the study could not explore. For instance, enhancement in the economic levels of the *vankars*, especially those who are commercially doing very well, seems to be leading to consumption patterns akin to those seen in the urban middle or upper middle classes. There does not appear to be any critical discussion among the *vankars* on this; we were unable to come to any conclusions on this aspect and thus it was not part of the study.

Within this overall change towards a greater ecological footprint, the attempt at some level of eco-friendliness and localization in the *kala* cotton production initiative, the gradual revival of natural dyes, and the absence of any significant entry of powerlooms and any larger textile industry in Kachchh are positive features from an ecological perspective.

Political sphere

There appears to be little or no linkage between the transformations in weaving as a livelihood and the political aspects of the *vankars*' lives. One striking phenomenon is the absence or very weak manifestation of collective mobilization relating to weaving in Kachchh, as compared to many other parts of India where weavers have collectivized as cooperatives, or as associations to take up advocacy on occupational policy issues. The last major mobilization appears to have been around the time of India's independence, when shortage of yarn was taken up as an issue affecting weaving. A cooperative of *vankars* did exist in the latter part of the twentieth century, but became non-functional reportedly due to internal dynamics and mismanagement, and has never been revived. Even on issues that had a significant impact on their lives, such as the imposition of high taxes under the Goods and Services Tax (GST) policy of the current central government, or the 'demonetization' of high-currency notes that took place in late 2016, there was no collective mobilization. There is also no sign of collectivization among the job workers to seek better working conditions (especially where still suffering from obvious exploitation), or to share resources to access markets directly and thereby become entrepreneurs themselves. This is partly due to the fact that the relationship between the two classes of weavers is not only of employer–employee, but also a complex of social bonds, loyalty, and feelings of being indebted due to entrepreneur weavers having helped out in times of social or personal crisis. This has traditionally also been a significant part of India's caste system (*jajmani* relations of patronage), which has elements of mutual benefit but from a class analysis could be said to be a subtle form of exploitation.

This absence of what could be called 'non-party political process' among the *vankars* could be explained as an outcome of their individualized business tradition (which is strong among many communities in Gujarat, who are well known for their entrepreneurial skills in many parts of the world!), the general sense that as businesspeople they will scrape through one way or the other, and a non-confrontational culture. There may also be a more structural socio-cultural reason, with the many years of social oppression and livelihood struggles reducing the capacity to organize (though there was no articulation to this effect during the study).

There is also little or no linkage between *vanaat* as an occupation and the electoral and party political process, from local to national level. A number of *vankars* have been in or currently occupy positions in local *panchayats* (village councils), district-level bodies, etc., but this does not seem to be an outcome of their status as *vankars* or their increased economic status, and conversely their position is not used to enhance the prospects of *vankars* as weavers in village or larger society. Some *panchayat* members and *sarpanches* (heads of village council) expressed that in their position, they seek to act for the benefit of all communities, not *vankars* in particular.

The linkages between state policy and the *vankars* are also weak. While in the late twentieth century government policies (such as some products being reserved to the handloom sector, with other industrial processes not being allowed to produce them) and programmes (such as subsidies and state procurement of handloom products) played a significant role, this appears to have diminished more recently. Indeed, the macro-policy environment has become less conducive in recent years to handloom weaving (as to handicrafts in general), with much less protection to these sectors against competition from the economically powerful corporate industrial sector. *Vankars* are accessing the market, with civil society organizations playing a key role in facilitating the value chain, and the state's role has receded into the background.

As an interesting offshoot of this study, and of a related visit to Himachal Pradesh by some members of the core study *vankar* team, discussion within the community to revive the weavers' cooperative has been strengthened. Additionally, the Kachchh Weavers Association, dormant for some time now, has been re-activated (see below under 'Knowledge, creativity and innovation'), which may enable greater political mobilization among *vankars* for livelihood issues.

Knowledge, creativity and innovation

Craft-based livelihoods have a distinct knowledge and skill base. Traditionally these have been passed down from parents to children, either by the latter simply being around the former when they were working and learning by observing, and/or being deliberately and systematically taught. In the case of handloom weaving, members of the younger generation have learned mostly by observing their parents, supplemented by deliberate imparting of skills only in the case of some exceptional techniques. Even today this form of transmission of knowledge and skill is predominant, though it is now also supplemented by more formal learning opportunities such as artisanal craft schools.

Strikingly, though, civil society interventions and the increasing market opportunities for the *vankars* have stimulated greater expression of creativity and innovation than in the past. The youth are not simply following the designs and techniques of their elders, but also innovating on these in trying to both create a market and also respond to it. The increasing incidence of women sitting at the loom has also created an opportunity for their creativity and innovation to be expressed. Remarkably, though, the younger generation of *vankars* continues to be committed to the traditional motifs and patterns and weaves (including the extra weft that characterizes the Kachchhi weave).

One concern is the absence of crafts learning in schools. Since children are increasingly going to school (a trend the *vankars* interviewed in our study have said they want to encourage), they are not learning weaving skills, and may be exposed to the predominant view of formal education that livelihoods such as crafts, farming and pastoralism are 'old-fashioned' and need to be abandoned in favour of modern skills. Recently Khamir and other organizations have proposed the introduction of craft learning in schools. Senior *vankars* also desire this to happen, noting that when they were in school there was at least a class on *udyog* ('industry') where traditional practical skills could be learned, which has now been replaced by learning on computers, with the arts and crafts relegated to textbooks.

One of the reasons for the vast range of design innovation today is an inherent feature of Kachchh weaving: the ability to weave very fine designs (extra weft) in relatively coarser yarns. Many weavers who had been weaving simpler designs in acrylic seem to have seamlessly transitioned to very fine cotton and silk weaving. Innovation stems from this innate understanding, ability and command over material and technique.

One potential source of tension relates to some initial signs of privatization of knowledge and skills among *vankars*. Traditionally, the *vanaat* knowledge, designs and motifs have been in the commons, according to the *vankar* elders. Some young people who are individually innovating, however, feel that they need to be acknowledged and that it is not fair for others to copy their innovations. The youth feel that with professionalism there is some preference for individualization of designs rather than general sharing. There is also a rising trend among youngsters to create their own brands and products. However, as yet there is no cut-throat competition, or serious resentment about being copied. The community by and large still considers weaving knowledge as being in the commons. A recent anomaly is the appropriation of the Kachchhi saris by one village (branding it the Bhujodi Sari), which several elders feel should not have happened; as Naranbhai Madan Siju of Bhujodi said, 'I gave my products the name Kachchh Carpets to enable other *vankars* anywhere in Kachchh to use the same brand'. The issue of privatization of knowledge is recognized as something the community needs to talk about, to avoid tensions in the future.

In 2012, the *vankars* formed a Kachchh Weavers Association (KWA), and obtained a Geographical Indication (GI) registration for the Kachchhi shawl, in a bid to stop cheaper industrial production. However, the GI recognition has so far not had any palpable effect or special benefit, nor have the KWA and the weavers

Figure 12.3 Prakash Naranbhai Vankar, continuing ancestral tradition of carpet weaving with new innovations

Source: Ashish Kothari.

in general pushed for its proper use. In late 2018, the KWA picked up the issue again, and proposed actions such as publicity about the GI and awareness among consumers of the need to buy the authentic Kachchhi product (Figure 12.3).

Key Findings, Lessons and Reflections

General findings

It is important to note that many of the changes and transformations seen during the course of the study are very recent, and no conclusion can be derived about their trajectory and sustainability. It will be interesting to see what directions they take, both for the *vankar* community and *vanaat* in Kachchh, as also for lessons relevant to weaving or craft elsewhere in India. Within this overall caveat, however, some broad findings can be stated as follows.

A general sense of well-being appears to have increased among the *vankars* in the fifteen villages studied, especially relative to the period immediately after the earthquake in 2001, and this is closely linked to the revival of *vanaat* as a livelihood. Several *vankars* (especially but not only entrepreneurs, and including youth) mentioned their preference for weaving due to the 'freedom' and autonomy it gives them, and the return of several young people into *vanaat* is a clear sign that the craft is doing well and attracting even those who could get other jobs. The ability and even preference of many *vankars* for staying (or returning to) their villages, counter to the dominant narrative that enjoins oppressed castes and classes and especially *Dalits* to head to the city, has been enabled by a combination of enhanced economic opportunities within weaving itself, improved social status and other circumstantial factors, with the availability of communications and other technologies enabling 'urban-like' access and facilities. Combined with a clear and consistent narrative on the reduction in casteism (especially its worst forms of untouchability), greater contact with the outside world, a visible sense of pride and dignity in their lives and livelihoods, and various other factors, an increase in well-being appears to be widespread in the villages studied.

The revival of handloom weaving, while reliant on the market, is seen as a phenomenon with interrelated economic, social, cultural, emotional and intellectual elements and meanings. It is not merely a 'job' that provides income, but also provides autonomy (control over means of production as a crucial element of this), freedom (in terms of aspects like when and how long to work), continued family and social connections (since production is at home and involves the whole household), space for innovation and expression of creativity, identity (as a community, as a distinct craft identified by a unique Kachchh design), comfort compared to many other occupations they had access to (such as industrial labour), a chance to express a spiritual responsibility (with *vanaat* skills being 'god's gift'), and other such non-economic aspects. Weaving does not appear to be an alienated form of labour for a substantial part of the community – though in the case of job workers who are producing for entrepreneurs, there may be elements of such alienation.

What is tangible here is the inherent self-confidence and creative growth that becomes possible when people flourish in practices that are in their heritage and conditioning. There is less alienation, possibilities for better integration with their own communities and an enhanced ability for problem-solving and navigating the complexities of the market.

The complex of above elements could be seen as comprising the unique identity of the Kachchhi *vankars*; it is an identity with strong continuations from the past but also significant aspects of the present, characterized by resilience, innovation and creativity, social and family bonds, the ability to hybridize the past with the present and future, and pride in the craft heritage as not only economic but also culturally important.

This narrative suggests that even though the occupation is linked inextricably to the market, it has not undergone the transformation into commodification and alienation as envisaged by Marx (Lukács 1972, Marx 1988). Or in other words, the craftsperson's labour has not been alienated from them. This has to do with many factors: the means of production, especially the loom, remains in the ownership or control of the producer family, even in the case of many job workers (though certainly not all – see below); even the product is at least partly in their control in that they can express creativity in making it, and there is negotiating power over its value in the market; the weaver is able to see and control most of the process towards the finished product within the family rather than being a small cog in an industrial mass production system with extreme division of labour; and additionally a worldview is prevalent that holds weaving as not only a commercial activity but also a cultural one, with important emotional, psychological and affective aspects. Indeed, the alienation that many youth felt when joining other industries (many expressed how they were not in control of their time or of production there) seems to have been a major reason for coming back to weaving where they have a sense of ownership and belonging. A similar point could be made with regard to the dignity of labour that is evidently present in weaving (compared to other jobs as labourers in industry, construction, etc.), which Gandhi (1959) emphasized as a crucial aspect of human fulfilment, and especially where the act of weaving encompasses both physical and intellectual labour. To quote Prakash Naran Vankar, a young weaver of Bhujodi, 'my loom is my computer; I have to continuously think, innovate, it is not only mechanical'. Finally, this also points to a worldview that focuses less on the cold 'efficiency' of modern industrial life and more on a multidimensional 'sufficiency' paradigm where workers are not necessarily seeking to (or forced to) maximize productivity but rather also take into account what is enjoyable, self-governing and creative (Bakshi 2017). The complex relationship between crafts (and the associated skills and knowledge), market, modernization and alienation (or the lack of it) has been well brought out in studies in other sectors, such as artisanal fisheries by Sundar (2018).

This is by no means universal. Several job workers expressed dissatisfaction with their economic and social life; some were clearly in distress, with signs of alienation and loss of dignity. There are marked geographic and class inequities among the *vankars*. Preliminary observations from the larger contextual survey being carried out across Kachchh suggest that a considerable section of the *vankar* community outside the studied fifteen villages is still facing multiple challenges and livelihood insecurity. But the transformations that have taken place in several families show the potential of such livelihoods for generating multiple benefits even in marginalized sections of society. Indeed, this and other stories across the world suggest that crafts of various kinds could be a major driver of economic, social, cultural and ecological transformation, especially when building on traditional skills and knowledge, linking to appropriate markets, and mindful of internal inequities in communities. Even in industrial societies, there seems to be an increasing movement to make things by hand or using physical labour and not only in a mechanized, automated way.

Along with the revival of *vanaat* has come a larger ecological footprint, in particular due to the significantly greater transportation of raw materials and of woven products to consumers in other parts of India and in Europe. This is a common trend in handloom clusters across the country. Interventions to try to make the craft more ecologically sensitive, e.g. through promotion of organic cotton and natural dyes, provide a counter-trend. The specific case of *kala* cotton is interesting as it highlights how in providing these contradictory trends, there may be a very low ecological impact of production but a higher impact on the consumption side.

A historical timeline of changes taking place (Figure 12.4) in the *vankar* community suggests that the economic sphere has primarily driven today's

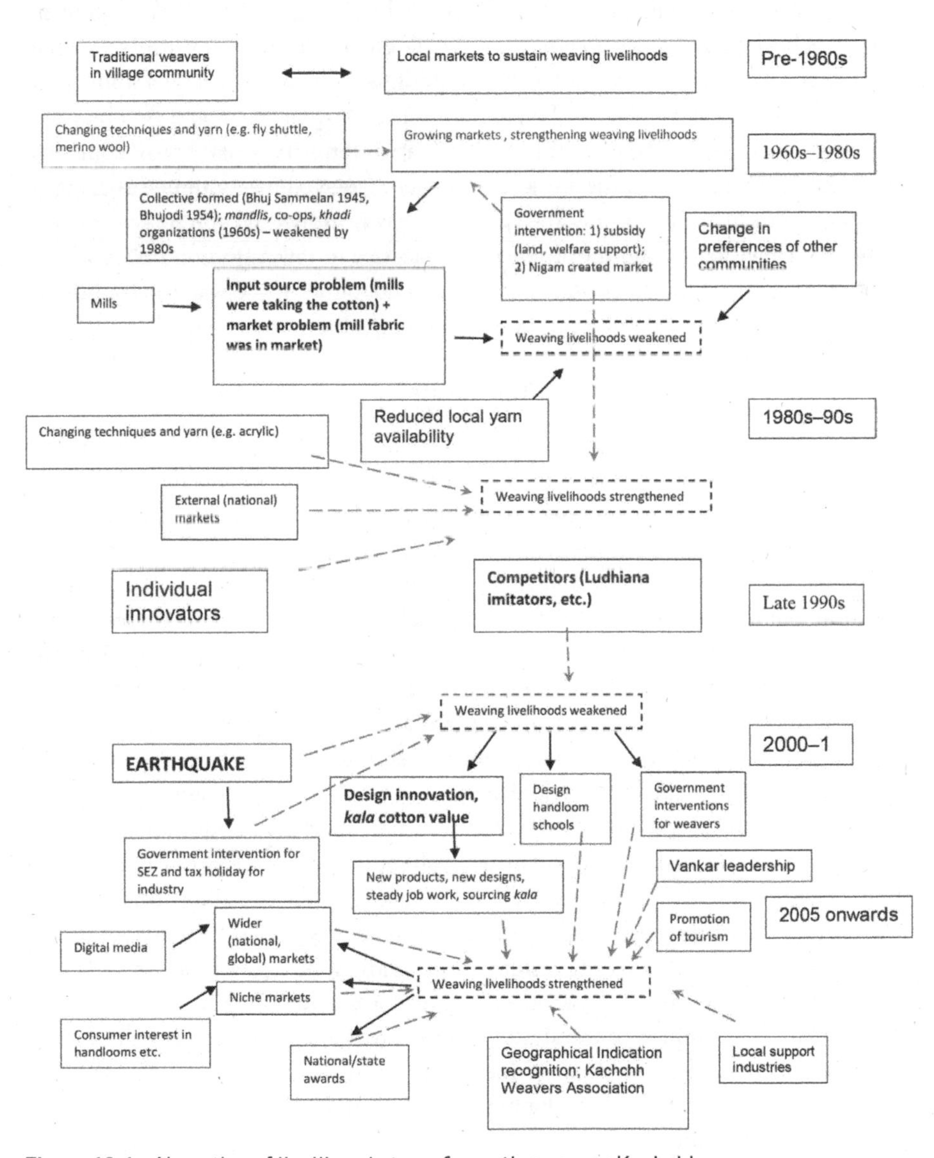

Figure 12.4 Narrative of livelihoods transformation among Kachchh weavers

transformations. The points of intervention and initiatives have focused primarily on economic and livelihood issues, i.e. to do with their occupation as weavers. It appears that transformations in other spheres of the lives of the community, particular a reduction in inequities in social spheres (such as in caste, gender and generational relations) are linked to this economic transformation, with socio-cultural factors also playing a significant role. However, this focus has led to an increase in economic inequities; and while some dimensions of it have been ecologically sensitive or positive, others are significantly not. Other aspects, such as political engagement and empowerment, which might address the resulting inequities have not been given attention. While there was an opportunity to delve into the ethical and spiritual values that underlie *vankar* society, only a glimpse of how these are changing due to the economic transformations could be obtained. Some articulations relating to this include an interesting balance of collective spirit and individualism in the work, the former displayed by continued sharing of design and product innovations, the latter in the way individual entrepreneurs have forged their own paths and the absence of collective mobilization on policy issues. Another glimpse of this, mentioned above, can be found in the way the traditional principle of limiting profit margins, to reduce inequalities, may be weakening.

The Future

The *vankar* community is concerned about and also hopeful for the future. Like any other primary sector industry activity dependent on an external market, *vanaat* remains subject to a number of macro-economic factors that are not in the control or even under the influence of producers. However, the *vankar* community is hopeful that its adaptability and resilience will help it to survive as it has in the past, and that the recent period of upswing could continue especially if *vankars* are able to continuously innovate and 'create' markets, and that institutions like Khamir and the artisanal craft schools will continue helping. Senior *vankars* realize that they will need to do more collective work (cooperatives, mutual help, advocacy), that special efforts are needed to encourage and incentivize the youth and women, and that issues like inequality and ecological impact need to be discussed and tackled. The *vankar* youth and women feel that given their increased empowerment, they can hope for continued enhancement in livelihood options. Many of these issues are new developments. This study itself, by initiating or encouraging conversations among the *vankars* on these issues, may act as a small trigger. For instance, at one of the final meetings to discuss the results of the study, some elder weavers proposed that the Kachchh Weavers Association organize a meeting with youth on visions for the future. As outputs of the study get circulated and read/seen/heard, further such conversations may take place. Based on all this learning, both the weavers themselves and institutions like Khamir could continue playing an enabling role in further transformation.

Notes

1 *Sandhani* (in Marathi) means 'to connect'. It is the first step of the weaving process on the loom, when two threads are tied together; but it is also used as a metaphor for connecting generations, connecting spiritually, connecting knowledge systems, etc. For the key findings of the study, see http://vikalpsangam.org/article/sandhani-weaving-transformations-in-kachchh-india-key-findings-and-analysis/#.XxfUDS2B2V4.

2 This study was coordinated by Ashish Kothari on behalf of Kalpavriksh and Durga Venkataswamy on behalf of Khamir, and involved in its core team Juhi Pandey and Ghatit Laheru of Khamir; Radhika Mulay, Kankana Trivedi and Arpita Lulla of Kalpavriksh; external consultants Meera Goradia and Arun Mani Dixit; and senior weavers Shamji Vishram Siju, Meghji Harji Vankar, Murji Hamir Vankar, Ramji Maheshwari and Naran Madan Siju. Suraj Jacob, as a member of the India advisory team for the ACKnowl-EJ project, provided valuable comments and input. Comments were also provided by several members of the project from within and outside India.

3 The study also gave rise to follow-up discussion with some sectors, like the youth, an example of which can be found here: https://vscoronatimes.blogspot.com/2020/04/vikalp-varta-2-youth-weave-new-story-in.html?fbclid=IwAR0BVFf2cSyn23SoT3_qJo_wnUklTOH-hwvWyYVu_BskoBu4M47uaHu8Ybc.

4 These are: Bhujodi, Awadhnagar, Kotay, Jamthada, Faradi, Siracha, Rampar Vekra, Godhra, Adhoi, Sanganara, Mathal, MotaVarnora, Dhanithar, Sarali and Ningal.

5 These films are available at: www.youtube.com/playlist?list=PLVGJfYVd8JMUYch7VAaCff-mcWhX0pQw4y.

6 *Dalits*, numbering about 300 million and with significant internal diversity, are among India's most oppressed and marginalized people. Dominant Hindu caste society holds them and their occupations to be the most 'impure'. Significant affirmative state action, including 'reservations' for jobs and educational institutions, social campaigns by government and civil society agencies, religious conversions to escape the shackles of Hindu casteism, and strident mobilization by *Dalits* themselves, has helped reduce extreme forms of discrimination, but for a majority, visible and invisible vestiges of marginalization and oppression continue. *Vankars* in Kachchh do not necessarily call themselves *Dalit*; several elders told us they feel this denotes 'downtrodden', and they prefer being called Meghwal or *vankar*.

7 The full report and associated films have several examples, with interviews of such youth.

8 Especially in the case of *kala* cotton products; there is still a substantial local market in acrylic products.

9 History is riddled with examples (vanilla, coffee, quinoa and many many more products) of the vulnerability of people dependent on external markets over which they have no control or even influence.

10 The study was not able to obtain the views of the global buyer or trader on this aspect, however.

11 The five 'units' studied are: four weavers making textile and different products from both the Kala and Bt cotton yarns, and Khamir which makes textile products only from Kala cotton yarn using only natural dyes.

References

Bakshi, R. (2017) Future Bazaar in India. In Kothari, A., Joy, K.J. (eds) *Alternative Futures: India Unshackled*. Delhi: Authors Upfront.

Gandhi, M.K. (1959) *Voice of Truth*. India: Shantilal H. Shah Navajivan Trust. http://gandhiashramsevagram.org/voice-of-truth/gandhiji-on-dignity-of-labour-bread-labour.php.

Lukács, G. (1972) *History and Class Consciousness: Studies in Marxist Dialectics*. Cambridge, MA: MIT Press.

Marx, K. (1988) *Economic and Philosophic Manuscripts of 1844* (trans. M. Milligan). Amherst, NY: Prometheus Books.

Sadgopal, A. (2016) An Agenda of Exclusion: 'Skill India' or Deskilling India. *Economic and Political Weekly*, 51(35), 27 August.

Shrivastava, A., Kothari, A. (2012) *Churning the Earth: The Making of Global India*. Delhi: Viking/Penguin.

Sundar, A. (2018) Skills for Work and the Work of Skills: Community, Labour and Technological Change in India's Artisanal Fisheries. *Journal of South Asian Development*, 13(3): 1–21.

PART III
Lessons from Ground-up Transformations

In this last part of the book, we present two final chapters in which we bring together the learnings from the in-depth case studies to develop a theory of 'transformation from the ground up', and share some key takeaways for our readers.

Dalit women farmers in Telangana, India, have formed their own collectives to revive millet farming (Photo Tejaswi Dantuluri)

13

Towards a Just Transformations Theory

Ashish Kothari, Leah Temper, Iokiñe Rodríguez, Mariana Walter, Begüm Özkaynak, Rania Masri, Mirna Inturias, Adrian Martin, Ethemcan Turhan, Neema Pathak Broome, Shrishtee Bajpai, Jen Gobby, Jérôme Pelenc, Meenal Tatpati and Shruti Ajit

Introduction

As communities and peoples across the world struggle to challenge the forces of inequality, exploitation and unsustainability, and achieve transformations towards a more just world, it is important to understand the processes and dynamics of such transformations. What motivates them, how are they achieved, what are the challenges they face, and how are such challenges overcome?

In this chapter, we share the results of the collective exercise carried out by the ACKnowl-EJ project core-team members to synthesize the key learnings from the project about how transformations to sustainability take place.[1] This was combined with knowledge and understanding of other environmental justice initiatives that team members have been involved with in the past.

As we explained in the introduction to this book, participants in the ACKnowl-EJ project were interested specifically in understanding processes of transformation that occur as a result of communities claiming justice for themselves and for nature, as part of what can be broadly called 'environmental justice', or 'just sustainabilities'.[2] This they do through resistance to ecologically damaging projects and/or situations of related deprivation and injustice, and/or the active construction of alternatives that reflect their ecological, social, cultural, economic, political and/or spiritual values.

Thus, our focus has been on deliberate collective action of communities towards a shared vision as the driver of such transformations (though this does not mean all members of the collective necessarily share such a vision; there are often internal differences). This focus stems from our understanding that movements from below hold the greatest promise for moving towards more just and sustainable futures, more than or often in opposition to state action and managed transitions, though the latter can also at times be a contribution to the drivers of overall transformation. As explained also in the introduction of the book and further expanded below, we have distinguished between 'transformations' that entail fundamental or systemic changes in the structures of injustice and unsustainability and towards structures of justice and sustainability, and 'transitions' or 'reforms' that do not challenge existing structures.

Such an understanding also necessitates a strong leaning towards research and action that is oriented towards transformation, both in the methodologies used (making them as participatory and community-led or community-based as possible) and in the analysis of the findings. This bias is made explicit and reflected in the various case studies and other outputs of the project.

Building on the conceptual framework we developed to carry out the ACKnowl-EJ project explained in the introduction of the book and our empirical findings from the case studies, in this chapter we provide key concepts about transformations to sustainability that we propose as a basis for developing a theory of just transformation to sustainability from the ground up. We first discuss what our understanding of 'transformation' is. Then in the following four sections we distinguish between how transformation happens (from conflicts to alternatives), how communities/movements seek transformation (strategies), what helps to make transformation happen (enablers), and what is transformed (scalar, temporal and spatial dynamics). Throughout, we illustrate the key points with examples of case studies presented in the rest of the book; these appear in italics.

What Is Transformation?

> Transformation
>
> *Etymology*: *trans* = across + *form* = mould, character
>
> *Dictionary definition*: marked or substantial change in form, structure, character or appearance

While the above dictionary and etymological understandings of the word 'transformation' are important, to elaborate and add nuance, we propose the following characteristics based on the learning of the ACKnowl-EJ project. Transformation from the ground up is:

1. A process in which conditions of injustice and unsustainability undergo profound changes towards situations of justice and sustainability. For instance, this may involve a move towards greater and more widespread participation in decision-making, greater economic security for everyone, greater power of local communities in relation to the state, greater gender, class, caste and other forms of equality within communities, legitimation of diverse knowledges and values, reduction of violence in all its forms (direct, cultural and structural), and so on. Transformation also calls for a proactive and collective vision of *possible* futures. For instance, in the Indian process Vikalp Sangam which brings together praxis of radical transformation from

across the country, there is a deliberative process of envisioning a more just, equitable and sustainable India (Kothari 2019), the framework of which was used as one basis of analysis in the ACKnowl-EJ project.[3] In the case of Boğaziçi University in Turkey, where in 2021 there was an attempt to take over the university through authoritarian politics, academics say that '*until the ideal of a free and independent university run based on participatory principles is realized in Turkey ... We do not accept, we will not give up*'. There is therefore a combination of objectives and intentions, practices and visions, and ethics and values guiding the actions of the relevant actors.

2. A process of emancipation that entails revealing, challenging and dealing with some or all root causes of oppressive conditions. These causes include structural and relational properties of the political, ecological, social, cultural and economic spheres of society, including prevalent forms of discrimination and domination such as (singly or in combination with) capitalism, colonialism, modernity,[4] patriarchy, racism, statism and speciesism or anthropocentrism.
3. Systemic and radical change, resulting in new (or revitalization of old) relations, structures and cultures (including narratives, knowledge, beliefs, institutions, norms, values, behaviour) that promote different forms of just and sustainable alternatives. It is in contrast with both business as usual approaches and reformist approaches, which seek gradual change within prevailing structural and relational conditions – though we recognize there is a distinction between those reforms that could, over time and building momentum, result in transformations such as progressive expansion of fundamental rights to humans and the rest of nature, and those that simply reinforce the structures of injustice such as the 'greening' of capitalism. As Andre Gorz (1967) said:

> A reformist reform is one which subordinates its objectives to the criteria of rationality and practicability of a given system and policy. Reformism rejects those objectives and demands – however deep the need for them – which are incompatible with the preservation of the system. On the other hand, a not necessarily reformist reform is one which is conceived not in terms of what is possible within the framework of a given system and administration, but in view of what should be made possible in terms of human needs and demands.

We come back to this later in the chapter.

To be clear, we do not claim here a comprehensive treatment or understanding of processes of transformation in all their diversity. There are significant gaps – for instance the aspect of technological transformations is not dealt with here. Environmental justice is itself a limited field of enquiry and frame of understanding, even though in the project it was very broadly understood. Nevertheless, and acknowledging fully the limitations of any single project and of the people involved in it, we feel that what has emerged from the project is significant enough to put out in this form.

How Transformation Happens: From Conflicts to Alternatives

Conflicts: Antagonistic relations between two or more sets of actors, in which the fundamental rights, values and well-being of one or more of them are threatened or under attack; 'actors' can include non-human species and other elements of nature. In this project we did not consider conflicts arising only out of specific activities such as extractivist projects, but also situations of deprivation and injustice such as real poverty, gender or caste or 'race' discrimination, etc.

Alternatives: Alternatives can be practical activities, policies, processes, technologies or concepts/frameworks that lead us to equity, justice and sustainability. They can be practised or proposed/propagated by communities, government agencies, civil society organizations, individuals and social enterprises, among others. They can simply be continuations from the past, reasserted in or modified for current times, or new ones; it is important to note that the term does not imply these are always 'marginal' or new, but that they are in contrast to the mainstream or dominant system.

Transformation is usually ***inseparable from conflict***, first because contradictions are a stimulant for change and second because movements for progressive change are resisted by those with vested interests in the status quo. This can include 'background' situations like structurally caused deprivation and inequality *(e.g. in the case of the weavers of Kachchh, India, macro-economic policies brought in by globalization in India, leading to unfair competition in the market from industrially produced textiles)*; it can be material/physical deprivation, dispossession or displacement, actual or proposed, often violent *(e.g. the enclosure of grazing commons for exclusionary wildlife conservation by the state, reducing the area available to the Raika pastoralists in Western India; the allocation of forest commons by the state for mining to external companies in the case of the Indigenous/*adivasi *villages of Korchi, Maharashtra, India; wood pirating by timber companies in Lomerío, Bolivia; the building of large-scale development projects in Indigenous peoples' territories, like the power line from Venezuela to Brazil in 1997; large-scale industrial investments and coal-fired power expansion in the Yeni Foça region of Turkey)*; it can be epistemic (forms of knowledge) impositions (*e.g. fire control policies in many parts of the world, such as Canaima National Park, Venezuela*) and displacement; and/or the hegemony of ontological systems (ways of being).

Conflict is rooted in situations that are perceived as violent and unjust and invite us to reflect on the opportunities that such clashes of interest and visions offer to produce transformation. Interestingly, at times a heightened conflict can even be a sign of imminent or ongoing (but not completely manifest) positive transformation, e.g. when the entrenched system hits back at a movement because it is actually being shaken and feeling the pressure.

Indeed, movements may deliberately exacerbate obvious conflicts to force a response, such as setting up blockades and other non-violent civil disobedience actions *(e.g. the anti-mining movement in Argentina)*, or even in violent ways. Conflicts also occur when authoritarian regimes perceive a threat from institutions promoting freedom of speech and independent academics, such as Turkish President Erdoğan's attempt to clamp down on Boğaziçi University (Kadıoğlu 2021). The presence of conflict can indicate the surfacing of the root causes of injustice and the rebalancing of power relations. It indicates a desire for change, and the development of a consciousness that change is possible and actionable despite obstacles, a necessary step towards transformation. Conversely, the absence of conflict can sometimes be a sign of co-optation by the system, which is very capable of 'tolerating' and even incorporating alternatives at scales and in forms that do not threaten it. This can also be a sign of great asymmetries in power that prevent the issues in conflict coming to the surface, often leading to more silent forms of resistance (Scott 1985). Deep understanding and analysis are needed to tease out the complex signals that conflicts or their absence are giving out.

Transformations often progress ***from resistance (conflict) to alternatives***. Resistance can arise in different contexts and different ways: as reactions to an externally imposed decision that forces those adversely impacted to act, or as proactive actions anticipating a situation of conflict. In a sense, resistance itself is an alternative, for it is an assertion of identity and power by the impacted, and usually includes implicit or explicit articulation of a different worldview from those being imposed by the dominant. Nevertheless, it can be considered as conceptually different from the construction of pathways of well-being that are alternative to the ones being imposed. In several case studies in this project, communities and citizens not only opposed a project or situation of conflict, but also proposed or implemented their own initiatives for livelihoods, knowledge, conservation and/or other aspects important for them.

Transformation involves the ***pursuit of justice***. This includes actions to reduce structures of discrimination such as patriarchy, capitalism, coloniality and racism; new forms of democracy *(such as the neighbourhood forums that flourished in Turkey after the Gezi movement, including the Yeni Foça Forum and the local assemblies in Argentina)* that empower previously marginalized voices to contend and change decisions across different spheres of transformation; or reassertion of traditional governance systems vis-à-vis the nation-state *(such as in the movements of Indigenous nations in Canada)* and more equitable distribution of rights and responsibilities to promote economic and other forms of security for all *(e.g. in the history of environmental conflict in Canada, resistance to extractivism has been led predominantly by First Nations like the Unist'ot'en, defending land and waters while fighting for self-determination in the face of ongoing colonialism)*.

Transformation relies on the ***creation of new meanings***, involving the dismantling of conventional understandings and categories and the breaking down of binaries and divisions, which could be present in both traditional and modern societies. For example, environmental justice is about reconceptualizing nature as the place where we live, work and play, especially for modern societies; this may well entail re-learning ancient ways of relating to the rest of nature. Recognizing the rights of nature, with its own agency and subjectivity, is one method. *The Haida Gwaii struggle against logging in Canada led to a rethinking of the categories of what a forest is.* Similar re-conceptualization is involved in the movements against plantations being equated to forests and movements across the world fighting (and winning) battles over rights to rivers.

Transformation requires ***co-evolutionary change across multiple spheres of society****:* the social, cultural, economic, political and ecological spheres of life and the multiple ethical and spiritual values that connect with these (see Figure 2.2 page 61, and Annex 1 page 68). The beginning point of transformation can be in any of these spheres, and we think that it is reductionist to assume that any one of these is always the key sphere. Transformations can begin within the economic dimension, but other dimensions can also be starting points *(in the cases of Canaima National Park in Venezuela and Lomerío in Bolivia, the important transformations started taking place first in the political sphere; movements of resistance to the authoritarian regime in Turkey are focused especially at the level of counter-hegemonic discourse).*

There is a dynamic relationship among the various spheres of society, with a major change in any of them likely to affect the others. At any given stage of the process, the transformation could be harmonious among these spheres and values, or arising from tensions, with spaces for change opening in the interstices produced by contradictions between spheres (e.g. economic and ecological). *The Kachchh weavers' case from India clearly displays these tensions and contradictions, with positive change in the livelihoods and economic situation of the weaver community and an attendant reduction in traditional discrimination on the basis of caste, gender and age, being tempered by an increase in economic inequality within the weaver community, and an overall larger ecological footprint of their products being sold globally rather than exchanged only locally. In the case of Korchi, India, significant transformation in many dimensions is taking place, but is (as in the case of the Kachchh weavers) partly dependent on an external, capitalist market. In Canaima, Venezuela, the Pemon Indigenous people have gained considerable agency and political recognition, but have had to move to extractivism to sustain livelihoods.*

A process of continuous or periodic self-reflection and self-assessment (with or without external facilitation) helps to generate awareness the relationships among the different spheres of society and enable possible corrective measures, though of course this would depend on the conditions available at a given time, including material conditions and what worldviews are prevailing. This conception of change across multiple spheres presents challenges for understanding, because case studies suggest uncertainty about whether transformation needs to occur across all (or most) spheres, or whether changes in different spheres might occur in sequence, over different timescales. *The Kachchh weavers study, for instance, brought out this complexity, posing challenges to the study team in trying to understand multiple dimensions of transformation (see below on this) with some depth within the given time and resource constraints of the project.*

Additionally, alternative transformations may not even be immediately visible in the midst of conflict scenarios; *at Boğaziçi University, continued resistance has created a credible narrative acknowledged by the public, which keeps the door open for alternative paths, paths that may effectively open up when movements continue believing in collective agency, stay resilient to institutional pressures, and keep proactively imagining and building alternative futures.*

How Communities/Movements Seek Transformation: Strategies

Transformation happens through ***confluences and alliances*** that build constituencies around progressive alternatives (though of course individual and isolated actions can also trigger processes towards transformation). This involves the creation of solidarities and cross-sectoral, cross-cultural and/or intercultural dialogues – the creation of physical, social and virtual spaces for sharing experiences, constructively challenging each other, engendering collaboration and collectively envisioning possible alternative futures. *The Lomerío case study in Bolivia is a very good case in point, as all the transformative strategies that the Monkoxi Indigenous people have put in place over the last four decades to advance their dream of self-government have been reliant on a wide range of alliances with NGOs, international cooperation, universities and key governmental actors. Similarly in Argentina, the ability of community assemblies to challenge large-scale mining activities, policies and discourses at local and national levels has been possible through the formation of a national network of assemblies and the creation of multiple alliances with institutional (local governments, church, unions) and professional actors (scientists).* Confluences increasingly occur across spheres, for example where movements involve solidarities between organizations that have traditionally sought to instigate change through single spheres (such as labour unions or environment groups). Such confluences can also be across space and time, and not necessarily even explicitly recognized or conscious, e.g. when news of resistance and alternative initiatives in one part of the world can inspire similar action in another part *(for instance Indigenous Life Plans, which started in Colombia in the 1990s and have become an inspiration for Indigenous peoples throughout Latin America and beyond)* or revived memories of historical incidents of transformation can inspire movements in the present. *For example, fighting against oil and gas pipelines, and working together to stem the expansion of the Alberta tar sands in Canada has brought together unprecedented coalitions of people, communities and organizations, serving to help break down 'issue silos' that have separated efforts for economic, social and environmental justice for decades. This emerging movement of movements forged in the fight against proposed pipeline after proposed pipeline has been opening up space in Canada to think together about just and sustainable alternatives to extractivism.*

Transformation takes place through a ***diversity of strategies.*** For example, urgent efforts to resist imminent threats require different strategies from longer-term efforts to build alternatives. Even longer-term efforts to transform power relations across different spheres require multiple strategies because different forms of power are amenable to different forms of action. As repeatedly demonstrated in the different chapters of this book, following the Conflict Transformation Framework (Rodríguez and Inturias 2018), such diversity of strategies may range from political mobilization, legal advocacy, lobbying, creating new institutional arrangements and modes of production, strengthening social and political organization, revitalizing local knowledge and histories, creatively using spaces within the system, recognizing that the system is not a monolith (e.g. rights-based laws enacted by the state), and others. *The analysis of multiple movements of Indigenous peoples in Canada and anti-mining movements in Argentina present in this book show a large diversity of such strategies; the campaign against mining and asserting local governance in Korchi, India, and the movement against the Brussels jail also reveal such diversity.* The strategy that is prioritized in a given moment in time largely depends on the actors that are leading the struggles, the alliances that can be built and where they seek to produce change, but also on the nature of the conflict (e.g. latent versus manifest). Conflicts over knowledge systems, which tend to be silent and involve invisible forms of power such as discourses and narratives of environmental change, require strategies that help create a new social consensus over meanings and values of nature. In this vein, the Argentina anti-mining movement has deployed diverse strategies and alliances that have successfully challenged extractive discourses, narratives and supporting actors and institutions.

Conflicts over extractive activities, which tend to be overt and are about an unfair distribution of harms and benefits or about lack of participation in decision-making, often require producing urgent changes in institutional and decision-making frameworks. This is why resistance strategies like political mobilization or creating new institutional decision-making structures are often privileged in these types of struggles. Yet long-term transformation requires also building new capabilities and networks, which is why working on issues such as strengthening local organization and sensitizing decision-makers or the business sector are also often part of the necessary strategies for change. *Similarly, the anti-coal movement in Aliağa, Turkey was always very busy with the continuous daily struggle and bureaucracies in the courts on the legal front. Hence, while trying to stop the coal-power plant projects, they had little time and resources to develop an alternative energy discourse that would offset or at least weaken the hegemonic modernist discourse of the state centred on looming energy scarcity from the 1980s onwards. Offering an alternative energy vision would surely be crucial for the long-term viability of the resistance as it would help them to break from a rather reactive and negative stance and enter a proactive one.*

As explored in the Argentinean and Canadian chapters (8 and 11 respectively), the EJAtlas documents how environmental justice groups have mobilized diverse strategies in their struggles, and how this diversification has been associated with their ability to succeed in their aims (stopping and suspending projects).

What Helps to Make Transformation Happen: Enablers

*Transformation requires a combination of **praxis and theory***, the hands-on and the conceptual, the local or grassroots and the wider scales (regional, national and/or global), the individual and the collective (behaviour, worldviews, etc.). Any one of these without the other will be incomplete (though at a given moment in time one may well be found without the other), and likely to be unsustainable.

Transformation requires a ***sense of community or collective***, something larger than an individual with which the individual can identify, something that is bigger than the sum of the individuals that comprise it, and something that produces and safeguards a set of commons (physical, material, intellectual, cultural, ethical). This can be a traditional village or urban community, in its old or modified form; or a completely new collective formation such as a new settlement, an educational institution, or others. A common purpose and identity, and norms that govern the commons, binds this community together. The collective strength that a community provides is a significant enabler of transformation, including at the level of individuals *(for instance, in Turkey, in both the Yeni Foça and Boğaziçi cases, cause-oriented activism brought a diverse set of actors to understand each other better, enabled collaboration despite differences, enhanced the sense of belonging and collective agency by strengthening the community culture, and reminded participants why they cared about localities and identities and how they relate to local and institutional histories).*

A community that is internally thriving, enabling spaces for meaningful participation to all members, providing a balance between individual freedom and identity and the health of the community, capable of intergenerational learning and respect (including enabling youth energy and innovation), is likely to be more successful in transformation than one that is ridden with internal inequities, divisions and rigidity. *An example of what happens when there is no conscious or proactive collective initiative is the case of the weavers of Kachchh, India, where significant positive transformation at the level of individual families is not necessarily translated into economic enhancement for the community as a whole (several weaver families are left behind), or does not enable an effective community response to negative state policies (such as taxation on handloom products that further disprivileges them vis-à-vis industrial products) ... though the absence of collective action is not the only cause of these problems or weaknesses.*

Within the sense of the collective, however, there will be complex dynamics between that collective and the individuals within it. Various participants of an initiative towards transformation will view it differently, will themselves be transformed (or not) in different (and not necessarily mutually complementary) ways, and will respond differently to various situations in the process. The process itself can often help resolve or reduce differences that can be conflictual; *in the case of the Boğaziçi resistance in Turkey, within participatory processes among academics, strategies were built in small increments, and while deciding on arenas of contestation to be mobilized, care was taken to respect the common wisdom emerging from discussions.*

Transformation has ***multiple enablers.*** These include:

- actual presence of and/or sense of injustice and the desire to change these;
- facilitation or leadership from within or outside the relevant actors/community – empowered local collectives or communities;
- horizontal (local to local or peer to peer) and vertical (local to wider scale) networks or alliances of learning and support;
- enabling policy frameworks;
- material conditions enabling action beyond survival; and
- cultural drivers including progressive worldviews and the intercultural exchange of such worldviews.

Some enablers or catalysts are only sparks, such as a sudden government move to displace people, or an earthquake; others are more sustained, such as leaders who sustain processes of transformation over a period of time. Different combinations of these will work in different situations; generalized 'recipes' for transformation need to be viewed with caution as they may not be sensitive to local conditions. The role of various forms of the state (including law), market and civil society, will differ in different contexts.

Additionally, some enablers may be direct and visible, and others indirect and invisible or 'behind the scenes', e.g. technological developments in one part of the world could have significant influence in another, enabling or disabling transformation in profound ways; or general changes in educational opportunities for marginalized people may create the social conditions for or against transformation. Another way of stating this is to see some enablers as having agency, while others are circumstantial.

Transformation requires or is likely to be ***more sustained when it includes a vision of alternatives***, a collective understanding of directions in which to head, based on shared ethical values that bind collective action and instil strength. This does not need to be fully cooked from the beginning, and indeed needs to constantly evolve and to encompass multiple perspectives and internal divergences, but without some such collective sense and agreement on direction and values (including individual ethical and spiritual persuasions) that are part of an evolving cultural ethos, transformations are likely to be short-lived or even counterproductive. *An example of such a vision of alternatives can be found in Bolivia and Venezuela, with the Monkoxi* *and Pemon Indigenous peoples respectively. In both, the longer-term vision of alternatives in their struggles involved making fundamental changes to the model of the nation-state in order to open up a space for the acknowledgement of pluricultural citizen rights in the national political and legal frameworks, including the right to territorial ownership, self-determination, political autonomy and defining their own forms of development, among others. Both cases involved intense social and political mobilizations to bring about these changes, which go well beyond the local scale of the conflicts. In the Brussels jail campaign, the resisters also came up with alternative strategies, which strengthened their position vis-à-vis the state. In the case of the Beirut resistance movement against waste, the absence of any alternative to how the city was handling garbage could have been a factor in its eventual collapse; possibly the same could be said of the Canaima case. On the other hand, the fact that a resistance movement comes up with an alternative does not guarantee its success; in the case of Canaima in Venezuela, for instance, despite this combination being strongly presented, the extractivist economy and state were too powerful.* Overall, though, the project was not able to go as deep into the connection between resistance and alternatives as desired.

Similarly, over the last few years in Canada there has been a new and powerful strategy for transformation emerging in Indigenous communities opposing ongoing oil and gas development; they are building solutions in the pathway of the problem. From the Healing Lodge and permaculture gardens at the Unist'ot'en camp in British Columbia, to the Treaty Truck House at the Mi'gmaq protest camp in Nova Scotia, to the Tiny House Warriors in Secwepemc territory, to the Watch House on Burnaby Mountain, Indigenous people are building low-carbon, beautiful, culturally grounded alternatives and placing these alternatives strategically to block the way of new oil and gas projects being pushed into their territories. These alternatives are offering inspiration by making clear that there are other ways to build economies. At the same time, they are enacting Indigenous sovereignty and lifeways.

Transformation requires movements that encompass ***plural forms of knowledge and values***. Epistemic and cognitive justice and the acknowledgement of other worlds and other ways of being are pre-conditions for abandoning previous ways and towards transformation. This may entail considerable decolonization of the mind in many parts of the world. It also requires that knowledge and its use is linked to ethical norms and values (such as those of equality, solidarity, reciprocity; see Annex page 328), if knowledge is to transform into wisdom, and contribute to transformation. The assertion of knowledge as a commons, enabling not only democratic sharing but also multiple points of synergistic innovation, is a crucial part of transformation. Where movements are built on confluences across scales and cultures, they will need to incorporate both universal narratives and contextual knowledge and values. For instance, a shared narrative of radical democratic decision-making processes and institutions in a Latin American Indigenous people may look different from equivalent narratives in a European city. *In Canada there* *are attempts at plural legal understandings where common law perspectives are joined with Indigenous law. On the ground this often entails the assertion of Indigenous law. For example, the Unist'ot'en are practising Anuk Nu'at'en (Wet'suwet'en law), based on a feast system of governance. Such legal approaches are based on an ecological wisdom and respect for nature lacking in colonial laws. Similarly, the Listuguj Mi'gmaq First Nation government took over the management of the salmon fishery by passing, implementing and enforcing their own Indigenous law. But the transformation may also be based on the merging of elements from various knowledge systems, as in the case of the Kachchh weavers in India where traditional skills and motifs have been combined with modern designs, products and technologies in innovative ways. Or the transformation may be entirely within modern contexts, as seen for instance in the emergence of counter-knowledge in the campaign against the Brussels jail proposal.*

What Is Transformed: Scalar, Temporal and Spatial Dynamics

Transformation occurs across ***spatial scales***, from the individual to humanity as a whole, from a single geographic unit to the entire landscape or seascape through a change in behaviour. This can happen to individuals, to social movements, communities or societal levels and the interrelations between them. We refer to this as the human or societal scale of transformation. The transformation of human behaviour is considered to be an essential part of transitions and transformations to global sustainability (Gifford 2011; Swim et al. 2011). The personal sphere considers the individual and collective beliefs, values and worldviews that shape the ways that the systems and structures (the political institutions) are perceived, and affects what types of solutions are considered 'possible'. *In the case of Lomerío, Bolivia, the movement created transformation from local to national levels, which in turn reflected back to further transformations at the local level.*[5] *Argentina's Union of Community Assemblies, a coalition of local assemblies, was able to transform local and national socio-environmental narratives and institutions.*

Even the smallest of transformations could be important; indeed macro-scale transformation may result from the confluence of smaller ones consciously or subconsciously linking up with each other until they reach critical mass. Conversely, such small initiatives, if scattered and unconnected and devoid of larger alliances, can also be more prone to being undermined or reversed by macro-forces. *In the case of Turkey, for instance, the primary source of tension seems to be the presence of an unquestioned commitment to rapid economic growth combined with energy scarcity and independence discourse on top of the absence of a deliberative planning process, a democratic scientific culture and a free press. While the movement was successful in stopping the coal-power plant at different periods in time over the forty years of struggle and surely made a difference, in each case, the state reacted to the success story and activism in Aliağa in typical hegemonic counter-movement fashion. The main challenge for the Aliağa resistance was to create synergies with other local environmental movements, and even an overarching national movement capable of sustained action. The anti-coal struggle in Aliağa marks a significant point in the history of environmental movements in Turkey in building politically conscious environmental resistance towards the emblematic Bergama gold mine case and beyond.*

Given the fact that there may be simultaneous transformations taking place across various spatial levels in (or because of) an initiative, it is very likely that these may not always be complementary. What is transformative for one community or set of individuals, or one section of society, may be regressive for another. *For instance, for some job workers in Kachchh, India, who have been left behind in the community's overall economic enhancement, the transformation may not be looking so positive. In the case of the Raika pastoralists in India, sheep-shearing as a communal activity is no longer commonly practised. This has created a livelihood opportunity for Muslim families who have had to give up their own herds due to a lack of state and policy support. While this has meant a forging of new bonds between the Raika and the Muslim shearers, both communities have experienced a loss of communal activities and livelihoods.*

Transformation involves different ***depths of change***, with a distinction between addressing superficial 'symptoms' and deeper, underlying 'causes'. While transformation must always involve deeper systemic or structural changes, it is not always possible to distinguish these from reformist changes, because movements typically require strategies to address both symptoms and causes, and sometimes a situation that is undergoing longer-term transformation may appear to be only reforming and it will depend on the strength of our analysis if we are able to discern this at a particular time and place. For instance, affirmative action for historically marginalized sections of society could remain non-transformative if it simply helps to absorb a few individuals from these sections into dominant society; however, it could lead to transformation over time by enabling sufficient numbers within these sections to challenge dominant power structures. Several movements engage with the state in order to expand spaces within the system, e.g. through rights-based legislations, even as they also struggle for more systemic change in the nature of democracy itself. *This is seen, for instance, in the movements of Indigenous peoples in Canada working to reclaim territory and gain autonomy, and by communities engaging with the Chavez government in Venezuela to seek constructive changes in law.*

Transformation involves multiple ***temporal dynamics***. It is a continuously evolving and dynamic, non-linear process, with continuities and ruptures and reversals, rather than a full and final end state. Pathways to the future will have roots in historical context but will not necessarily repeat history. Some forms of activity and change can be spontaneous, abrupt and episodic, such as protests to oppose individual projects. Other activities are long-wave, slower and more continuous, such as the construction of constituencies and alternatives. *The Beirut (Lebanon)* *waste protests built on similar movements over a decade. The movement against mining and towards self-determination in Korchi, India, builds on a long history of resistance movements, but also includes specific episodic actions that appear as flashpoints. Cumulative and sudden episodic events are also the story of the Raika pastoralists in India.*

Some changes can be reversed in an instant or over time – *the adoption of extractive activities by Indigenous peoples who first fought them in Canaima, Venezuela, and the co-option of the Lebanon waste movement by the state are cases in point.* Others are more resistant to violence and co-option, including values, knowledge and subjectivities (things we produce in common). For instance, a process of community mobilization against the arbitrary power of the state regarding what to do with forest commons, and the creation of institutional processes to govern these commons, could well be reversed in future if the state hits back with greater force or if internal power hierarchies undermine the institution, but for the moment, this does constitute transformation. Indeed, transformation requires the development of these long-wave resources in order to build the capacity to respond and adapt to unexpected challenges and setbacks, and a faith in the possibility of transformation even in the midst of such setbacks – in sum, a culture and capacity for resilience. Even slower, longer-wave changes evolve in non-linear ways characterized by waves (ups and downs but still an overall progression towards goals), cycles (coming back to more or less original states over time), spirals (slow progression towards goals within what seems to be a cyclical pattern) and tipping points (when a critical mass is reached that produces a state-change in the system). Some transformations may even emerge from actions that do not seem to be immediately transformative.

These multiple and complex dynamics contribute to the difficulty of observing transformation at any one point in time. This is why efforts at systematizing and learning from transformations with the protagonists of resistances and alternatives over a sustained period of time, as we have done with the ACKnowl-EJ project, are so important. We hope the lesson we have learned in this process will inspire others to continue adding to this 'just transformations theory' in the making.

Annex: Ethical Values of Transformation

(Adapted from Vikalp Sangam n.d.)

Practical and conceptual alternatives vary widely, and none are replicable in precise form from one place to the other, given the diversity of local situations. Search for such alternatives is perennial. New circumstances will demand new responses – hence, alternatives will have to keep evolving and changing.

The way alternative transformations are attempted by the actors concerned, and observed by others, is based very much on their worldviews. These encompass spiritual and/or ethical positions on one's place in the universe, relations with other humans and the rest of nature, identity and other aspects. Initiatives towards alternatives espouse or are based on many values and principles that emanate from or are encompassed in such worldviews, keeping in mind also that even within single communities there may be more than one worldview, with differences emanating from how members are placed regarding gender, class, caste, ethnicity, age and other considerations.

It is possible to derive the crucial, commonly held principles underlying alternative initiatives. Given below is an initial list of such values/principles; these are not necessarily distinct from each other, but rather interrelated and overlapping.

We note here that there is a list of even more fundamental human ethical values that should be the bedrock of the principles below, including compassion, empathy, honesty, integrity and truthfulness, tolerance, generosity, caring, and others. These are espoused by most spiritual traditions and secular ethical systems, and are worth keeping central to a discussion of the values/principles described below.

Ecological integrity and the rights of nature

The functional integrity of the ecological and ecoregenerative processes (especially the global freshwater cycle), ecosystems and biological diversity that are the basis of all life on Earth.

The right of nature and all species (wild and domesticated) to survive and thrive in the conditions in which they have evolved, and respect for and celebration of the 'community of life' as a whole (while keeping in mind natural evolutionary processes of extinction and replacement, and that human use of the rest of nature is not necessarily antithetical to treating it with respect).

Equity, justice, inclusion and access

Equitable access and inclusion of all human beings in current and future generations (intergenerational) of decision-making and participation, to the conditions needed for human well-being (socio-cultural, economic, political, ecological and psychological), without endangering any other person's access; and social, economic and environmental justice for all regardless of gender, class, caste, ethnicity, race and other attributes, (including a special focus on including those currently left out for reasons of physical/mental/social 'disability'). There is also a need to acknowledge unjust and unfair dynamics within families and try to address them.

Right to and responsibility of meaningful participation

The right of each citizen and community to have *agency*, to meaningfully participate in crucial decisions affecting their life, and the right to the conditions that provide the ability for such participation, as part of a radical, participatory democracy.

Corresponding to such rights, the responsibility of each citizen and community to ensure meaningful decision-making that is based on the twin principles of ecological sustainability and socio-economic equity.

Diversity and pluralism

Respect for the diversity of environments and ecologies, species and genes (wild and domesticated), cultures, ways of living, knowledge systems, values, livelihoods, perspectives and polities (including those of Indigenous peoples and local communities, and of youth), in so far as they are in consonance with the principles of sustainability and equity.

Collective commons and solidarity, in balance with individual freedoms

Collective and cooperative thinking and working founded on the socio-cultural, economic and ecological commons (moving away from private property), respecting both common custodianship and individual freedoms and choices (including the right to be 'different', such as in sexual orientation) and innovations within such collectivities, with inter-personal and inter-community solidarity, relationships of caring and sharing, and common responsibilities, as fulcrums.

Resilience and adaptability

The ability of communities and humanity as a whole to respond, adapt and sustain the resilience needed to maintain ecological sustainability and equity in the face of external and internal forces of change, including through respecting the conditions enabling the resilience of nature. Sustaining initiatives in the midst of changing generational values/priorities, larger economic and political systems.

Subsidiarity, self-reliance and ecoregionalism

Local rural and urban communities (small enough for all members to take part in decision-making) as the fundamental unit of governance, self-reliant for basic needs,[6] linked with each other at bioregional and ecoregional levels into landscape, regional, national and international institutions that are answerable to these basic units. (The term 'self-reliant' here means self-sufficiency for basic needs as far as possible, and the right to access what is not possible to meet locally, from more centralized systems guaranteed by the state.)

Autonomy and sovereignty

Collective rights and capacities to self-govern or self-rule and be self-reliant, as peoples and communities, including custodianship of territories and elements of nature they live within or amidst; including mechanisms of direct or radical democracy.

Simplicity and/or sufficiency – need over greed

The ethic of living on and being satisfied with what is adequate for life and livelihood, in tune with what is ecologically sustainable and equitable. There is a need to elaborate and distinguish between need and want.

Dignity and creativity of labour and work/innovation

Respect for all kinds of labour, physical and intellectual, with no occupation or work being inherently superior to another; giving manual labour and family/women's 'unpaid' work and processes of sharing/caring their rightful place, but with no inherent attachment of any occupation with particular castes or genders; the need for all work to be dignified, safe and free from exploitation (requiring toxic/hazardous processes to be stopped); reducing work hours; and moving towards removing the artificial dichotomy between 'work' and 'leisure' by enabling more creative and enjoyable engagement; encouraging a spirit of enquiry and inquisitiveness.

Non-violence, harmony, peace, co-existence and interdependence

Attitudes and behaviour towards others that respect their physical, psychological and spiritual well-being; the motivation not to harm others; conditions that engender harmony and peace among and between peoples. Wasting and irresponsible use of resources – food, water, energy – is a type of violence against the unprivileged.

Efficiency in production and consumption

Efficiency in the use of elements of nature and natural (including human) resources, in terms of eliminating or minimizing waste (and *not* in modern industrial terms of narrow productivity).

Dignity and trust

Respect of every person's entitlement to be treated with dignity and trust, regardless of sexual, ethnic, class, caste, age or other identity, and without being 'judged' as a person on moral grounds.

Fun

Inculcating the spirit and practice of enjoyment, fun and lightness of being in all aspects of life, without causing harm to others.

Notes

1 Unless otherwise indicated, throughout this chapter we use the term 'transformations' to denote positive moves towards sustainability and justice, noting here that transformations could also be regressive.

2 We use these terms to mean processes leading to situations of increasing equality and equity among peoples, and between people and the rest of nature, and an active respect of the ecological limits that humanity lives within on Planet Earth. See also Bennett et al.'s (2019) proposal of a Just Transformation to Sustainability approach that includes recognitional, procedural and distributional considerations.

3 www.vikalpsangam.org, see also Vikalp Sangam (2017).

4 We distinguish between 'modernity' as a colonizing project of cultural and knowledge homogenization and a unilinear view of progress, and 'modern values' that may have positive features such as an emphasis on equality.
5 We use the term 'local' in relation to 'global' in terms of spatial scale, or place, not in terms of the scope of knowledge or visions being generated, where the local–global dichotomy is somewhat artificial.
6 Food, water, shelter, sanitation, clothing, personal security, learning/education, health and livelihood. This does not for the moment incorporate some of the other non-material needs such as those identified by Manfred Max-Neef et al. (1989).

References

Bennett, N.J., Blythe, J., Cisneros-Montemayor, A.M., Singh, G.G., Rashid Sumaila, U. (2019) Just Transformations to Sustainability. *Sustainability*, 11(14): 3881. https://doi.org/10.3390/su11143881.

Gifford, R. (2011) The Dragons of Inaction: Psychological Barriers That Limit Climate Change Mitigation and Adaptation. *American Psychologist*, 66(4): 290–302.

Gorz, A. (1967) *Strategy for Labour: A Radical Proposal*. Boston: Beacon Press.

Kadıoğlu, A. (2021) Autocratic Legalism in New Turkey. *Social Research: An International Quarterly*, 88(2): 445–71.

Kothari, A. (2019) Collective dreaming: Democratic Visioning in the Vikalp Sangam Process. *Economic and Political Weekly*, 54(34), 24 August.

Max-Neef, M., Elizalde, A., Hopenhayn, M. (1989) Human Scale Development: An Option for the Future. *Development Dialogue*, 1989(1). www.daghammarskjold.se/wp-content/uploads/1989/05/89_1.pdf.

Rodríguez, I., Inturias, M. (2018) Conflict Transformation in Indigenous Peoples' Territories: Doing Environmental Justice with a 'Decolonial Turn'. *Development Studies Research*, 5(1): 90–105. https://doi.org/10.1080/21665095.2018.1486220.

Scott, J. (1985) *Weapons of the Weak: Everyday Forms of Peasant Resistance*. New Haven: Yale University Press.

Swim, J.K., Stern, P.C., Doherty, T.J., Clayton, S., Reser, J.P., Weber, E.U., Gifford, R., Howard, G.S. (2011) Psychology's Contributions to Understanding and Addressing Global Climate Change. *American Psychologist*, 66(4): 241–50.

Vikalp Sangam (n.d.) The Search for Alternatives: Key Aspects and Principles. http://vikalpsangam.org/about/the-search-for-alternatives-key-aspects-and-principles/.

—— (2017) Alternatives Transformation Format. https://vikalpsangam.org/wp-content/uploads/migrate/Resources/alternatives_transformation_format_revised_20.2.2017.pdf.

14
Takeaways for Environmental Justice Movements

Leah Temper, Mariana Walter and Iokiñe Rodríguez

Our intention with this book has been to capture, understand and support transformations to alternative futures that are born from community resistance, often on the margins. We have wanted to see how communities 'on the commodity frontiers' suffering directly from the impacts and injustices of extractive economies have managed to organize and intentionally transform these conditions towards situations of justice and sustainability.

While the question of transformation and how to enable, support and accelerate the necessary changes to avert ecological planetary disaster has become a key driving question in sustainability research, the bulk of this research still focuses on managed transitions from above and the necessary technologies needed to accompany these. Our contribution in contrast has been a vision of transformation at the worm's eye level from the grassroots up, and of the social innovations and the tools/technologies born from local knowledge and imagination that we believe are necessary for achieving deep system-level change.

We have tried to dig deep to understand the grassroots 'radicalis' of transformative futures from the messy, courageous and never-ending work of communities and environmental justice movements building strategies and overcoming obstacles at the local level. As we noted in the previous chapter, transformations are not linear and are enmeshed in histories that span decades, if not centuries. What we have captured in this book is therefore just a snapshot of the ongoing labour of imagining, conceptualizing and putting into place these alternatives from below. We have not focused on the important question of how these local processes can lead to world-scale change and how to scale them up and out. However, these local movements radiate outwards and are part of broader movements for decolonization, environmental justice and global policy transformations. While these experiences are ingrained locally, they can become seeds of inspiration and transformation across space and time (as seen in Chapter 8 on Argentina). We believe that the process of activist scholarship and documentation, the cases we engaged with, and the methodological challenges addressed during the ACKnowl-EJ project and shared in this book hold learnings for activists, researchers and social changemakers working with communities to transform the world.

To this end, we would like to close this book with three key takeaways, reflections and learnings from these ground-up struggles that we hope will be useful for all those engaged in trying to make just transformations to sustainability:

- How to best defend against the intent of hegemonic forces on stifling transformation towards injustice, and how to sustain collective power in the face of institutional blocks and cultural resistance to change.
- How the co-production of knowledge can help.
- How the global transformation spurred by the pandemic in the past few years has deepened our understanding of transformation across scales.

First Takeaway: Handbrakes to Destruction

Across all the case studies of this book we have seen diverse ways in which the state and corporate interests divide, conquer, repress and empty out movements to maintain the status quo and advance diverse, unsustainable and unjust development practices. Communities fight back in myriad ways, but perhaps the key learning that emerges on how to resist this type of atomization points to careful attendance to both the internal work of building governance and cultural revitalization, at the same time as engaging in the outer work of activism and the need for creating, sustaining and mobilizing shared counter-narratives about the alternative futures the movement aims to move towards. Community-led transformative processes entail a sustained engagement with inner and outer transformative and discursive labour.

As Rodríguez and Aguilar showed for the Pemon case in Venezuela (Chapter 4), as with many other Indigenous peoples of the world, they are increasingly experiencing a disconnection from nature and the local environment because of rapid processes of cultural change and assimilation policies. They note that their efforts into reshaping the relationship between Indigenous peoples and the state led to a neglect of strengthening their own decision-making and governance mechanisms. This eventually allowed the state to exploit these internal weaknesses to advance its extractivist agenda.

According to Rodríguez and Aguilar, development counter-narratives cannot emerge or be sustained over time unless long-term and continual endogenous processes of cultural revitalization are put in place to help strengthen Indigenous peoples' own knowledge and value systems and cultural identities.

Ways to do this include the creation of discursive, cultural and territorial revitalization strategies. In Bolivia, the Monkoxi show a more promising evolution in their work on cultural revitalization, born from their own analysis of their transformations over the last four decades. According to Inturias et al. (Chapter 10), in recent times, the Indigenous Union of Lomerío (CICOL) has been much more actively engaged in developing strategies to strengthen its cultural power and finding ways for the younger generations, in particular, to stay connected with their identity, culture and territory. Examples of this include the Monkoxi's recent work developing a PhotoVoice project with youth and the development of an intercultural education strategy to be used in the schools of Lomerío.

In the case of the emergence and spread of Argentina's anti-mining movement, Walter and Wagner (Chapter 8) posed that in order to understand the shift in environmental and mining discourses and policies in the country, it is crucial to observe the cultural transformations that occurred within local and inter-local organizations. Local assemblies became a space of individual and collective empowerment, where local actors – many of whom had no previous mobilization experience – could express their views and feel listened to and recognized. Assemblies made it possible for marginalized actors (e.g. women, Indigenous communities) to become protagonists in the construction and dissemination of counter-hegemonic narratives and worldviews, weaving networks with diverse actors and successfully transforming institutions.

It should be noted that the creation and elaboration of development counter-narratives inevitably entails internal conflicts within the movements. Such narratives will not be uniform or universally shared. But the work of their elaboration and formalization, such as through community life plans ('*planes de vida*'), serves to make the alternatives that movements are fighting for tangible. As Pelenc argues (Chapter 6), these internal conflicts are normal because movements operate as an 'oppositional public space' where different practices and political discourses can be confronted and discussed. While he notes that they can generate a great fatigue among the resisters, the work of creating such shared documents, records and moments for movement reflexiveness is invaluable when confronted with actors who aim to sow division and discord to conquer. Our work as engaged researchers sometimes entails supporting these processes.

In this vein, this book is built on the idea that documenting and sharing stories of transformation holds value, as a process to highlight local knowledge, cultures and lessons, and as a way to foster inspiration, hope and visioning for communities and researchers on the ground and beyond.

In Lebanon, the key weakness that Masri identifies as stifling the transformative potential of the movement is precisely the lack of political imagination – the inability to look forward and outwards enough to truly envision an alternative political system beyond the sectarian state (Chapter 5). This is another way in which becoming mired in technical solutions and technocratic politics framed in narrow terms can lead to demobilization and fragmentation of the movement into the 'art of the possible'.

Finally, of course, the use of violence and criminalization remains a major tool of the state and corporate powers. Different forms of violence have been documented across the cases presented in this book, especially but not only against activists. One learning is that while we are not all equally exposed and Indigenous and racialized communities face the brunt of this violence, no one trying to transform the system is immune, whether they are in a democracy such as so-called Canada or whether they are academics, as the experience of our colleagues in Turkey shows. Violence blurs territories, actors and learnings, hindering the capabilities for transformation as well as the conditions to document, support and learn from local trajectories and experiences. Not all cases initially envisioned for this book were finalized, as the safety of local researchers, communities and organizations was put at risk. This occurred in India, but it is occurring in many other regions and countries in the world where violence is isolating and making invisible territories, people and their knowledges. In this vein, while a great deal of interest has been shown in transformation processes and enablers, much more attention should be placed on the increasing and diverse forms of direct and indirect violence against communities, organizations, journalists and researchers that are hindering the consolidation, emergence and diffusion of just and sustainable processes of transformations.

Second Takeaway: On Doing Co-produced Transformative Research

The ACKnowl-EJ project was a particular collective endeavour, not only focused on exploring cases of grassroots-led transformation, but also reflecting on the role of active and engaged research in these processes. This book has explored different transformation elements, their actors, their complex dynamics, scales, times, advances and setbacks, pointing to key processes and lessons. Moreover, the book has addressed the value of research in processes of transformation, highlighting the relevance of reflective research processes, co-production alliances, intercultural dialogues and spaces of learning and sharing.

ACKnowl-EJ researchers developed ongoing activities to document and reflect on our research practice, our approach to co-production, our ethics and processes, as well as our relationships with ourselves and others, and the processes of knowledge production over the course of the project. As we discussed in Chapter 1, engaging in co-production process entails diverse tensions with ourselves (challenging our lenses), our jobs (as researchers expected to provide certain outputs in certain timeframes; career pressures), our institutions (precarity, short-term funding and long-term processes) and our work with communities (expectations, aims, narratives, etc.) that play out in different ways as we engage in co-production processes.

Moreover, as we have mentioned throughout the book, we used and developed different methodological approaches such as the EJAtlas, a collaborative tool to document EJ struggles and support activists; the Alternatives Transformation Format, to support self-evaluation methods for just and sustainable transformations; the Tarot Activity, which used collage materials to reflect on researcher identities, positionalities and ethics; power analysis tools for activist empowerment and conflict transformation; back-casting and scenario-building for strategizing with mobilized groups; and future visioning methods to pattern hope. These are among the tools that allowed researchers and local groups to co-examine, learn about and co-produce transformative processes.

The Conflict Transformation and Alternatives Transformation frameworks (Chapter 2) were developed to learn from transformations brought about by resistance movements, but also to enhance and inform struggles and processes of change. The use of the frameworks sparked conversations in the communities that could be both uncomfortable and productive. Conversations explored how to create solidarity across different interests in other contexts. As already discussed, bringing people to the table and helping voices to be heard that otherwise would not be given space can change consciousness but also political dynamics. One example is ACKnowl-EJ's work with Raika women (Chapter 7), where the research team recorded, for the first time, the voices and worldviews of women in the community, which initiated discursive change, challenging the patriarchal decision processes. Work in Turkey led to discussions and rethinking of the concept of development and to a broadening of what people there understood as 'environmental'; this significantly contributed to future visioning and enabled the movement to evolve from being defensive to taking a more propositional position.

Looking back from the vantage point of a few years later, some of the fruits of our engaged approach can be discerned. In Bolivia, the conflict transformation work helped elucidate some of the gaps in the transformation strategies. The joint research clearly showed that the Monkoxi had focused their mobilization and resistance strategies much more on structural change than cultural revitalization. Because of this, in recent years, and as mentioned above, the Monkoxi have started working much more on cultural transformations, strengthening their traditional language (Besiro) and developing strategies to reconnect the youth to their territory.

As this book was being written we also learned more about the powerful impact of co-produced forms of communication, where researchers and communities co-design communication outputs that are attentive to local communication languages, channels and interests. For instance, in India, youth and women groups were key to fostering the production of short videos, documentaries and graphic novels that allowed them to express research results in a way that was relevant to them and their communities. These outputs were used by and for the communities and to help counter the effect of authoritarian regimes. Nevertheless, further research and attention should be targeted on studying how to include and reach vulnerable groups, such as those that do not read or write; and the relevance of developing diverse outputs that go beyond academic papers, narratives and languages to explore visual, musical, artistic and written forms of communication with local groups.

Third Takeaway: Writing about Transformations in Times of Global Crisis: Moving Lenses and Images

Since 2019, when the drafting of this book began, many things have happened with deep implications for us, as mothers, fathers and researchers, as well as for the grassroots-led transformative processes that we have been working with.

Some transformations have been negative, while others became opportunities, and many remain to be seen. The Covid-19 global pandemic at the outset captured the imagination of many transformative social movements who in the words of Arundhati Roy saw 'an opening to a world otherwise'. Calls for 'No return to normal' envisioned transformative paths towards coping with the global pandemic that could strengthen mutual aid, refocus the reproductive care system in our economies, and be resilient to transformative changes in our lifestyles and systems of consumption and production.

As the pandemic progressed, its regressive qualities became clearer. In many of our case study sites, it bolstered authoritarianism, leading to surveillance and militarization, intensifying local conflicts and injustices as it led to measures that slowed, paused or regressed ongoing counter-hegemonic transformative processes (forbidding local protest, making invisible repression). Amidst this, creative forms of protest, organization and networking also emerged. While the impact of the shift in values, expectations and trauma unleashed by the pandemic are still unfolding processes, one learning that became evident was how communities with common and robust webs of relations forged through their joint struggles were able to confront in a more resilient way some of its challenges.

In India, the Vikalp Sangam process gave birth to an initiative called Vikalp Sutra which examines and supports the struggles for dignified livelihoods in the face of Covid-19. For example, in Korchi *taluka gram sabha* leaders came together to ease the distress caused by the lockdown by distributing essentials using the *gram sabha* funds. A study on the impact of the pandemic on the Raika noted how many pastoralists mentioned that they might have experienced fewer negative impacts resulting from the lockdown than many others with rural livelihoods. They attributed this to the fact that pastoral communities have historically needed to adapt to climatic, political and other changes and have adapted resilience to system-level transformations.

The network of the ACKnowl-EJ project itself became an agent for solidarity in these difficult times, by being part of a series of worldview dialogues led by the Global Tapestry of Alternatives (www.globaltapestryoflaternatives.org). This series aimed to facilitate sessions with activists, scholars, researchers, mobilizers and practitioners across the world who have started exploring systemic alternatives to dominant regimes. One example was a webinar series called 'Dialogue on Alternatives in the Time of Global Crises', which offered an opportunity for an encounter between Monkoxi communities in Bolivia and communities from India and other parts of the world, to share experiences and strategies to cope with Covid-19.[1]

Covid-19 made visible and allowed an appreciation of the value of some features and dynamics of communities. For instance, in Raika, factors like the diversification of jobs have really helped local communities adapt to the pandemic. People had to leave their jobs in the cities and return to the communities, as it was the animals that kept them going. Covid made visible some key elements for local resilience and transformation.

Exchanges produced during Covid between communities with whom we were working show how empowered grassroots communities in control of their local means of production often fared much better than their urban counterparts, who were more exposed to the failures of national governments to protect citizens from both Covid and its economic impacts. Those who had already advanced on developing local self-reliance, community production and democratic governance had increased capacity to guard against the vagaries of global markets. Such grassroots governance initiatives thus demonstrate how those empowered by alternative politics and means of livelihood can better withstand system-level disruptions such as the pandemic.

But we cannot close this book without voicing our grave concern about the fact that Covid-19 has increased extractivist pressures worldwide, but especially in the Global South, fostering new and exacerbating ongoing socio-environmental struggles. Covid-19 has furthered public indebtment, increasing state and corporative pressures to extract (and export) natural resources to level trade balance accounts, at a great social and environmental cost. Moreover, climate emergency, global energy security and economic crisis narratives are accelerating projects to extract coal, gas, oil, uranium, lithium, graphite, nickel and copper among many other materials considered strategic for energy security and energy transition worldwide, but again especially in the Global South. On the ground, we have seen how these projects are violently put forward, criminalizing and repressing communities, neglecting consultation rights with opaque decision-making processes, and irreversibly transforming territories and transformative paths. In the name of multiple crises and emergencies, top-down approaches are prioritized, bulldozing bottom-up alternative knowledge and processes of transformation that, as we have seen, were key to navigate the pandemic crisis.

In the preparation of this book, we also witnessed impacts in our lives and places of study due to shifting political regimes, such as the authoritarian regimes in Turkey and India, and Lebanon's socio-political crisis after the catastrophic harbour explosion. Such processes showed that even within academia we are not safe from processes of oppression and politicization. This was a reminder that academia is not a safe space, and we need to defend the spaces we do have. As learned by our Boğaziçi colleagues from Turkey, building internal and external networks of support and collaboration within and outside academia across time has been relevant to defend these spaces. Moreover, as external pressures grow for both academics and activists in Turkey, sustaining interpersonal relations and having face-to-face gatherings to meet and share emotional challenges has been central to protecting people and maintaining struggles. As experienced with Covid-19, structural control is easier when people are isolated.

In different places, the pandemic has slowed processes of transformation through alienation, increased authoritarianism as a counter-transformative force, and the inability to organize face to face. While some of the dreams of transformation expressed at the outset of this earth-shattering event may have dimmed, we find hope in the non-linear, multi-scalar, cascading, recursive nature of such processes and take heart from Özkaynak et al.'s (Chapter 3) exhortation not to dichotomize such processes into victories and failures, but to continue to build strategies as handbrakes to authoritarian environmental destruction and towards transformative alternatives.

Note

1 Recording here: https://globaltapestryofalternatives.org/webinars:2020:13?redirect=1.

Notes on Contributors

Vladimir Aguilar is a lawyer specializing in Indigenous law in Venezuela and Latin America. In addition to being a political scientist and doctor in Development Studies, he is director of the Working Group on Indigenous Affairs (GTAI) at the Universidad de los Andes, Venezuela.

Shruti Ajit is currently Programme Officer at Women4Biodiversity, working on gender justice and environmental rights research and advocacy. She is currently focusing on women-led restoration initiatives and their relevance in the larger biodiversity conservation context. Since 2015, with Kalpavriksh Environment Action Group, she has been working on documentation of community-led conservation initiatives and through that informing local, national and global policies for inclusive and community-led forms of conservation.

Miguel Aragón is a lawyer and specialist in Indigenous agrarian issues, a legal advisor to CICOL, and a university professor (maragon@nur.edu.bo).

Cem İskender Aydın is an assistant professor at the Institute of Environmental Sciences at Boğaziçi University. His research focuses on energy and climate policy/politics, environmental justice and mapping environmental conflicts, and environmental governance. His work has appeared in journals such as *Energy Policy*, *PLoS One* and *Frontiers in Energy*.

Shrishtee Bajpai is a researcher-activist-writer from India, networking and researching on systemic alternatives to dominant extractive systems. She writes about topics including environmental justice, pluriversal alternatives, more-than-human governance, Indigenous worldviews and rights of nature. She is a member of Kalpavriksh and helps coordinate the Vikalp Sangam process which networks and researches on systemic alternative initiatives in India. She is a core team member of Global Tapestry of Alternatives and serves on the executive committee of Global Alliance for the rights of nature. She is an avid birder and photographer, and loves walking.

Jen Gobby is an affiliate assistant professor at Concordia University and holds a PhD in Renewable Resources from McGill University. She is the founder of the MudGirls Natural Building Collective and currently works as coordinator with Research for the Front Lines, a network that supports the research needs of movements and communities fighting for climate and environmental justice across so-called Canada. She collaborates with Indigenous Climate Action on their Decolonizing Climate Policy project and is the author of *More Powerful Together: Conversations with Climate Activists and Indigenous Land Defenders* (Fernwood Press).

Ashish Kothari is a founder-member of Kalpavriksh, and a member of many people's movements. He has taught at the Indian Institute of Public Administration, coordinated India's National Biodiversity Strategy & Action Plan, served on the boards of Greenpeace International & India, ICCA Consortium, and been a judge at the International Rights of Nature Tribunal. He helps coordinate Vikalp Sangam (www.vikalpsangam.org), Global Tapestry of Alternatives (www.globaltapestryofalternatives.org) and Radical Ecological Democracy (www.radicalecologicaldemocracy.org). He is co-author/co-editor of *Churning the Earth*, *Alternative Futures* and *Pluriverse: A Post-Development Dictionary*.

Mirna Inturias is a social researcher, specializing in Indigenous issues, identity and interculturality, Indigenous education, and environmental conflict transformation. She has conducted research on environmental conflicts in protected areas and Indigenous territories of the Bolivian Oriente, Chaco and the Amazon. She is a founding member of the Confluencias Group and is part of numerous Latin American reflection and research networks. She is professor and main researcher at the Social Research Center of Nur University, Santa Cruz, Bolivia. She is author, co-author or editor of ten books, among them *Capacity Building in Socio-environmental Conflict Transformation*.

Arpita Lulla is a researcher, filmmaker and consultant. A member of Kalpavriksh, her current work revolves around understanding and networking sustainable initiatives, which are alternatives to the current extractive development ideology. She is also a freelance digital marketer, plays ultimate frisbee, and is on a quest for mindful and holistic living practices.

Adrian Martin is a professor in Environment and Development at the School of International Development (DEV), University of East Anglia, UK. He is a social scientist who specializes in interdisciplinary research to inform the management of natural resources in developing countries, particularly in relation to governance of protected areas, integrated conservation and development, participatory forestry and agricultural intensification.

Elmar Masay is former chief general and technical advisor of the Indigenous Union of Native Communities of Lomerío (CICOL).

Rania Masri is currently a co-director at the North Carolina Environmental Justice Network. Previously, she was the founding Associate Director of the Asfari Institute for Civil Society and Citizenship at the American University of Beirut, and a professor at the Department of Environmental Sciences at the University of Balamand in Lebanon. She is the author of numerous book chapters, including 'Assault on Iraq's Environment: Radioactive Waste and Disease' (in *Iraq: Its History, People and Politics*), 'The Media's Deadly Spin on Iraq' (in *Iraq Under Siege: The Deadly Impact of Sanctions and War*) and 'The Al-Aqsa Intifada: A Natural Consequence of the Military Occupation, the Oslo Accords, and the "Peace Process"' (in *The Struggle for Palestine*).

Begüm Özkaynak is a professor in the Department of Economics at Boğaziçi University. She specializes in research related to sustainability economics, environmental conflicts and justice, and energy governance. She is a member of the International Society of Ecological Economics and has been serving as co-editor-in-chief of the *Ecological Economics* journal since 2022.

Jérôme Pelenc is lecturer at the Université Toulouse 2 Jean Jaurès and researcher at the UMR LISST-Dynamiques Rurales. He has worked on the resistance movements against mega-infrastructure projects, and is co-author of the book *Résister aux grands projets inutiles et imposés* (Textuel, 2018).

Anacleto Peña is the current Chief General of the Indigenous Union of Native Communities of Lomerío (CICOL).

Neema Pathak Broome is a member of Kalpavriksh and is ICCA's Regional Coordinator for South Asia. She has helped the Ministry of Environment and Forests (MoEF) draft a set of guidelines for identifying and supporting Community Conserved Areas in India. She has authored and co-authored a number of publications on protected areas, community-based conservation, and the relationship between decentralization and conservation.

Iokiñe Rodríguez is an associate professor in the School of International Development (DEV), University of East Anglia, UK Her work focuses on environmental conflict transformations, local knowledge, power, environmental justice, equity and intercultural dialogue in Latin America. As a researcher in the ACKnowl-EJ project, she contributed to the development of conceptual frameworks to evaluate alternative transformations, to the co-production of knowledge strategy and to two case studies.

Mukesh Shende is a member of Amhi Amchya Arogyasathi, based in Nagpur, India.

Meenal Tatpati is a senior research and advocacy associate with Kalpavriksh. Her work focuses on areas of law and environmental jurisprudence of forests in India, pastoralism studies, Indigenous and local worldviews, developmental projects, protected areas and people interface. She is also a member of the IUCN World Commissions on Protected Areas and Environmental Law.

Leah Temper is the Campaign Director of the Fossil Fuel Ad Ban Campaign at the Canadian Association of Physicians for the Environment. She is a co-founder of the Global Atlas of Environmental Justice and the co-coordinator of ACKnowl-EJ (Activist-Academic Co-production of Knowledge for Environmental Justice). She is the author, co-author or editor of over fifty articles, chapters and reports including *Ecological Economics from the Ground Up* (Earthscan).

Ethemcan Turhan is an assistant professor of environmental planning at the Department of Spatial Planning and Environment, University of Groningen. His research is situated in the broadly defined field of political ecology with empirical attention to climate justice and energy democracy. He co-edited *Transforming Socio-Natures in Turkey: Landscapes, State and Environmental Movements* (Routledge, 2019) and *Urban Movements and Climate Change: Loss, Damage and Radical Adaptation* (Amsterdam University Press, forthcoming).

Lucrecia Wagner is a post-doctoral researcher at the National Scientific and Technical Research Council (CONICET, Argentina), and a member of the Environmental History Group at the Argentinean Institute of Snow, Glaciology and Environmental Sciences (IANIGLA). She is professor of the PhD in Social Sciences at the National University of Cuyo (UNCuyo), and the Master's in Environmental and Territorial Policies. Her main research topics are environmental conflicts, especially related to extractive activities, with emphasis on environmental legislation and social participation.

Mariana Walter is a political ecologist and ecological economist engaged researcher. She was the Scientific Coordinator for ACKnowl-EJ, based at the Institute of Sciences and Technologies of the Autonomous University of Barcelona (ICTA-UAB). She is currently a visiting postdoctoral researcher at the JHU-UPF Public Policy Center, Department of Political and Social Sciences and UPF Barcelona School of Management (UPF-BSM) at the Universitat Pompeu Fabra. She is a member of the Coordination and Direction Group of the Environmental Justice Atlas.

Lena Weber is an artist, writer and researcher who completed her doctorate at the Autonomous University of Barcelona. Her work examines how activist-academic researchers use critical and transgressive methodologies for socio-ecological transformation, and how we can teach and learn research in a way that breathes life into ourselves and the world around us.

Index

Thanks to our Patreon subscriber:

Ciaran Kane

Who has shown generosity and comradeship in support of our publishing.